PEARSON
Math Makes Sense 8

WNCP Edition

Lorraine Baron
Garry Davis
Susan Ludwig
Kanwal Neel
Robert Sidley

Trevor Brown
Sharon Jeroski
Elizabeth Milne
John Pusic
David Sufrin

With Contributions from

Michael Davis
Cathy Heideman
A. J. Keene
Antonietta Lenjosek
Margaret Sinclair

Craig Featherstone
Jason Johnston
Bryn Keyes
Peggy Morrow
Elizabeth A. Wood

Marc Garneau
Don Jones
Georgia Konis-Chatzis
Carole Saundry

PEARSON
Education Canada

Publisher
Mike Czukar

Research and Communications Manager
Barbara Vogt

Publishing Team
Enid Haley
Alison Rieger
Ioana Gagea
Lynne Gulliver
Stephanie Cox
Judy Wilson
Claire Burnett
Lesley Haynes
Ellen Davidson
Marina Djokic
Jane Schell

Photo Research
Karen Hunter
Jeanne McKane

Design
Word & Image Design Studio Inc.

Copyright © 2008 Pearson Education Canada, a division of Pearson Canada Inc.

All Rights Reserved. This publication is protected by copyright, and permission should be obtained from the publisher prior to any prohibited reproduction, storage in a retrieval system, or transmission in any form or by any means, electronic, mechanical, photocopying, recording, or likewise. For information regarding permission, write to the Permissions Department.

ISBN-13 978-0-321-54803-0
ISBN-10   0-321-54803-5

Printed and bound in the United States.

1 2 3 4 5 – 12 11 10 09 08

The information and activities presented in this book have been carefully edited and reviewed. However, the publisher shall not be liable for any damages resulting, in whole or in part, from the reader's use of this material.

Brand names that appear in photographs of products in this textbook are intended to provide students with a sense of the real-world applications of mathematics and are in no way intended to endorse specific products.

The publisher wishes to thank the staff and students of St. Stephen School and D. A. Morrison Middle School for their assistance with photography.

Statistics Canada information is used with the permission of Statistics Canada. Users are forbidden to copy this material and/or redisseminate the data in an original or modified form, for commercial purposes, without the express permission of Statistics Canada. Information on the availability of the wide range of data from Statistics Canada can be obtained from Statistics Canada's Regional Offices, its World Wide Web site at http://www.statscan.ca and its toll-free access number 1-800-263-1136.

Consultants, Advisers, and Reviewers

Series Consultants
Trevor Brown
Maggie Martin Connell
Mignonne Wood
Craig Featherstone
John A. Van de Walle

Assessment Consultant
Sharon Jeroski

Aboriginal Content Consultants
Susan Hopkins, Tlicho Community Services Agency, Behchoko, NT
Jed Prest, Inclusive Schooling Coordinator, Tlicho Community Services Agency, Rae-Edzo, NT

Advisers and Reviewers

Pearson Education thanks its advisers and reviewers, who helped shape the vision for *Pearson Mathematics Makes Sense* through discussions and reviews of prototype materials and manuscript.

Alberta

Joanne Adomeit
Calgary Board of Education

Bob Berglind
Formerly Calgary Board of Education

Jacquie Bouck
Lloydminster Public School Division 99

Auriana Burns
Edmonton Public School Board

Daryl Chichak
Edmonton Catholic School District

Derek Christensen
Edmonton Public Schools

Lissa D'Amour
Medicine Hat School District 76

Jaime Didychuk
Prairie Rose School Division

Brenda Foster
Calgary R.C.S.S.D. 1

Florence Glanfield
University of Alberta

Mary-Elizabeth Kaiser
Calgary Board of Education

Dona Kutryk
Edmonton Public Schools

Jodi Mackie
Edmonton Public School Board

David S. Moss
Edmonton Catholic Schools

Lisa Spicer
Red Deer Public School Board

Jeffrey Tang
Calgary R.C.S.S.D. 1

Cathy Turner
Calgary Public Board of Education

C. D. Wiersma
Calgary Catholic School District

British Columbia

Lorraine Baron
Central Okanagan School District 23

Donna Beaumont
Burnaby School District 41

Bob Belcher
School District 62 (Sooke)

Mark Chow
School District 36 (Surrey)

Dean Coder
School District 73 (Kamloops/Thompson)

David Ellis
School District 39 (Vancouver)

Marc Garneau
Surrey School District 36

Catherine A. R. Gillion
School District 35 (Langley)

Karen Greaux
School District 75 (Mission)

Hardeep Grewal
School District 75 (Mission)

Barry Gruntman
School District 53 (Okanagan/Similkameen)

Jane M. Koleba
School District 61 (Greater Victoria)

Selina Millar
Surrey School District 36

Camille Quigley
School District 75 (Mission)

Shannon Sharp
School District 44 (North Vancouver)

Kirsten Urdahl-Serr
School District 42 (Maple Ridge-Pitt Meadows)

Chris Van Bergeyk
Central Okanagan School District 23

Christine VanderRee
Comox Valley School District 71

Denise Vuignier
Burnaby School District 41

Mignonne Wood
Formerly Burnaby School District 41

Manitoba

Rosanne Ashley
Winnipeg School Division

Ralph Mason
University of Manitoba

Christine Ottawa
Mathematics Consultant, Winnipeg

Gretha Pallen
Formerly Manitoba Education

D. E. Saj
Interlake School Division

Judy Strick
Lakeshore School District

Gay Sul
Frontier School Division

Saskatchewan

Susan Beaudin
File Hills Qu'Appelle Tribal Council

Gisèle Carlson
Regina School District 4

Edward Doolittle
First Nations University, University of Regina

Lori Jane Hantelmann
Regina School Division 4

Angie Harding
Regina R.C.S.S.D. 81

Jeff Rambow
Regina R.C.S.S.D. 81

Cheryl Shields
Spirit School Division

Shawna D. Stangel
Regina Public Schools

Table of Contents

Investigation: Triangle, Triangle, Triangle	2

Unit 1: Square Roots and the Pythagorean Theorem

Launch	4
1.1 Square Numbers and Area Models	6
1.2 Squares and Square Roots	11
1.3 Measuring Line Segments	17
1.4 Estimating Square Roots	22
Game: Fitting In	28
Technology: Investigating Square Roots with a Calculator	29
Mid-Unit Review	30
1.5 The Pythagorean Theorem	31
Technology: Verifying the Pythagorean Theorem	37
1.6 Exploring the Pythagorean Theorem	39
1.7 Applying the Pythagorean Theorem	46
Strategies for Success: Getting Unstuck	52
Unit Review	54
Practice Test	58
Unit Problem: The Locker Problem	60

Unit 2: Integers

Launch	62
2.1 Using Models to Multiply Integers	64
2.2 Developing Rules to Multiply Integers	70
Game: What's My Product?	76
2.3 Using Models to Divide Integers	77
Mid-Unit Review	83
2.4 Developing Rules to Divide Integers	84
2.5 Order of Operations with Integers	90
Strategies for Success: Understanding the Problem	94
Unit Review	96
Practice Test	99
Unit Problem: Charity Golf Tournament	100

v

UNIT 3: Operations with Fractions

Launch	102
3.1 Using Models to Multiply Fractions and Whole Numbers	104
3.2 Using Models to Multiply Fractions	110
3.3 Multiplying Fractions	115
3.4 Multiplying Mixed Numbers	121
Game: Spinning Fractions	127
Mid-Unit Review	128
3.5 Dividing Whole Numbers and Fractions	129
3.6 Dividing Fractions	135
3.7 Dividing Mixed Numbers	141
3.8 Solving Problems with Fractions	147
3.9 Order of Operations with Fractions	153
Strategies for Success: Checking and Reflecting	156
Unit Review	158
Practice Test	162
Unit Problem: Sierpinski Triangle	164
Cumulative Review Units 1–3	166

UNIT 4: Measuring Prisms and Cylinders

Launch	168
4.1 Exploring Nets	170
4.2 Creating Objects from Nets	177
4.3 Surface Area of a Right Rectangular Prism	183
4.4 Surface Area of a Right Triangular Prism	188
Mid-Unit Review	194
4.5 Volume of a Right Rectangular Prism	195
Game: Largest Box Problem	201
4.6 Volume of a Right Triangular Prism	202
4.7 Surface Area of a Right Cylinder	209
4.8 Volume of a Right Cylinder	215
Strategies for Success: Choosing the Correct Answer	220
Unit Review	222
Practice Test	226
Unit Problem: Prism Diorama	228
Investigation: Pack It Up!	230

UNIT 5: Percent, Ratio, and Rate

Launch	232
5.1 Relating Fractions, Decimals, and Percents	234
5.2 Calculating Percents	242
5.3 Solving Percent Problems	248
5.4 Sales Tax and Discount	256
Mid-Unit Review	263
5.5 Exploring Ratios	264
5.6 Equivalent Ratios	269
Strategies for Success: Explaining Your Thinking	276
Game: Triple Play	278
5.7 Comparing Ratios	279
5.8 Solving Ratio Problems	287
5.9 Exploring Rates	294
5.10 Comparing Rates	300
Unit Review	307
Practice Test	312
Unit Problem: What Is the Smartest, Fastest, Oldest?	314

UNIT 6: Linear Equations and Graphing

Launch	316
6.1 Solving Equations Using Models	318
6.2 Solving Equations Using Algebra	327
6.3 Solving Equations Involving Fractions	333
6.4 The Distributive Property	338
6.5 Solving Equations Involving the Distributive Property	344
Game: Make the Number	349
Mid-Unit Review	350
6.6 Creating a Table of Values	351
6.7 Graphing Linear Relations	359
Technology: Using Spreadsheets to Graph Linear Relations	366
Strategies for Success: Choosing a Strategy	368
Unit Review	370
Practice Test	374
Unit Problem: Planning a Ski Trip	376
Cumulative Review Units 1–6	378

vii

UNIT 7 — Data Analysis and Probability

Launch	380
7.1 Choosing an Appropriate Graph	382
Technology: Using Spreadsheets to Record and Graph Data	391
7.2 Misrepresenting Data	394
Technology: Using Spreadsheets to Investigate Formatting	403
Mid-Unit Review	406
7.3 Probability of Independent Events	407
Strategies for Success: Doing Your Best on a Test	414
Game: Empty the Rectangles	416
7.4 Solving Problems Involving Independent Events	417
Technology: Using Technology to Investigate Probability	423
Unit Review	424
Practice Test	428
Unit Problem: Promoting Your Cereal	430

UNIT 8 — Geometry

Launch	432
8.1 Sketching Views of Objects	434
Technology: Using a Computer to Draw Views of Objects	440
8.2 Drawing Views of Rotated Objects	441
8.3 Building Objects from Their Views	447
Technology: Using a Computer to Construct Objects from Their Views	454
Mid-Unit Review	455
8.4 Identifying Transformations	456
8.5 Constructing Tessellations	462
Game: Target Tessellations	470
8.6 Identifying Transformations in Tessellations	471
Technology: Using a Computer to Create Tessellations	479
Strategies for Success: Explaining Your Answer	480
Unit Review	482
Practice Test	486
Unit Problem: Creating Tessellating Designs	488
Investigation: A Population Simulation	490
Cumulative Review Units 1–8	492
Answers	496
Illustrated Glossary	538
Index	544
Acknowledgments	547

Welcome to
Pearson Math Makes Sense 8

Math helps you understand your world.

This book will help you improve your problem-solving skills and show you how you can use your math now, and in your future career.

The opening pages of **each unit** are designed to help you prepare for success.

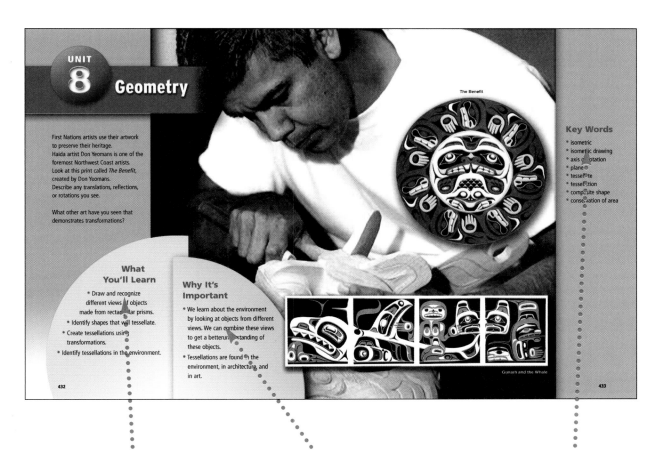

Find out **What You'll Learn** and **Why It's Important**. Check the list of **Key Words**.

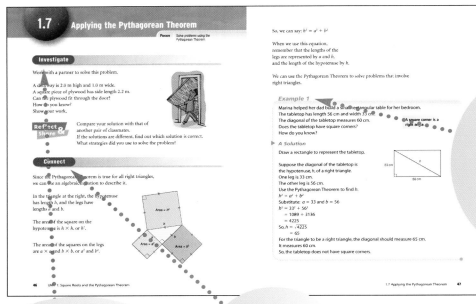

Examples show you how to use the ideas and that there may be different ways to approach the question.

Investigate an idea or problem, usually with a partner, and often using materials.

Connect summarizes the math.

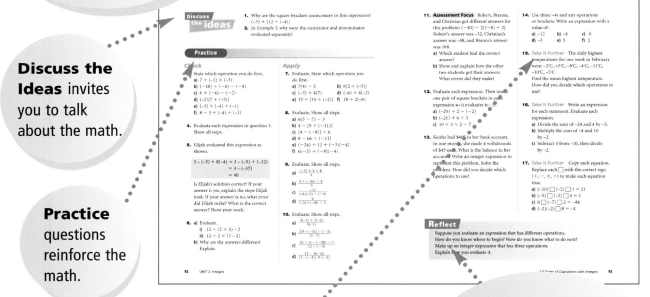

Discuss the Ideas invites you to talk about the math.

Practice questions reinforce the math.

Take It Further questions offer enrichment and extension.

Reflect on the big ideas of the lesson. Think about your learning style and strategies.

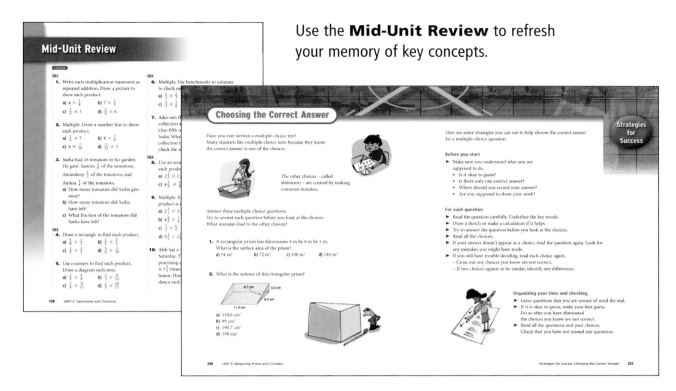

Use the **Mid-Unit Review** to refresh your memory of key concepts.

Strategies for Success sections suggest ways you can help yourself to show your best performance.

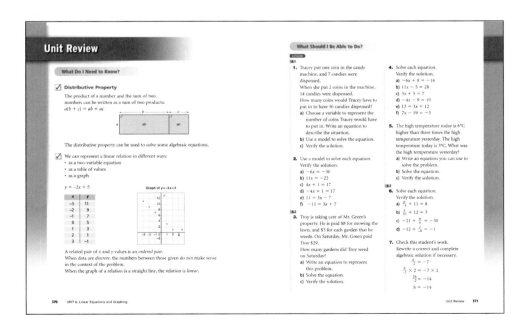

What Do I Need to Know? summarizes key ideas from the unit.

What Should I Be Able to Do? allows you to find out if you are ready to move on. The *Practice and Homework Book* provides additional support.

xi

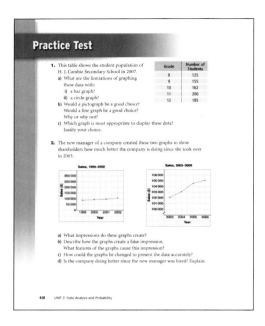

The **Practice Test** models the kind of test your teacher might give.

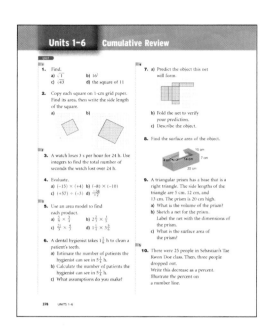

Keep your skills sharp with **Cumulative Review.**

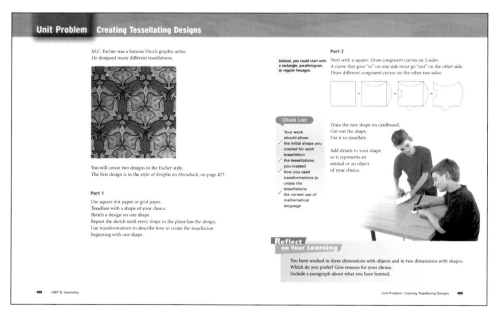

The **Unit Problem** presents problems to solve, or a project to do, using the math of the unit.

Explore some interesting math when you do the **Investigations**.

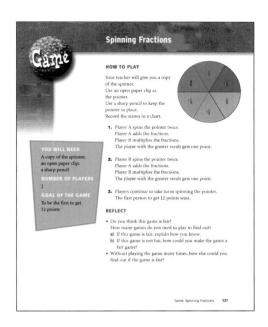

Play a **Game** with your classmates or at home to reinforce your skills.

Icons remind you to use **technology**. Follow the instructions for using a computer or calculator to do math.

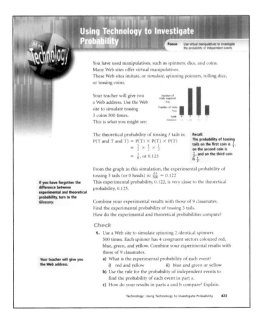

The **Illustrated Glossary** is a dictionary of important math words.

xiii

Investigation

Triangle, Triangle, Triangle

Work with a partner.

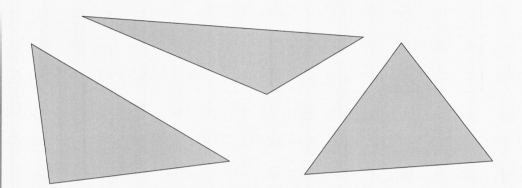

Materials:
- centimetre ruler
- 1-cm grid paper
- scissors

Part 1

➤ On grid paper, draw a large right triangle. Make sure its base is along a grid line and the third vertex is at a grid point.
Estimate the area of the triangle.
➤ On another sheet of grid paper, draw a congruent triangle.
➤ Cut out both triangles.
Place the triangles edge to edge to make a quadrilateral.
➤ Describe the quadrilateral. How many different quadrilaterals can you make? Sketch all your quadrilaterals.
➤ Calculate the area of each quadrilateral you made.
Compare the area of each quadrilateral to the area of the triangle. What do you notice?

Part 2

➤ Draw a large acute triangle with its base along a grid line and the third vertex at a grid point.
Estimate the area of the triangle.
➤ Draw a congruent triangle.
➤ Cut out both triangles.
Then, cut along a grid line on each triangle to make 2 triangles.

➤ Arrange the 4 triangles edge to edge to make a quadrilateral with no gaps or overlaps.
➤ Describe the quadrilateral. How many different quadrilaterals can you make? Sketch all your quadrilaterals.
➤ Calculate the area of each quadrilateral you made.
Calculate the area of the original acute triangle. Compare the areas. What do you notice?

Part 3

Create a summary of your work.
Show all your calculations. Explain your thinking.

Take It Further

➤ Draw an obtuse triangle on grid paper.
Predict its area.
➤ How did you use what you learned about acute and right triangles to make your prediction?
➤ Find a way to check your prediction.

UNIT 1
Square Roots and the Pythagorean Theorem

Some of the greatest builders are also great mathematicians. They use concepts of geometry, measurement, and patterning.

Look at the architecture on these pages.
What aspects of mathematics do you see?

In this unit, you will develop strategies to describe distances that cannot be measured directly.

What You'll Learn

- Determine the square of a number.
- Determine the square root of a perfect square.
- Determine the approximate square root of a non-perfect square.
- Develop and apply the Pythagorean Theorem.

Why It's Important

The Pythagorean Theorem enables us to describe lengths that would be difficult to measure using a ruler. It enables a construction worker to make a square corner without using a protractor.

Key Words

- square number
- perfect square
- square root
- irrational number
- leg
- hypotenuse
- Pythagorean Theorem
- Pythagorean triple

1.1 Square Numbers and Area Models

Focus Relate the area of a square and square numbers.

A rectangle is a quadrilateral with 4 right angles.
A square also has 4 right angles.

A rectangle with base 4 cm and height 1 cm
is the same as a rectangle with base 1 cm and height 4 cm.

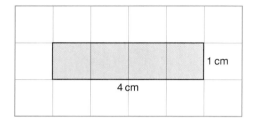

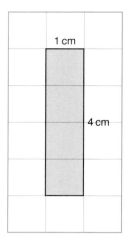

These two rectangles are *congruent*.
Is every square a rectangle?
Is every rectangle a square?

Investigate

Work with a partner.
You will need grid paper and 20 square tiles like this:
Use the tiles to make as many different rectangles
as you can with each area.

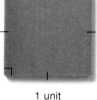

1 unit

- 4 square units
- 6 square units
- 8 square units
- 9 square units
- 12 square units
- 16 square units
- 20 square units

Draw the rectangles on grid paper.
➤ For how many areas above were you able to make a square?
➤ What is the side length of each square you made?
➤ How is the side length of a square related to its area?

Compare your strategies and results with those
of another pair of classmates.
Find two areas greater than 20 square units for
which you could use tiles to make a square.
How do you know you could make a square for
each of these areas?

6 UNIT 1: Square Roots and the Pythagorean Theorem

Connect

When we multiply a number by itself, we *square* the number.
For example: The square of 4 is 4 × 4 = 16.
We write: 4 × 4 = 4^2
So, 4^2 = 4 × 4 = 16
We say: Four squared is sixteen.
16 is a **square number**, or a **perfect square**.

One way to model a square number is to draw a square whose area is equal to the square number.

Example 1

Show that 49 is a square number.
Use a diagram, symbols, and words.

▶ **A Solution**

Draw a square with area 49 square units.
The side length of the square is 7 units.
Then, 49 = 7 × 7 = 7^2
We say: Forty-nine is seven squared.

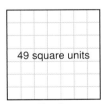

49 square units   7 units

7 units

Example 2

A square picture has area 169 cm².
Find the perimeter of the picture.

▶ **A Solution**

The picture is a square with area 169 cm².
Find the side length of the square:

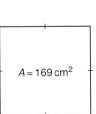

A = 169 cm²

Find a number which, when multiplied by itself, gives 169.
13 × 13 = 169
So, the picture has side length 13 cm.

Perimeter is the distance around the picture.
So, P = 13 cm + 13 cm + 13 cm + 13 cm
       = 52 cm
The perimeter of the picture is 52 cm.

1. Is 1 a square number? How can you tell?
2. Suppose you know the area of a square. How can you find its perimeter?
3. Suppose you know the perimeter of a square. How can you find its area?

Practice

Check

4. Match each square below to its area.

   a)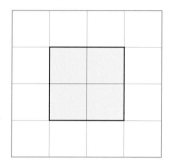

   b)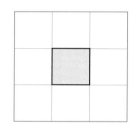

   c)

   i) 1 unit × 1 unit = 1 square unit
   ii) 2 units × 2 units = 4 square units
   iii) 3 units × 3 units = 9 square units

5. Find the area of a square with each side length.
   a) 8 units   b) 10 units   c) 3 units

6. Use square tiles.
   Make as many different rectangles as you can with area 36 square units.
   Draw your rectangles on grid paper.
   Is 36 a perfect square?
   Justify your answer.

Apply

7. Use square tiles.
   Make as many different rectangles as you can with area 28 square units.
   Draw your rectangles on grid paper.
   Is 28 a perfect square?
   Justify your answer.

8. Show that 25 is a square number. Use a diagram, symbols, and words.

9. Show that 12 is not a square number. Use a diagram, symbols, and words.

8 UNIT 1: Square Roots and the Pythagorean Theorem

10. Use a diagram to show that each number below is a square number.
   a) 1   b) 144
   c) 121   d) 900

11. Find the side length of a square with each area.
   a) 100 m²   b) 64 cm²
   c) 81 m²   d) 400 cm²

12. Which of these numbers is a perfect square?
   How do you know?
   a) 10   b) 50
   c) 81   d) 20

13. Use 1-cm grid paper.
   Draw as many different rectangles as you can with area 64 cm².
   Find the base and height of each rectangle.
   Record the results in a table.

Base (cm)	Height (cm)	Perimeter (cm)

   Which rectangle has the least perimeter?
   What can you say about this rectangle?

14. I am a square number.
   The sum of my digits is 9.
   What square numbers might I be?

15. These numbers are not square numbers. Which two consecutive square numbers is each number between?
   Describe the strategy you used.
   a) 12   b) 40
   c) 75   d) 200

16. The floor of a large square room has area 144 m².
   a) Find the length of a side of the room.
   b) How much baseboard is needed to go around the room?
   c) Each piece of baseboard is 2.5 m long.
   How many pieces of baseboard are needed?
   What assumptions do you make?

17. A garden has area 400 m².
   The garden is divided into 16 congruent square plots.
   Sketch a diagram of the garden.
   What is the side length of each plot?

1.1 Square Numbers and Area Models

18. **Assessment Focus** Which whole numbers between 50 and 200 are perfect squares?
Explain how you know.

19. Lee is planning to fence a square kennel for her dog.
Its area must be less than 60 m².
   a) Sketch a diagram of the kennel.
   b) What is the kennel's greatest possible area?
   c) Find the side length of the kennel.
   d) How much fencing is needed?
   e) One metre of fencing costs $10.00. What is the cost of the fencing? What assumptions do you make?

20. **Take It Further** Devon has a piece of poster board 45 cm by 20 cm.
His teacher challenges him to cut the board into parts, then rearrange the parts to form a square.
   a) What is the side length of the square?
   b) What are the fewest cuts Devon could have made? Explain.

21. **Take It Further** The digital root of a number is the result of adding the digits of the number until a single-digit number is reached. For example, to find the digital root of 147:
$1 + 4 + 7 = 12$ and $1 + 2 = 3$
   a) Find the digital roots of the first 15 square numbers.
      What do you notice?
   b) What can you say about the digital root of a square number?
   c) Use your results in part b. Which of these numbers might be square numbers?
      i) 440   ii) 2809   iii) 3008
      iv) 4225   v) 625

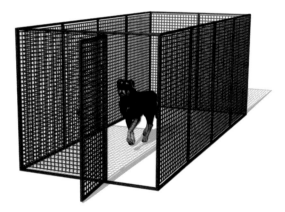

Reflect

Use diagrams to explain why 24 is not a square number but 25 is a square number.

1.2 Squares and Square Roots

Focus Find the squares and square roots of whole numbers.

A factor is a number that divides exactly into another number.
For example, 1, 2, 3, and 6 are factors of 6.
What are the factors of 10?

Investigate

Work with a partner.
Your teacher will give you a copy of this chart.
It shows the factors of each whole number from 1 to 8.
Complete your copy of the chart.

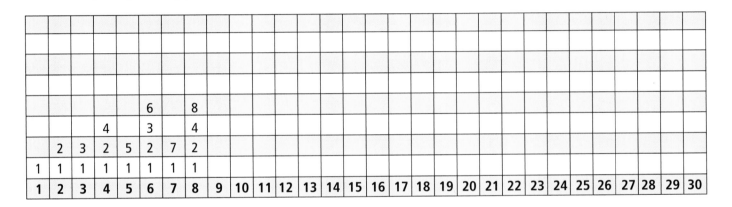

- Which numbers have only two factors?
  What do you notice about these numbers?
- Which numbers have an even number of factors, but more than 2 factors?
- Which numbers have an odd number of factors?

Compare your chart and results with those of another pair of classmates.
Which numbers in the chart are square numbers? How do you know?
What seems to be true about the factors of a square number?

Connect

➤ Here are some ways to tell whether a number is a square number.
If we can find a division sentence for a number so that the quotient is equal to the divisor, the number is a square number.
For example, 16 ÷ 4 = 4, so 16 is a square number.

dividend   divisor   quotient

➤ We can also use factoring.
Factors of a number occur in pairs.
These are the dimensions of a rectangle.

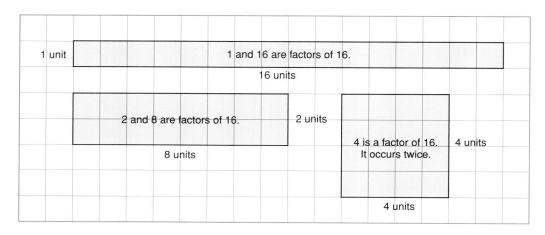

Sixteen has 5 factors: 1, 2, 4, 8, 16

Since there is an odd number of factors, one rectangle is a square.
The square has side length 4 units.
We say that 4 is a **square root** of 16.
We write: $4 = \sqrt{16}$

A factor that occurs twice is only written once in the list of factors.

When a number has an odd number of factors, it is a square number.

When we multiply a number by itself, we square the number.
Squaring and taking the square root are inverse operations. That is, they undo each other.

$4 \times 4 = 16$ $\sqrt{16} = \sqrt{4 \times 4} = \sqrt{4^2}$
so, $4^2 = 16$ $= 4$

12   UNIT 1: Square Roots and the Pythagorean Theorem

Example 1

Find the square of each number.
a) 5   b) 15

▶ A Solution

a) The square of 5 is: $5^2 = 5 \times 5$
$ = 25$

b) The square of 15 is: $15^2 = 15 \times 15$
$ = 225$

Example 2

Find a square root of 64.

▶ A Solution

Use grid paper.
Draw a square with area 64 square units.
The side length of the square is 8 units.
So, $\sqrt{64} = 8$

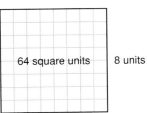

Example 2
Another Solution

Find pairs of factors of 64.
Use division facts.

$64 \div 1 = 64$ 1 and 64 are factors.
$64 \div 2 = 32$ 2 and 32 are factors.
$64 \div 4 = 16$ 4 and 16 are factors.
$64 \div 8 = 8$ 8 is a factor. It occurs twice.

The factors of 64 are: 1, 2, 4, **8**, 16, 32, 64
A square root of 64 is 8, the factor that occurs twice.

Example 3

The factors of 136 are listed in ascending order.
136: 1, 2, 4, 8, 17, 34, 68, 136
Is 136 a square number?
How do you know?

Numbers are in ascending order when they are written from least to greatest number.

▶ *A Solution*

A square number has an odd number of factors.
One hundred thirty-six has 8 factors.
Eight is an even number.
So, 136 is not a square number.

▶ *Example 3*
Another Solution

List the factors of 136 in a column,
in ascending order.
Beside this column, list the factors
in descending order.
Multiply the numbers in each row.

$$1 \times 136 = 136$$
$$2 \times 68 = 136$$
$$4 \times 34 = 136$$
$$8 \times 17 = 136$$
$$17 \times 8 = 136$$
$$34 \times 4 = 136$$
$$68 \times 2 = 136$$
$$136 \times 1 = 136$$

The same factor does not occur
in the same place in both columns.
So, 136 cannot be written as the product
of 2 equal numbers.

So, 136 is not a square number.

Discuss the ideas

1. Squaring a number and taking a square root are inverse operations. What other inverse operations do you know?
2. When the factors of a perfect square are written in order from least to greatest, what do you notice?
3. Why do you think numbers such as 4, 9, 16, … are called perfect squares?
4. Suppose you list the factors of a perfect square. Why is one factor a square root and not the other factors?

Practice

Check

5. Find the square of each number.
   a) 4   b) 6
   c) 2   d) 9

6. Find.
   a) 8^2 b) 3^2
   c) 1^2 d) 7^2

7. Find a square root of each number.
   a) 25   b) 81
   c) 64   d) 169

8. a) Find the square of each number.
   i) 1   ii) 10
   iii) 100   iv) 1000
   b) Use the patterns in part a. Predict the square of each number.
   i) 10 000   ii) 1 000 000

9. a) Use a table like this.

Number = 50	
Factor Pairs	
1	50
2	

List the factor pairs of each number. Which numbers are square numbers? How do you know?
   i) 50   ii) 100
   iii) 144   iv) 85
   b) Find a square root of each square number in part a.

Apply

10. List the factors of each number in ascending order.
Find a square root of each number.
   a) 256   b) 625   c) 121

11. The factors of each number are listed in ascending order.
Which numbers are square numbers? How do you know?
   a) 225: 1, 3, 5, 9, 15, 25, 45, 75, 225
   b) 500: 1, 2, 4, 5, 10, 20, 25, 50, 100, 125, 250, 500
   c) 324: 1, 2, 3, 4, 6, 9, 12, 18, 27, 36, 54, 81, 108, 162, 324
   d) 160: 1, 2, 4, 5, 8, 10, 16, 20, 32, 40, 80, 160

12. a) List the factors of each number in ascending order.
   i) 96   ii) 484
   iii) 240   iv) 152
   v) 441   vi) 54
   b) Which numbers in part a are square numbers?
   How can you tell?

13. Find each square root.
   a) $\sqrt{1}$ b) $\sqrt{49}$
   c) $\sqrt{144}$ d) $\sqrt{9}$
   e) $\sqrt{16}$ f) $\sqrt{100}$
   g) $\sqrt{625}$ h) $\sqrt{225}$

1.2 Squares and Square Roots

14. Find a square root of each number.
   a) 3^2 b) 6^2
   c) 10^2 d) 117^2

15. Find the square of each number.
   a) $\sqrt{4}$ b) $\sqrt{121}$
   c) $\sqrt{225}$ d) $\sqrt{676}$

16. **Assessment Focus** Find each square root. Use a table, list, or diagram to support your answer.
   a) $\sqrt{169}$
   b) $\sqrt{36}$
   c) $\sqrt{196}$

17. Find the number whose square root is 23. Explain your strategy.

18. Use your results from questions 6b and 13d. Explain why squaring and taking the square root are inverse operations.

19. Order from least to greatest.
   a) $\sqrt{36}$, 36, 4, $\sqrt{9}$
   b) $\sqrt{400}$, $\sqrt{100}$, 19, 15
   c) $\sqrt{81}$, 81, $\sqrt{100}$, 11
   d) $\sqrt{49}$, $\sqrt{64}$, $\sqrt{36}$, 9

20. Which perfect squares have square roots between 1 and 20? How do you know?

21. Take It Further
   a) Find the square root of each palindromic number.

   > A palindromic number is a number that reads the same forward and backward.

   i) $\sqrt{121}$
   ii) $\sqrt{12\ 321}$
   iii) $\sqrt{1\ 234\ 321}$
   iv) $\sqrt{123\ 454\ 321}$

   b) Continue the pattern. Write the next 4 palindromic numbers in the pattern and their square roots.

22. Take It Further
   a) Find.
      i) 2^2 ii) 3^2
      iii) 4^2 iv) 5^2
   b) Use the results from part a. Find each sum.
      i) $2^2 + 3^2$ ii) $3^2 + 4^2$
      iii) $2^2 + 4^2$ iv) $3^2 + 5^2$
   c) Which sums in part b are square numbers? What can you say about the sum of two square numbers?

Reflect

Which method of determining square numbers are you most comfortable with? Justify your choice.

1.3 Measuring Line Segments

Focus Use the area of a square to find the length of a line segment.

Investigate

Work with a partner. You will need 1-cm grid paper.
Copy the squares below.
Without using a ruler, find the area and side length of each square.

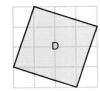

What other squares can you draw on a 4 by 4 grid?
Find the area and side length of each square.
Write all your measurements in a table.

Reflect & Share

How many squares did you draw?
Describe any patterns in your measurements.
How did you find the area and side length of each square?
How did you write the side lengths of squares C and D?

Connect

We can use the properties of a square to find its area or side length.
Area of a square = length × length = (length)^2
When the side length is ℓ, the area is ℓ^2.
When the area is A, the side length is $\sqrt{A}$.
We can calculate the length of any line segment on a grid by thinking of it as the side length of a square.

1.3 Measuring Line Segments

Example 1

Find the length of line segment PQ.

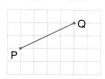

▶ **A Solution**

Use a straightedge and protractor to construct a square on line segment PQ. Then, the length of the line segment is the square root of the area.

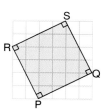

Cut the square into 4 congruent triangles and a smaller square.

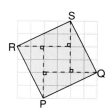

The area of each triangle is: $A = \frac{bh}{2}$

Substitute: $b = 4$ and $h = 2$

$A = \frac{4 \times 2}{2}$

$= 4$

The area of each triangle is 4 square units.
The area of the 4 triangles is:
4×4 square units $= 16$ square units

The area of the small square is: $A = \ell^2$
Substitute: $\ell = 2$
$A = 2^2$
$= 4$
The area of the small square is: 4 square units

The area of PQRS = Area of triangles + Area of small square
$\qquad$ = 16 square units + 4 square units
$\qquad$ = 20 square units
So, the side length of the square, PQ = $\sqrt{20}$ units

Since 20 is not a square number, we cannot write $\sqrt{20}$ as a whole number.
A number like $\sqrt{20}$ is called an **irrational number**.
In Lesson 1.4, you will learn how to find an approximate value
for $\sqrt{20}$ as a decimal.

Example 2 illustrates another way to find the area of a square with its vertices where grid lines meet.

Example 2

a) Find the area of square ABCD.
b) What is the side length AB of the square?

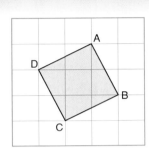

▶ A Solution

a) Draw an enclosing square JKLM.
The area of JKLM = 3^2 square units
= 9 square units

The triangles formed by the enclosing
square are congruent.
Each triangle has area:
$\frac{1}{2}$ × 1 unit × 2 units = 1 square unit

So, the 4 triangles have area:
4 × 1 square unit = 4 square units
The area of ABCD = Area of JKLM − Area of triangles
= 9 square units − 4 square units
= 5 square units

b) So, the side length AB = $\sqrt{5}$ units

1. Is the area of every square a square number?
2. When will the length of a line segment be an irrational number?

Practice

Check

3. Simplify.
  a) 3^2 b) 4^2 c) 7^2
  d) 10^2 e) 6^2 f) 12^2

4. Find.
  a) $\sqrt{1}$ b) $\sqrt{64}$ c) $\sqrt{144}$
  d) $\sqrt{169}$ e) $\sqrt{121}$ f) $\sqrt{625}$

5. The area A of a square is given. Find its side length.
  Which side lengths are whole numbers?
  a) $A = 36$ cm^2 b) $A = 49$ m^2
  c) $A = 95$ cm^2 d) $A = 108$ m^2

6. The side length s of a square is given. Find its area.
  a) $s = 8$ cm   b) $s = \sqrt{44}$ cm
  c) $s = \sqrt{7}$ m   d) $s = 13$ m

Apply

7. Copy each square on grid paper.
  Find its area.
  Then write the side length of the square.
  a)

  b)
  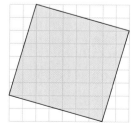

  c)
  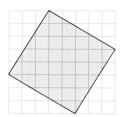

8. Copy each square on grid paper.
  Which square in each pair has the greater area?
  Show your work.
  a)
  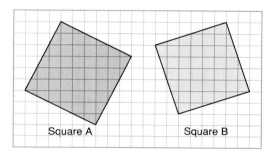
  Square A   Square B

  b)
  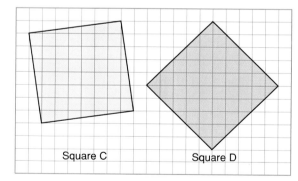
  Square C   Square D

9. Copy each square on grid paper.
  Find its area.
  Order the squares from least to greatest area.
  Then write the side length of each square.

a) 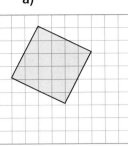   b)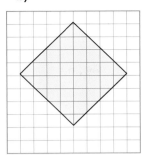

c) 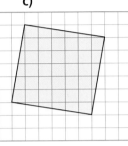   d)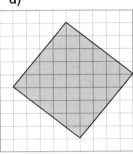

10. Copy each line segment on grid paper. Draw a square on each line segment. Find the area of the square and the length of the line segment.

a)   b)

c)    d)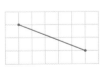

11. Assessment Focus Without measuring, determine which line segment is shorter. Explain how you know.

a) 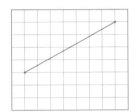   b)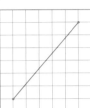

12. Take It Further Plot each pair of points on a coordinate grid. Join the points to form a line segment. Find the length of the line segment.
a) P(1, 3), Q(5, 5)
b) R(−3, 2), S(1, −3)
c) T(−4, 2), U(−1, 3)
d) W(6, 0), X(8, 2)

13. Take It Further On square dot paper, draw a square with area 2 square units. Write to explain how you know the square has this area.

14. Take It Further Use a 7-cm by 7-cm grid. Construct a square with side length 5 cm.
No side can be horizontal or vertical.
Explain your strategy.

Reflect

How is the area of a square related to its side length?
How can we use this relationship to find the length of a line segment?
Include an example in your explanation.

1.4 Estimating Square Roots

Focus Develop strategies for estimating a square root.

You know that a square root of a given number is a number which, when multiplied by itself, results in the given number.
For example, $\sqrt{9} = \sqrt{3 \times 3}$
$= 3$

You also know that the square root of a number is the side length of a square with area that is equal to that number.
For example, $\sqrt{9} = 3$

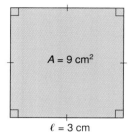

$A = 9 \text{ cm}^2$
$\ell = 3 \text{ cm}$

Investigate

Work with a partner.
Use a copy of the number line below.
Place each square root on the number line to show its approximate value: $\sqrt{2}, \sqrt{5}, \sqrt{11}, \sqrt{18}, \sqrt{24}$
Write each estimated square root as a decimal.
Use grid paper if it helps.

Compare your answers with those of another pair of classmates. What strategies did you use to estimate the square roots? How could you use a calculator to check your estimates?

22   UNIT 1: Square Roots and the Pythagorean Theorem

Connect

Here is one way to estimate the value of $\sqrt{20}$:
- 25 is the square number closest to 20, but greater than 20.
  On grid paper, draw a square with area 25.
  Its side length is: $\sqrt{25} = 5$
- 16 is the square number closest to 20, but less than 20.
  Draw a square with area 16.
  Its side length is: $\sqrt{16} = 4$

Draw the squares so they overlap.

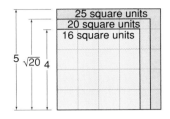

A square with area 20 lies between these two squares.
Its side length is $\sqrt{20}$.
20 is between 16 and 25, but closer to 16.
So, $\sqrt{20}$ is between $\sqrt{16}$ and $\sqrt{25}$, but closer to $\sqrt{16}$.
So, $\sqrt{20}$ is between 4 and 5, but closer to 4.
An estimate of $\sqrt{20}$ is 4.4 to one decimal place.

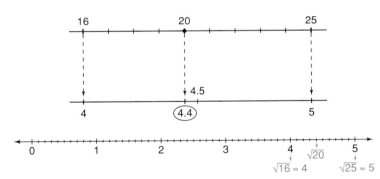

Example 1

Which whole number is $\sqrt{96}$ closer to?
How do you know?

▶ **A Solution**

$81 < 96 < 100$
So, $\sqrt{81} < \sqrt{96} < \sqrt{100}$
$\quad 9 < \sqrt{96} < 10$
$\sqrt{96}$ is between 9 and 10.
96 is closer to 100 than to 81.
So, $\sqrt{96}$ is closer to $\sqrt{100}$, or 10.

1.4 Estimating Square Roots

Example 1
Another Solution

Use number lines.

96 is between 81 and 100, but closer to 100.
So, $\sqrt{96}$ is between $\sqrt{81}$ and $\sqrt{100}$,
but closer to $\sqrt{100}$, or 10.

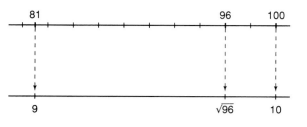

Example 2

A square garden has area 139 m².
a) What are the approximate dimensions of the garden to two decimal places?
b) Net-wire fencing is needed to keep out coyotes. About how much fencing would be needed around the garden?

A Solution

a) Draw a square to represent the garden.

The side length of the square is: $\sqrt{139}$
Estimate:
$121 < 139 < 144$
So, $11 < \sqrt{139} < 12$
With a calculator, use guess and test to refine the estimate.
Try 11.5: $11.5 \times 11.5 = 132.25$ (too small)
Try 11.8: $11.8 \times 11.8 = 139.24$ (too large, but close)
Try 11.78: $11.78 \times 11.78 = 138.7684$ (close)
Try 11.79: $11.79 \times 11.79 = 139.0041$ (very close)

The side length of the garden is 11.79 m, to two decimal places.

b) To find how much fencing is needed, find the perimeter of the garden.
The perimeter of the garden is about:
4×11.79 m $= 47.16$ m

To be sure there is enough fencing, round up.
About 48 m of fencing are needed to go around the garden.

Discuss the ideas

1. Which type of number has an exact square root?
2. Which type of number has an approximate square root?
3. How can you use perfect squares to estimate a square root, such as $\sqrt{8}$?

Practice

Check

4. Find.
   a) $\sqrt{15 \times 15}$
   b) $\sqrt{22 \times 22}$
   c) $\sqrt{3 \times 3}$
   d) $\sqrt{1 \times 1}$

5. Between which two consecutive whole numbers is each square root? How do you know?
   a) $\sqrt{5}$
   b) $\sqrt{11}$
   c) $\sqrt{57}$
   d) $\sqrt{38}$
   e) $\sqrt{171}$
   f) $\sqrt{115}$

6. Copy this diagram on grid paper. Then estimate the value of $\sqrt{7}$ to one decimal place.

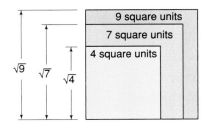

7. Use the number line below.
   a) Which placements are good estimates of the square roots? Explain your reasoning.

   b) Use the number line to estimate the value of each square root that is incorrectly placed.

Apply

8. Use a copy of this number line. Place each square root on the number line to show its approximate value.
   a) $\sqrt{11}$
   b) $\sqrt{40}$
   c) $\sqrt{30}$
   d) $\sqrt{55}$

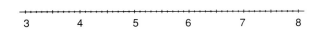

9. In each pair, is the given whole number greater than, less than, or equal to the square root? Justify your answer.
   a) 7, $\sqrt{14}$
   b) 8, $\sqrt{60}$
   c) 11, $\sqrt{121}$
   d) 12, $\sqrt{150}$

10. Which whole number is each square root closer to? How do you know?
    a) $\sqrt{58}$
    b) $\sqrt{70}$
    c) $\sqrt{90}$
    d) $\sqrt{151}$

11. Is each statement true or false? Explain.
   a) $\sqrt{17}$ is between 16 and 18.
   b) $\sqrt{5} + \sqrt{5}$ is equal to $\sqrt{10}$.
   c) $\sqrt{131}$ is between 11 and 12.

12. Use guess and test to estimate each square root to two decimal places. Record each trial.
   a) $\sqrt{23}$ b) $\sqrt{13}$ c) $\sqrt{78}$
   d) $\sqrt{135}$ e) $\sqrt{62}$ f) $\sqrt{45}$

13. Find the approximate side length of the square with each area.
   Give each answer to one decimal place.
   a) 92 cm²   b) 430 m²
   c) 150 cm²   d) 29 m²

14. Which estimates are good estimates of the square roots?
   Explain your reasoning.
   a) $\sqrt{17}$ is about 8.50.
   b) $\sqrt{20}$ is about 4.30.
   c) $\sqrt{8}$ is about 2.83.
   d) $\sqrt{34}$ is about 5.83.

15. **Assessment Focus** A student uses a square canvas for her painting. The canvas has area 5 m². She wants to frame her artwork.
   a) What are the dimensions of the square frame to two decimal places?
   b) The framing can be purchased in 5-m or 10-m lengths. Which length of framing should she purchase? Justify your choice.

16. A square lawn is to be reseeded. The lawn has area 152 m².
   a) What are the approximate dimensions of the lawn to two decimal places?
   b) A barrier of yellow tape is placed around the lawn to keep people off. About how much tape is needed?

17. Which is the closer estimate of $\sqrt{54}$: 7.34 or 7.35?
   How did you find out?

18. Most classrooms are rectangles. Measure the dimensions of your classroom. Calculate its area. Suppose your classroom was a square with the same area. What would its dimensions be?

19. **Take It Further** A square carpet covers 75% of the area of a floor. The floor is 8 m by 8 m.

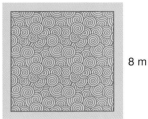

a) What are the dimensions of the carpet? Give your answer to two decimal places.
b) What area of the floor is not covered by the carpet?

20. **Take It Further** Is the product of two perfect squares always, sometimes, or never a perfect square? Investigate to find out. Write about your findings.

21. **Take It Further** An approximate square root of a whole number is 7.67. Is the whole number closer to 49 or 64? How do you know?

22. **Take It Further** Write five numbers whose square roots are between 9 and 10. Explain your strategy.

23. **Take It Further** Simplify each expression. Give your answer to two decimal places when necessary.
a) $\sqrt{81} + \sqrt{16}$
b) $\sqrt{81 + 16}$
c) $\sqrt{\sqrt{81} + 16}$
d) $\sqrt{81 + \sqrt{16}}$
e) $\sqrt{\sqrt{81} + \sqrt{16}}$

24. a) Estimate each square root to two decimal places.
   i) $\sqrt{2}$ ii) $\sqrt{200}$ iii) $\sqrt{20\,000}$
b) Look at your results in part a. What patterns do you see?
c) Use the patterns in part b to estimate.

Reflect

What is your favourite method for estimating a square root of a number that is not a perfect square? Use an example to explain how you would use your method.

Fitting In

HOW TO PLAY

Your teacher will give you 3 sheets of game cards.
Cut out the 54 cards.

1. Place the 1, 5, and 9 cards on the table.
   Spread them out so there is room for
   several cards between them.
   Shuffle the remaining cards.
   Give each player 6 cards.

2. All cards laid on the table must be arranged
   from least to greatest.
   Take turns to place a card so it touches another card
   on the table.
   - It can be placed to the right of the card if its value
     is greater.
   - It can be placed to the left of the card if its value is less.
   - It can be placed on top of the card if its value is equal.
   - However, it cannot be placed between two cards that
     are already touching.

In this example, the $\sqrt{16}$ card cannot be placed because
the 3.5 and the 5 cards are touching.
The player cannot play that card in this round.

YOU WILL NEED

1 set of *Fitting In* game cards; scissors

NUMBER OF PLAYERS

2 to 4

GOAL OF THE GAME

To get the lowest score

What other games could you play with these cards? Try out your ideas.

 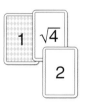 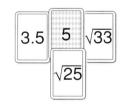 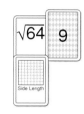

3. Place as many of your cards as you can. When no player
   can place any more cards, the round is over.
   Your score is the number of cards left in your hand.
   At the end of five rounds, the player with the lowest
   score wins.

Investigating Square Roots with a Calculator

Focus Use a calculator to investigate square roots.

 We can use a calculator to calculate a square root.

➤ Use the square root key to find the square root of 16.
A square root of 16 is 4.
Check by multiplying or by squaring: $4 \times 4 = 16$

➤ Many square roots are not whole numbers.
Use a calculator to find a square root of 20.
4.472135955 should be displayed.
A square root of 20 is 4.5 to one decimal place.

Investigate to compare what happens when we use a calculator with a key versus a 4-function calculator.

➤ Find $\sqrt{20} \times \sqrt{20}$ using each calculator.
Record what you see in the display each time.
Which display is accurate? How do you know?

➤ Check what happens when you enter $4.472135955 \times 4.472135955$ into each calculator.
Suppose you multiplied using pencil and paper.
Would you expect a whole number or a decimal? Explain.

➤ $\sqrt{20}$ cannot be described exactly by a decimal.
The decimal for $\sqrt{20}$ never repeats and never terminates.
We can write $\sqrt{20}$ with different levels of accuracy.
For example, $\sqrt{20}$ is 4.5 to one decimal place, 4.47 to two decimal places, and 4.472 to three decimal places.

The calculator rounds or truncates the decimal for $\sqrt{20}$. So, the number in the display is only an approximate square root.

Check

Find each square root.
Which square roots are approximate?
Justify your answer.

a) $\sqrt{441}$ b) $\sqrt{19}$ c) $\sqrt{63}$ d) $\sqrt{529}$

Mid-Unit Review

LESSON

1.1
1. Which numbers below are perfect squares? Draw diagrams to support your answers.
   a) 15   b) 26   c) 65   d) 100

1.2
2. Find a square root of each number.
   a) 16   b) 49   c) 196   d) 400

3. Find.
   a) 11^2 b) $\sqrt{64}$ c) $\sqrt{169}$ d) $\sqrt{225}$

1.1
1.2
4. Copy each square onto 1-cm grid paper.
   i) Find the area of each square.
   ii) Write the side length of each square as a square root.
   a)   b)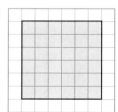

5. List the factors of each number below in order from least to greatest. Which of the numbers are square numbers? How do you know? For each square number below, write a square root.
   a) 216   b) 364   c) 729

1.3
6. If you know a square number, how can you find its square root? Use diagrams, symbols, and words.

7. a) The area of a square is 24 cm². What is its side length? Why is the side length not a whole number?
   b) The side length of a square is 9 cm. What is its area?

1.3
1.4
8. Copy this square onto 1-cm grid paper.

   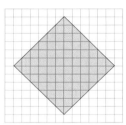

   a) What is the area of the square?
   b) Write the side length of the square as a square root.
   c) Estimate the side length to one decimal place.

9. Find.
   a) $\sqrt{12 \times 12}$ b) $\sqrt{34 \times 34}$

10. Between which two consecutive whole numbers does each square root lie? How do you know? Sketch a number line to show your answers.
    a) $\sqrt{3}$ b) $\sqrt{65}$ c) $\sqrt{72}$ d) $\sqrt{50}$

11. Use guess and test to estimate each square root to two decimal places. Record each trial.
    a) $\sqrt{17}$ b) $\sqrt{108}$ c) $\sqrt{33}$ d) $\sqrt{79}$

1.5 The Pythagorean Theorem

Focus Discover a relationship among the side lengths of a right triangle.

We can use the properties of a right triangle to find the length of a line segment. A right triangle has two sides that form the right angle. The third side of the right triangle is called the **hypotenuse**. The two shorter sides are called the **legs**.

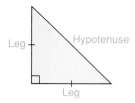
Isosceles right triangle

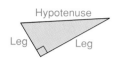
Scalene right triangle

Investigate

Work on your own.
You will need grid paper, centimetre cubes, and a protractor.

➤ Copy line segment AB.
Draw right triangle ABC that has segment AB as its hypotenuse.
Draw a square on each side of △ABC.
Find the area and side length of each square.

➤ Draw 3 different right triangles, with a square on each side.
Find the area and side length of each square.
Record your results in a table.

> See page 18 if you have forgotten how to find the area of a square on a line segment.

	Area of Square on Leg 1	Length of Leg 1	Area of Square on Leg 2	Length of Leg 2	Area of Square on Hypotenuse	Length of Hypotenuse
Triangle ABC						
Triangle 1						
Triangle 2						
Triangle 3						

Compare your results with those of another classmate.
What relationship do you see among the areas of the squares on the sides of a right triangle? How could this relationship help you find the length of a side of a right triangle?

Connect

Here is a right triangle, with a square drawn on each side.

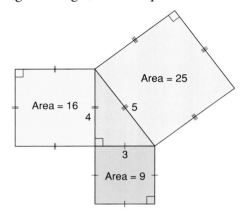

The area of the square on the hypotenuse is 25.
The areas of the squares on the legs are 9 and 16.

The Pythagorean Theorem is named for the Greek mathematician, Pythagoras.

Notice that: $25 = 9 + 16$
A similar relationship is true for all right triangles.

In a right triangle, the area of the square on the hypotenuse is equal to the sum of the areas of the squares on the legs.

This relationship is called the **Pythagorean Theorem**.

We can use this relationship to find the length of any side of a right triangle, when we know the lengths of the other two sides.

Example 1

Find the length of the hypotenuse.
Give the length to one decimal place.

▶ A Solution

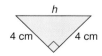

Label the hypotenuse h.
The area of the square on the hypotenuse is h^2.
The areas of the squares on the legs are
$4 \times 4 = 16$ and $4 \times 4 = 16$.
So, $h^2 = 16 + 16$
$ = 32$

The area of the square on the hypotenuse is 32.
So, the side length of the square is: $h = \sqrt{32}$
Use a calculator.
$h \doteq 5.6569$
So, the hypotenuse is 5.7 cm to one decimal place.

Example 2

Find the unknown length to one decimal place.

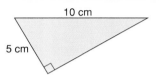

▶ **A Solution**

Label the leg g.
The area of the square on the hypotenuse is $10 \times 10 = 100$.
The areas of the squares on the legs are
g^2 and $5 \times 5 = 25$.
So, $100 = g^2 + 25$
To solve this equation, subtract 25 from each side.
$100 - 25 = g^2 + 25 - 25$
$75 = g^2$
The area of the square on the leg is 75.
So, the side length of the square is: $g = \sqrt{75}$
Use a calculator.
$g \doteq 8.66025$
So, the leg is 8.7 cm to one decimal place.

1. How can you identify the hypotenuse of a right triangle?
2. How can you use the Pythagorean Theorem to find the length of the diagonal in a rectangle?

Practice

Check

3. Find the area of the indicated square.
   a)
   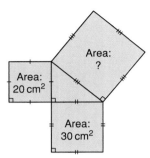

   b)
   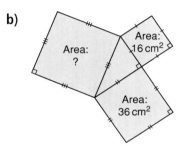

4. Find the area of the indicated square.
   a)
   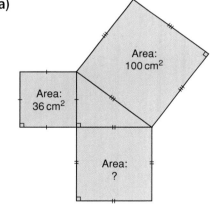

   b)
   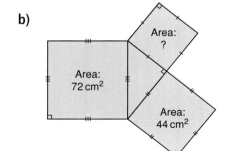

5. Find the length of each hypotenuse. Give your answers to one decimal place where needed.

6. Find the length of each leg labelled ℓ. Give your answers to one decimal place where needed.

   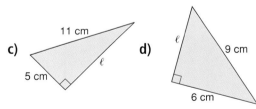

Apply

7. Find the length of each side labelled with a variable. Give your answers to one decimal place where needed.

   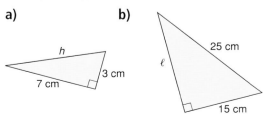

c)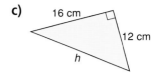

8. Find the length of the diagonal, d, in each rectangle.
   Give your answers to two decimal places where needed.

   a) b) c)
   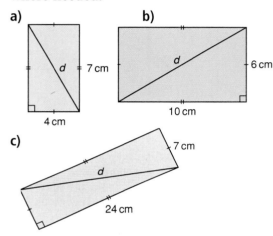

9. Find the length of the diagonal, d, in each rectangle.
   What patterns do you notice?
   Write to explain.
   Use your patterns to draw the next rectangle in the pattern.

   a) b) c)
   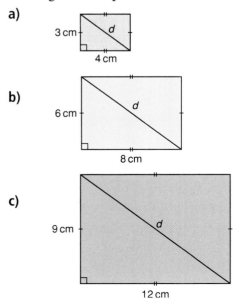

10. Suppose you are given the side lengths of a right triangle.
    Which length is the length of the hypotenuse?
    Explain how you know.

11. Use the rectangle on the right. Explain why the diagonals of a rectangle have the same length.

    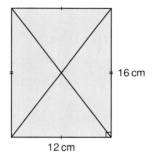

12. **Assessment Focus** The hypotenuse of a right triangle is $\sqrt{18}$ units.
    What are the lengths of the legs of the triangle?
    How many different answers can you find?
    Sketch a triangle for each answer.
    Explain your strategies.

13. Find the length of each side labelled with a variable.

    a) b) c)
    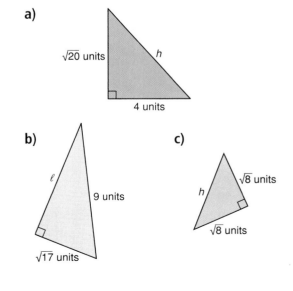

1.5 The Pythagorean Theorem   35

14. Use what you know about the Pythagorean Theorem. On grid paper, draw a line segment with each length. Explain how you did it.
a) $\sqrt{5}$ b) $\sqrt{10}$
c) $\sqrt{13}$ d) $\sqrt{17}$

15. Take It Further The length of the hypotenuse of a right triangle is 15 cm. The lengths of the legs are whole numbers of centimetres.
Find the sum of the areas of the squares on the legs of the triangle.
What are the lengths of the legs? Show your work.

16. Take It Further An artist designed this logo. It is a right triangle with a semicircle drawn on each side of the triangle. Calculate the area of each semicircle. What do you notice? Explain.

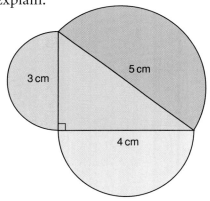

17. Take It Further Use grid paper. Draw a right triangle with a hypotenuse with each length.
a) $\sqrt{20}$ units b) $\sqrt{89}$ units c) $\sqrt{52}$ units

18. Take It Further Your teacher will give you a copy of part of the Wheel of Theodorus.

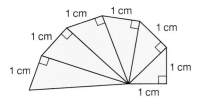

Theodorus was born about 100 years after Pythagoras. Theodorus used right triangles to create a spiral. Today the spiral is known as the Wheel of Theodorus.

a) Find the length of the hypotenuse in each right triangle.
Label each hypotenuse with its length as a square root.
What pattern do you see?
b) Use a calculator. Write the value of each square root in part a to one decimal place.
c) Use a ruler. Measure the length of each hypotenuse to one decimal place.
d) Compare your answers in parts b and c. What do you notice?

Reflect

Suppose your classmate missed today's lesson. Use an example to show your classmate how to find the length of the third side of a right triangle when you know the lengths of the other two sides.

36 UNIT 1: Square Roots and the Pythagorean Theorem

Verifying the Pythagorean Theorem

Focus Use a computer to investigate the Pythagorean relationship.

Geometry software can be used to create and transform shapes. Use available geometry software.

Open a new sketch.
Display a coordinate grid.

> Should you need help at any time, use the software's Help menu.

➤ Construct points A(0, 0), B(0, 4) and C(3,0).
Join the points to form right △ABC.

➤ Construct a square on each side of the triangle.
To form a square on BC, rotate line segment BC 90° clockwise about C,
then 90° counterclockwise about B.

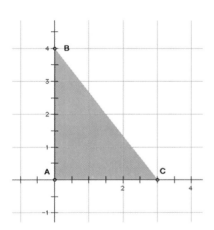

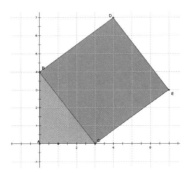

➤ Join the points at the ends of the rotated segments to form a square.
Label the vertices of the square.
This is what you should see on the screen.

Technology: Verifying the Pythagorean Theorem

➤ Repeat the previous step, with appropriate rotations, to form squares on AB and AC.

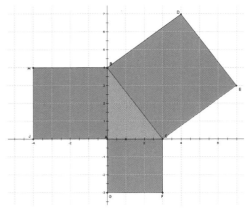

➤ Use the software to find the area of each square. What relationship is shown?

➤ Drag a vertex of the triangle and observe what happens to the area measurements of the squares.
How does the geometry software verify the Pythagorean Theorem?

Use the software to investigate "what if" questions.

➤ What if the triangle was an acute triangle? Is the relationship still true? Explain.

➤ What if the triangle was an obtuse triangle? Is the relationship still true? Explain.

1.6 Exploring the Pythagorean Theorem

Focus Use the Pythagorean Theorem to identify right triangles.

Look at these triangles.
Which triangle is a right triangle? How do you know?
Which triangle is an obtuse triangle? An acute triangle?
How did you decide?

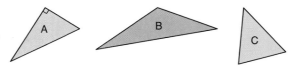

Investigate

Work on your own.
You will need grid paper and a protractor.

➤ Draw an obtuse triangle.
Draw a square on each side.
Find the area and side length of each square.

➤ Draw an acute triangle.
Draw a square on each side.
Find the area and side length of each square.

➤ Record your results in a table.

	Area of Square on Shortest Side	Length of Shortest Side	Area of Square on Second Side	Length of Second Side	Area of Square on Longest Side	Length of Longest Side
Obtuse Triangle						
Acute Triangle						

Compare your results with those of 3 other classmates.
What relationship do you see among the areas of the squares on the sides of an obtuse triangle?
What relationship do you see among the areas of the squares on the sides of an acute triangle?
How could these relationships help you identify the type of triangle when you know the lengths of its sides?

Connect

Here are an acute triangle, a right triangle, and an obtuse triangle, with squares drawn on the sides of each triangle.

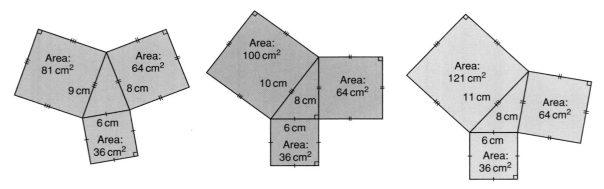

For each triangle, compare the sum of the areas of the two smaller squares to the area of the largest square.

Triangle	Sum of Areas of Two Smaller Squares	Area of Largest Square
Acute	36 + 64 = 100	81
Right	36 + 64 = 100	100
Obtuse	36 + 64 = 100	121

Notice that the Pythagorean Theorem is true for the right triangle only.

We can use these results to identify whether a triangle is a right triangle.
If Area of square A + Area of square B
= Area of square C,
then the triangle is a right triangle.

If Area of square A + Area of square B
≠ Area of square C,
then the triangle is not a right triangle.

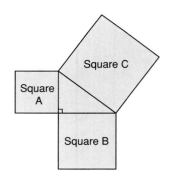

The symbol ≠ means "does not equal."

40   UNIT 1: Square Roots and the Pythagorean Theorem

Example 1

Determine whether each triangle with the given side lengths is a right triangle.
a) 6 cm, 6 cm, 9 cm   b) 7 cm, 24 cm, 25 cm

▶ **A Solution**

a) Sketch a triangle with a square on each side.
The square with the longest side is square C.
Area of square A = 6 cm × 6 cm
   = 36 cm²
Area of square B = 6 cm × 6 cm
   = 36 cm²
Area of square C = 9 cm × 9 cm
   = 81 cm²
Area of square A + Area of square B = 36 cm² + 36 cm²
   = 72 cm²
Since 72 cm² ≠ 81 cm², the triangle is not a right triangle.

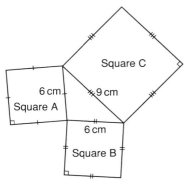

b) Sketch a triangle with a square on each side.
The square with the longest side is square C.
Area of square A = 7 cm × 7 cm
   = 49 cm²
Area of square B = 24 cm × 24 cm
   = 576 cm²
Area of square C = 25 cm × 25 cm
   = 625 cm²
Area of square A + Area of square B = 49 cm² + 576 cm²
   = 625 cm²
Since 625 cm² = 625 cm², the triangle is a right triangle.

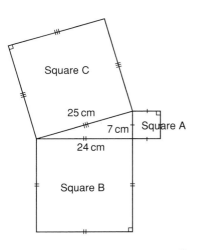

A set of 3 whole numbers that satisfies the Pythagorean Theorem is called a **Pythagorean triple**.
For example, 3-4-5 is a Pythagorean triple because $3^2 + 4^2 = 5^2$.

From part b of *Example 1*, a triangle with side lengths 7 cm, 24 cm, and 25 cm is a right triangle.
So, 7-24-25 is a Pythagorean triple.

Example 2

Which of these sets of numbers is a Pythagorean triple?
How do you know?
a) 8, 15, 18   b) 11, 60, 61

▶ **A Solution**

Suppose each set of numbers represents the side lengths of a triangle.
When the set of numbers satisfies the Pythagorean Theorem,
the set is a Pythagorean triple.

a) Check:
Does $8^2 + 15^2 = 18^2$?
L.S. = $8^2 + 15^2$ R.S. = 18^2
     = 64 + 225   = 324
     = 289
Since 289 ≠ 324, 8-15-18 is not a Pythagorean triple.

b) Check:
Does $11^2 + 60^2 = 61^2$?
L.S. = $11^2 + 60^2$ R.S. = 61^2
     = 121 + 3600   = 3721
     = 3721
Since 3721 = 3721, 11-60-61 is a Pythagorean triple.

Discuss the ideas

1. Suppose a square has been drawn on each side of a triangle.
   How can you tell whether the triangle is obtuse?
   How can you tell whether the triangle is acute?
2. Can a right triangle have an hypotenuse of length $\sqrt{65}$ units?

History

"Numbers Rule the Universe!" That was the belief held by a group of mathematicians called the Brotherhood of Pythagoreans. Their power and influence became so strong that fearful politicians forced them to disband. Nevertheless, they continued to meet in secret and teach Pythagoras' ideas.

UNIT 1: Square Roots and the Pythagorean Theorem

Practice

Check

3. The area of the square on each side of a triangle is given. Is the triangle a right triangle? How do you know?

   a)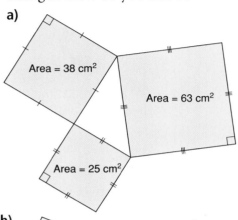

   b)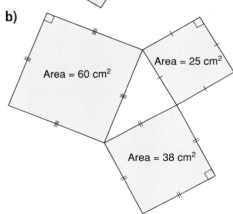

   c)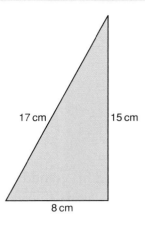

4. Which of these triangles appears to be a right triangle? Determine whether each triangle is a right triangle. Justify your answers.

   a) 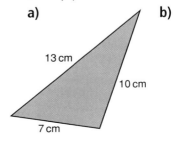   b)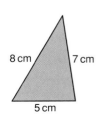

5. Look at the triangle below. Can the Pythagorean Theorem be used to find the length of the side labelled with a variable? Why or why not?

   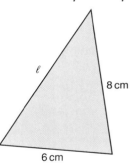

Apply

6. Determine whether a triangle with each set of side lengths is a right triangle. Justify your answers.
   a) 16 cm, 30 cm, 34 cm
   b) 8 cm, 10 cm, 12 cm
   c) 20 m, 25 m, 15 m
   d) 28 m, 53 m, 45 m
   e) 17 mm, 14 mm, 5 mm
   f) 30 mm, 9 mm, 25 mm
   g) 9 cm, 9 cm, 15 cm
   h) 10 cm, 26 cm, 24 cm

1.6 Exploring the Pythagorean Theorem

7. Which sets of numbers below are Pythagorean triples? How did you decide?
   a) 16, 30, 34
   b) 6, 8, 9
   c) 15, 39, 36
   d) 16, 65, 63
   e) 9, 30, 35
   f) 40, 42, 58

8. An elder and his granddaughter, Kashala, are laying a plywood floor in a cabin.
   The floor is rectangular, with side lengths 9 m and 12 m.
   Kashala measures the diagonal of the floor as 15 m.
   Is the angle between the two sides a right angle?
   Justify your answer.

9. A triangle has side lengths 6 cm, 7 cm, and $\sqrt{13}$ cm.
   Is this triangle a right triangle?
   Do these side lengths form a Pythagorean triple? Explain.

10. **Assessment Focus**
    May Lin uses a ruler and compass to construct a triangle with side lengths 3 cm, 5 cm, and 7 cm. Before May Lin constructs the triangle, how can she tell if the triangle will be a right triangle? Explain.

11. Look at the Pythagorean triples below.
    3, 4, 5   6, 8, 10
    12, 16, 20   15, 20, 25
    21, 28, 35
    a) Each set of numbers represents the side lengths of a right triangle. What are the lengths of the legs? What is the length of the hypotenuse?
    b) Describe any pattern you see in the Pythagorean triples.
    c) Use a pattern similar to the one you found in part b. Generate 4 more Pythagorean triples from the triple 5, 12, 13.

12. Two numbers in a Pythagorean triple are given.
    Find the third number.
    How did you find out?
    a) 14, 48, ☐   b) 32, 24, ☐
    c) 12, 37, ☐   d) 20, 101, ☐

13. In Ancient Egypt, the Nile River overflowed every year and destroyed property boundaries. Because the land plots were rectangular, the Egyptians needed a way to mark a right angle. The Egyptians tied 12 evenly spaced knots along a piece of rope and made a triangle. Explain how this rope could have been used to mark a right angle.

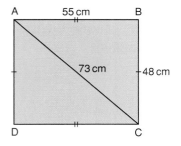

14. Is quadrilateral ABCD a rectangle? Justify your answer.

15. **Take It Further** The perimeter of a right triangle is 90 m.
    The length of the longest side of the triangle is 41 m.
    Find the lengths of the other two sides.
    How did you find out?

16. **Take It Further** Use your data and those of your classmates from *Investigate*, on page 39.
    a) Use the results for the obtuse triangles. How does the area of the square on the longest side compare to the sum of the areas of the squares on the other two sides?
    b) Use the results for the acute triangles. How does the area of the square on the longest side compare to the sum of the areas of the squares on the other two sides?
    c) Use the patterns in parts a and b to classify all the triangles in question 6.

17. **Take It Further** You can use expressions to generate the numbers in a Pythagorean triple.
    Choose a number, then choose a greater number.
    Use these expressions to find the numbers in a Pythagorean triple:
    - 2(lesser number)(greater number)
    - (greater number)^2 − (lesser number)^2
    - (greater number)^2 + (lesser number)^2

    If a spreadsheet is available, enter the formulas. Change the numbers you start with to generate 15 Pythagorean triples.

Reflect

When you know the side lengths of a triangle, how can you tell whether it is a right triangle?
What other condition must be satisfied for the numbers to be a Pythagorean triple?
Use examples in your explanation.

1.7 Applying the Pythagorean Theorem

Focus Solve problems using the Pythagorean Theorem.

Investigate

Work with a partner to solve this problem.

A doorway is 2.0 m high and 1.0 m wide.
A square piece of plywood has side length 2.2 m.
Can the plywood fit through the door?
How do you know?
Show your work.

Compare your solution with that of another pair of classmates.
If the solutions are different, find out which solution is correct.
What strategies did you use to solve the problem?

Connect

Since the Pythagorean Theorem is true for all right triangles, we can use an algebraic equation to describe it.

In the triangle at the right, the hypotenuse has length h, and the legs have lengths a and b.

The area of the square on the hypotenuse is $h \times h$, or h^2.

The areas of the squares on the legs are $a \times a$ and $b \times b$, or a^2 and b^2.

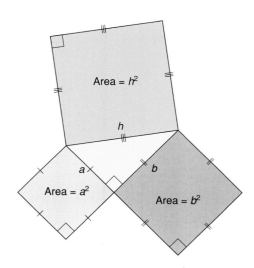

46   UNIT 1: Square Roots and the Pythagorean Theorem

So, we can say: $h^2 = a^2 + b^2$

When we use this equation, remember that the lengths of the legs are represented by a and b, and the length of the hypotenuse by h.

We can use the Pythagorean Theorem to solve problems that involve right triangles.

Example 1

Marina helped her dad build a small rectangular table for her bedroom. The tabletop has length 56 cm and width 33 cm. The diagonal of the tabletop measures 60 cm. Does the tabletop have square corners? How do you know?

A square corner is a right angle.

▶ A Solution

Draw a rectangle to represent the tabletop.

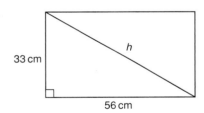

Suppose the diagonal of the tabletop is the hypotenuse, h, of a right triangle.
One leg is 33 cm.
The other leg is 56 cm.
Use the Pythagorean Theorem to find h.
$h^2 = a^2 + b^2$
Substitute: $a = 33$ and $b = 56$
$h^2 = 33^2 + 56^2$
$ = 1089 + 3136$
$ = 4225$
So, $h = \sqrt{4225}$
$ = 65$

For the triangle to be a right triangle, the diagonal should measure 65 cm.
It measures 60 cm.
So, the tabletop does not have square corners.

Example 2

A ramp is used to load a snow machine onto a trailer.
The ramp has horizontal length 168 cm
and sloping length 175 cm.
The side view is a right triangle.
How high is the ramp?

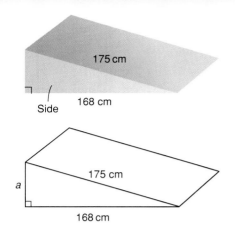

▶ **A Solution**

The side face of the ramp is a
right triangle with hypotenuse 175 cm.
One leg is 168 cm.
The other leg is the height.
Label it a.

Use the Pythagorean Theorem.
$h^2 = a^2 + b^2$
Substitute: $h = 175$ and $b = 168$
$175^2 = a^2 + 168^2$ Use a calculator.
$30\,625 = a^2 + 28\,224$
Subtract 28 224 from each side to isolate a^2.
$30\,625 - 28\,224 = a^2 + 28\,224 - 28\,224$
$2401 = a^2$
The area of the square with side length a is 2401 cm².
$a = \sqrt{2401}$ Use a calculator.
$ = 49$
The ramp is 49 cm high.

Discuss the ideas

1. What do you need to know to apply the Pythagorean Theorem to calculate a distance?
2. Does it matter which two sides of the triangle you know?
3. When you use the Pythagorean Theorem, how can you tell whether to add or subtract the areas?

Practice

Check

4. Find the length of each hypotenuse. Give your answers to one decimal place where needed.

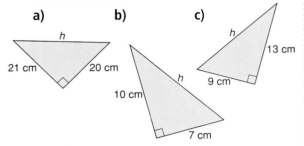

a) 21 cm, 20 cm, h
b) 10 cm, 7 cm, h
c) 13 cm, 9 cm, h

5. Find the length of each leg labelled with a variable. Give your answers to one decimal place where needed.

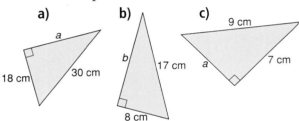

a) 18 cm, 30 cm, a
b) 17 cm, 8 cm, b
c) 9 cm, 7 cm, a

6. A 5-m ladder leans against a house. It is 3 m from the base of the wall. How high does the ladder reach?

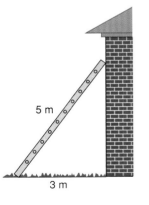

Apply

7. As part of a design for a book cover, Brandon constructed a right triangle with sides 10 cm and 24 cm.
   a) How long is the third side?
   b) Why are two answers possible to part a?

8. Find the length of each line segment. Give your answers to one decimal place.
   a)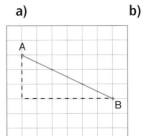
   b)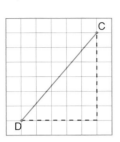

9. Alyssa has made a picture frame for the painting she just finished. The frame is 60 cm long and 25 cm wide. To check that the frame has square corners, Alyssa measures a diagonal. How long should the diagonal be? Sketch a diagram to illustrate your answer.

10. A boat is 35 m due south of a dock. Another boat is 84 m due east of the dock. How far apart are the boats?

11. A baseball diamond is a square with side length about 27 m. The player throws the ball from second base to home plate. How far did the player throw the ball? Give your answer to two decimal places.

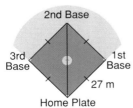

1.7 Applying the Pythagorean Theorem 49

12. Copy each diagram on grid paper. Explain how each diagram can be used to illustrate the Pythagorean Theorem.

a)
b)

13. The size of a TV set is described by the length of a diagonal of the screen. One TV is labelled 27 inches, which is about 70 cm. The screen is 40 cm high. What is the width of the screen? Give your answer to one decimal place. Draw a diagram to illustrate your answer.

14. **Assessment Focus** Look at the grid. Without measuring, find another point that is the same distance from A as B is. Explain your strategy. Show your work.

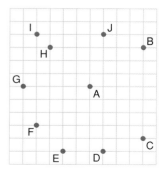

15. A line segment joins points P(−1, 2) and Q(4, 5). Calculate the length of line segment PQ. Give your answer to one decimal place.

16. Joanna usually uses the sidewalk to walk home from school. Today she is late, so she cuts through the field. How much shorter is Joanna's shortcut?

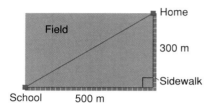

17. Felix and Travis started at the same point in the campground. They walked in different directions. Felix walked 650 m and Travis walked 720 m. They were then 970 m apart. Were they travelling along paths that were at right angles to each other? Explain your thinking.

18. A plane takes off from a local airport. It travels due north at a speed of 400 km/h. The wind blows the plane due east at a speed of 50 km/h. How far is the plane from the airport after 1 h? Give your answer to one decimal place.

19. Take It Further A Powwow is a traditional practice in some First Nations cultures. Women dancers have small cone-shaped tin jingles sewn onto their dresses, one for each day of the year. A typical jingle has a triangular cross-section. Suppose the triangle has base 6 cm and height 7 cm. Use the diagram to help you find the slant height, s, of the jingle. Give your answer to one decimal place.

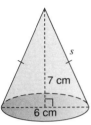

20. Take It Further What is the length of the diagonal in this rectangular prism?

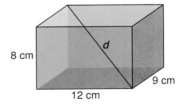

21. Take It Further Rashad is flying a kite. His hand is 1.3 m above the ground. Use the picture below. How high is the kite above the ground?

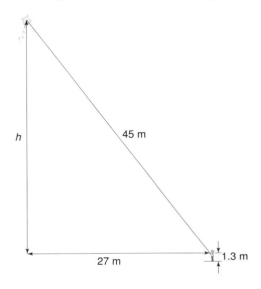

22. Take It Further Two cars meet at an intersection.
One travels north at an average speed of 80 km/h.
The other travels east at an average speed of 55 km/h.
How far apart are the cars after 3 h?
Give your answer to one decimal place.

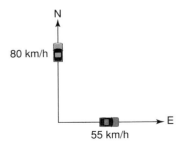

Reflect

When can you use the Pythagorean Theorem to solve a problem?
Use examples in your explanation.

Getting Unstuck

Have you ever got stuck trying to solve a math problem? Almost everybody has at one time or another.

To be successful at math, you should know what to do when you get stuck trying to solve a problem.

Consider this problem:

Use square dot paper.
Draw a line segment with length $\sqrt{5}$ cm.

Suppose you get stuck.
Here are a few things you can try.

- Explain the problem to someone else in your own words.
- Make a simpler problem.
  For example, how could you draw a line segment with length $\sqrt{2}$ cm?
- Draw a picture to represent the problem.
- Think about what you already know.
  For example, what do you know about square roots? How can you use dot paper to draw line segments of certain lengths?
- Try to approach the problem a different way.
- Talk to someone else about the problem.

Strategies for Success

Using More Than One Strategy

There is almost always more than one way to solve a problem. Often, the more ways you can use to solve a problem, the better your understanding of math and problem solving.

Use at least two different ways to show each statement below is true.

1. $\sqrt{11}$ is between 3 and 4.
2. This line segment has length $\sqrt{13}$ units.

3. The number 25 is a perfect square.
4. A triangle with side lengths 5 cm, 6 cm, and $\sqrt{11}$ cm is a right triangle.
5. The area of this square is 25 square units.

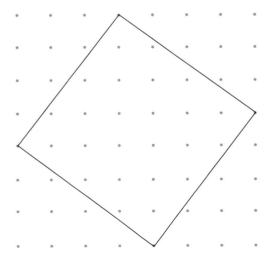

Strategies for Success: Getting Unstuck

Unit Review

What Do I Need to Know?

✓ **Side Length and Area of a Square**

The side length of a square is equal
to the square root of its area.
Length = $\sqrt{\text{Area}}$
Area = (Length)^2

✓ **The Approximate Square Root of a Number**

For numbers that are not perfect squares, we can determine the
approximate square root using estimation or a calculator.

✓ **The Pythagorean Theorem**

In a right triangle, the area of the square
on the hypotenuse is equal to the sum of
the areas of the squares on the two legs.
$h^2 = a^2 + b^2$
Use the Pythagorean Theorem to find
the length of a side in a right triangle,
when two other sides are known.

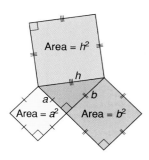

What Should I Be Able to Do?

LESSON
1.1

1. Use square tiles.
   Make as many different rectangles as
   you can with area 24 square units.
   Draw your rectangles on grid paper.
   Is 24 a perfect square?
   Justify your answer.

2. Which of these numbers is
   a perfect square?
   Use a diagram to support your
   answer.
   a) 18   b) 25   c) 44   d) 80

54   UNIT 1: Square Roots and the Pythagorean Theorem

LESSON

3. I am a square number.
   The sum of my digits is 7.
   What square number might I be?
   How many different numbers can you find?

1.2

4. Find the square of each number.
   a) 5   b) 7   c) 9   d) 13

5. Find a square root.
   a) 7^2 b) $\sqrt{289}$ c) $\sqrt{400}$

6. a) List the factors of each number in ascending order.
      i) 108   ii) 361   iii) 150
      iv) 286   v) 324   vi) 56
   b) Which numbers in part a are square numbers? How can you tell?

7. The area of a square is 121 cm². What is the perimeter of the square? How did you find out?

1.3

8. Copy this square onto grid paper. Find its area. Then write the side length of the square.

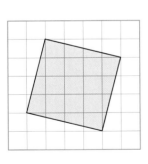

9. The area of each square is given. Find its side length. Which side lengths are whole numbers?
   a) 75 cm²   b) 96 cm²   c) 81 cm²

10. Without measuring, which line segment is longer? How can you tell?
    a) 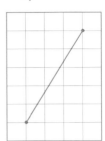   b)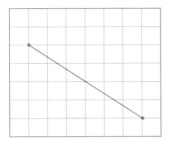

11. Find.
    a) $\sqrt{26 \times 26}$
    b) $\sqrt{5 \times 5}$
    c) $\sqrt{50 \times 50}$
    d) $\sqrt{13 \times 13}$

12. Between which two consecutive whole numbers is each square root? How did you find out?
    a) $\sqrt{46}$ b) $\sqrt{84}$
    c) $\sqrt{120}$ d) $\sqrt{1200}$

1.4

13. Without using a calculator, estimate each square root to the nearest whole number.
    a) $\sqrt{6}$ b) $\sqrt{11}$
    c) $\sqrt{26}$ d) $\sqrt{35}$
    e) $\sqrt{66}$ f) $\sqrt{86}$

14. Estimate each square root to one decimal place. Show your work.
    a) $\sqrt{55}$ b) $\sqrt{75}$
    c) $\sqrt{95}$ d) $\sqrt{105}$
    e) $\sqrt{46}$ f) $\sqrt{114}$

LESSON

15. Which is the better estimate of $\sqrt{72}$: 8.48 or 8.49?
How do you know?

16. This First Nations quilt is a square, with area 16 900 cm². How long is each side of the quilt?

17. Is each statement true or false? Justify your answers.
a) $\sqrt{2} + \sqrt{2} = 2$
b) $\sqrt{29}$ is between 5 and 6.
c) $\sqrt{9} + \sqrt{25} = \sqrt{64}$

1.5
18. Find the length of each side labelled with a variable. Give your answers to one decimal place where needed.
a)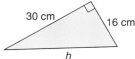
b)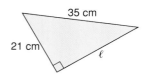

c)

19. Find the length of the diagonal, d, in each rectangle. Give your answers to one decimal place where needed.
a) 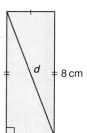 b)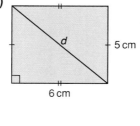

20. The area of the square on each side of a triangle is given.
Is the triangle a right triangle? How do you know?

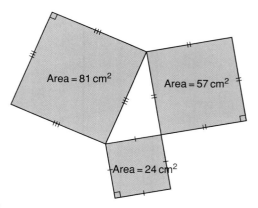

1.6
21. A triangle has side lengths 7 cm, 12 cm, and 15 cm.
Is the triangle a right triangle? Justify your answer.

LESSON

22. Identify the sets of numbers that are Pythagorean triples.
How did you decide?
a) 24, 32, 40
b) 11, 15, 24
c) 25, 60, 65
d) 5, 8, 9

23. Two numbers in a Pythagorean triple are 20 and 29.
Find the third number.
How many solutions are possible?
Justify your answer.

1.7

24. Look at the map below. The side length of each grid square is 10 km.
How much farther is it to travel from Jonestown to Skene by car than by helicopter?

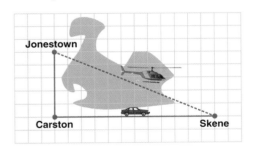

25. Find the perimeter of △ABC.

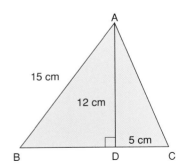

26. There is a buried treasure at one of the points of intersection of the grid lines below.
Copy the grid.

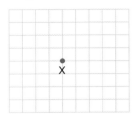

The treasure is $\sqrt{13}$ units from the point marked X.
a) Where might the treasure be? Explain how you located it.
b) Could there be more than one position? Explain.

27. Two boats started at the same point. After 2 h, one boat had travelled 20 km due east. The other boat had travelled 24 km due north. How far apart are the boats?
Explain your thinking.
Give your answer to one decimal place.

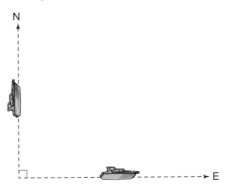

Practice Test

1. Find.
   Give your answers to two decimal places where needed.
   a) $\sqrt{121}$
   b) 14^2
   c) $\sqrt{40}$
   d) the square of 9

2. Explain why $\sqrt{1} = 1$.

3. A square tabletop has perimeter 32 cm.
   What is the area of the tabletop? Explain your thinking.
   Include a diagram.

4. a) What is the area of square ABCD?
   b) What is the length of line segment AB?
   Explain your reasoning.

   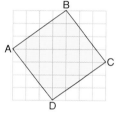

5. The area of a square on each side of a triangle is given.
   Is the triangle a right triangle?
   How do you know?
   a) 15 cm², 9 cm², 24 cm²
   b) 11 cm², 7 cm², 20 cm²

6. Find the length of each side labelled with a variable.
   Give your answers to one decimal place where needed.
   a)
   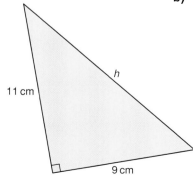
   b)
   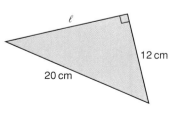

7. Which of the sets of numbers below is a Pythagorean triple?
How did you find out?
a) 20, 48, 54
b) 18, 24, 30

8. A parking garage in a shopping mall has ramps from one level to the next.
a) How long is each ramp?
b) What is the total length of the ramps?

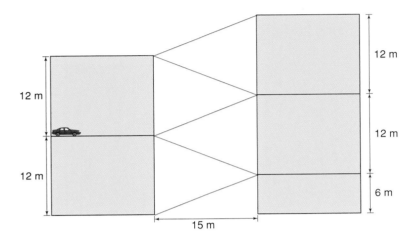

9. Draw these 3 line segments on 1-cm grid paper.
a) Find the length of each line segment to one decimal place.
b) Could these line segments be arranged to form a triangle?
If your answer is no, explain why not.
If your answer is yes, could they form a right triangle?
Explain.

10. Rocco runs diagonally across a square field.
The side of the field has length 38 m.
How many times will Rocco have to run diagonally across the field
to run a total distance of 1 km?
Give your answer to the nearest whole number.

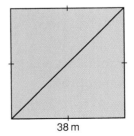

Unit Problem The Locker Problem

Part A

In a school, there is a row of 25 lockers, numbered 1 to 25.
A student goes down the row and opens every locker.
A second student goes down the row and closes every other locker, beginning with locker 2.
A third student changes every 3rd locker, beginning with locker 3.
 • If the locker is closed, the student opens it.
 • If the locker is open, the student closes it.
A fourth student changes every 4th locker, beginning with locker 4.
This continues until 25 students have gone down the row.
Which lockers will be open after the 25th student has gone down the row?

1. Draw a chart similar to the one below.
   Extend the chart for 25 lockers and 25 students.
   The first student opens every locker.
   Write O for each opened locker.
   The second student closes every second locker, beginning with locker 2.
   Write C for each locker that was closed.
   The third student changes every third locker, beginning with locker 3. Write C or O for each locker that was changed.
   Complete the chart for 25 students.

Student \ Locker	1	2	3	4	5	6	7	8	9	10	11	12	13	14	15
1	O	O	O	O	O	O	O	O	O	O	O	O	O	O	O
2		C		C		C		C		C		C		C	

2. Which lockers remain open after 25 students have gone down the row? What pattern do you notice?
   How can you use this pattern to tell which lockers will be open?
3. Suppose there were 100 lockers and 100 students. Which lockers remain open after 100 students have gone down the row?

60   UNIT 1: Square Roots and the Pythagorean Theorem

4. Suppose there were 400 lockers and 400 students. Which lockers remain open after 400 students have gone won the row?
5. What is the rule for any number of lockers and students? Why do you think your rule works?

Part B

6. Copy and complete a chart similar to the one below for the square numbers to 900. What patterns do you see?

Number	Square Numbers	Difference Between Square Numbers
1	1	
2	4	3
3	9	5
4	16	7
5	25	9

7. We can use the completed chart to find some Pythagorean triples. For example, the difference between 25 (5^2) and 16 (4^2) is 9, or 3^2. So, 3-4-5 is a Pythagorean triple.
   a) How does it help to have the differences between square numbers in numerical order to identify a Pythagorean triple?
   b) Will the difference between 2 consecutive square numbers ever be 16? Explain.
   c) Which difference between square numbers will indicate the next Pythagorean triple?
   d) Use your chart to identify 2 more Pythagorean triples.
   e) Extend the chart to find 1 more Pythagorean triple.

Check List

Your work should show:
- ✓ completed charts, clearly labelled
- ✓ clear descriptions of the patterns observed
- ✓ reasonable explanations of your thinking and your conclusions

Reflect on Your Learning

How do you think the Pythagorean Theorem could be used by a carpenter, a forester, a surveyor, a wildlife conservation officer, or the captain of a ship?

UNIT 2
Integers

As of 2007, Mike Weir is the only Canadian to ever win the US Masters Golf Championship. Weir defeated Len Mattiace on the first extra hole of a playoff to win the 2003 Masters.

Here are seven players, in alphabetical order, and their leaderboard entries.

- What was Weir's leaderboard entry?
- Order the entries from least to greatest.
- Why do you think golf is scored using integers?
- What other uses of integers do you know?

Player	Over/Under Par
Jim Furyk	−4
Retief Goosen	+1
Jeff Maggert	−2
Phil Mickelson	−5
Vijay Singh	−1
Mike Weir	−7
Tiger Woods	+2

Par for the tournament is 288. Jim Furyk shot 284. His score in relation to par is 4 under, or −4.

What You'll Learn

- Multiply and divide integers.
- Use the order of operations with integers.
- Solve problems involving integers.

Why It's Important

We use integers in everyday life, when we deal with weather, finances, sports, geography, and science.

Key Words

- positive integer
- negative integer
- zero pair
- opposite integers
- zero property
- distributive property
- commutative property
- product
- quotient
- grouping symbol
- order of operations

2.1 Using Models to Multiply Integers

Focus Use a model to multiply two integers.

We can think of multiplication as repeated addition.
5 × 3 is the same as adding five 3s: 3 + 3 + 3 + 3 + 3
As a sum: 3 + 3 + 3 + 3 + 3 = 15
As a product: 5 × 3 = 15

How can we think of 3 × 5?
One way is to use a number line.
Take 3 steps each of size 5.
So, 3 × 5 = 15

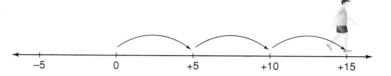

Investigate

Work with a partner.
You will need masking tape and a metre stick.
Make a large number line across the floor.
Divide the line into intervals of 15 cm.
Label the line from −15 to +15.

➤ Walk the line to multiply integers.
  − Start at 0.
  − For negative numbers of steps, face the negative end of the line before walking.
  − For negative step sizes, walk backward.
  For example:

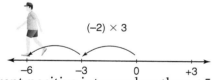

➤ Choose 2 different positive integers less than +5.

➤ Find all possible products of the integers and their opposites.
  Take turns. One partner walks the line to find each product.
  The other partner records on a number line and writes the
  multiplication equation each time.

Reflect & Share
Share your results with the class. What patterns do you notice?
How can you predict the product of two integers?

64   UNIT 2: Integers

Connect

Recall that we can use coloured tiles to model integers.
One yellow tile models +1, and one red tile models −1.

They combine to form a zero pair: (+1) + (−1) = 0

We can extend our use of coloured tiles to model the multiplication of two integers.
Let a circle represent the "bank." Start with the bank having zero value.
The first integer tells us to deposit (put in) or to withdraw (take out).
When the first integer is positive, put tiles in.
When the first integer is negative, take tiles out.
The second integer tells us what to put in or take out.

▶ Multiply: (+4) × (+3)
+4 is a positive integer.
+3 is modelled with 3 yellow tiles.
So, *put* 4 sets of 3 yellow tiles *into* the circle.

There are 12 yellow tiles in the circle.
They represent +12.
So, (+4) × (+3) = +12

(+4) × (+3) = (+3) + (+3) + (+3) + (+3)
Make 4 deposits of +3.

▶ Multiply: (+4) × (−3)
+4 is a positive integer.
−3 is modelled with 3 red tiles.
So, *put* 4 sets of 3 red tiles *into* the circle.

There are 12 red tiles in the circle.
They represent −12.
So, (+4) × (−3) = −12

(+4) × (−3) = (−3) + (−3) + (−3) + (−3)
Make 4 deposits of −3.

▶ Multiply: (−4) × (−3)
−4 is a negative integer.
−3 is modelled with 3 red tiles.
So, *take* 4 sets of 3 red tiles *out* of the circle.
There are no red tiles to take out.

Make 4 withdrawals of −3.

2.1 Using Models to Multiply Integers

So, add zero pairs until there are enough red tiles to remove.
Add 12 zero pairs.
Take out 4 sets of 3 red tiles.

There are 12 yellow tiles left in the circle.
They represent +12.
So, (−4) × (−3) = +12

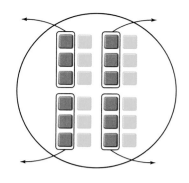

➤ Multiply: (−4) × (+3)
−4 is a negative integer.
+3 is modelled with 3 yellow tiles.
So, take 4 sets of 3 yellow tiles out of the circle.
There are no yellow tiles to take out.
So, add zero pairs until there are enough yellow tiles to remove.
Add 12 zero pairs.
Take out 4 sets of 3 yellow tiles.

Make 4 withdrawals of +3.

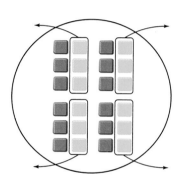

There are 12 red tiles left in the circle.
They represent −12.
So, (−4) × (+3) = −12

Example 1

Use tiles to find: (−3) × (+2)

➤ **A Solution**

−3 is a negative integer.
+2 is modelled with 2 yellow tiles.
So, take 3 sets of 2 yellow tiles out of the circle.
Since there are no yellow tiles to take out,
add 6 zero pairs.
Take out 3 sets of 2 yellow tiles.
6 red tiles remain.
They represent −6.
So, (−3) × (+2) = −6

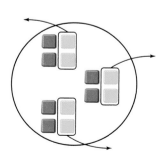

Example 2

The temperature fell 3°C each hour for 6 h.
Use an integer number line to find the total change in temperature.

▶ **A Solution**

−3 represents a fall of 3°C.
+6 represents 6 h.
Using integers, we need to find: $(+6) \times (-3)$
$(+6) \times (-3) = (-3) + (-3) + (-3) + (-3) + (-3) + (-3)$
Use a number line. Start at 0.
Move 3 units left, 6 times.

To add a negative integer, move left on a number line.

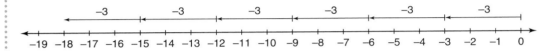

So, $(+6) \times (-3) = -18$
The total change in temperature is −18°C.

Discuss the ideas

1. Which model do you find easiest to use to multiply integers?
2. What other models can you think of?
3. What do you notice about the effect of the order of the integers on the product?
4. Use a model to explain multiplication of an integer by 0.

Math Link

Sports

In hockey, each player has a plus/minus statistic. A player's plus/minus statistic increases by 1 when his team scores a goal while he is on the ice. A player's plus/minus statistic decreases by 1 when his team is scored against while he is on the ice. For example, a player begins a game with a plus/minus statistic of −7. During the game, his team scores 3 goals while he is on the ice and the opposing team scores 1 goal. What is the player's new plus/minus statistic?

Practice

Check

5. Write a multiplication expression for each repeated addition.
   a) (−1) + (−1) + (−1)
   b) (−2) + (−2) + (−2) + (−2) + (−2)
   c) (+11) + (+11) + (+11) + (+11)

6. Write each multiplication expression as a repeated addition. Use coloured tiles to find each sum.
   a) (+7) × (−4)
   b) (+6) × (+3)
   c) (+4) × (+6)
   d) (+5) × (−6)

7. Which integer multiplication does each number line represent? Find each product.
   a)
   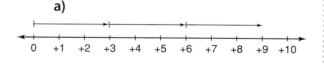
   b)
   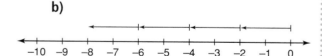

8. Use a number line. Find each product.
   a) (+6) × (−1)
   b) (+3) × (+9)
   c) (+2) × (+6)
   d) (+4) × (−5)

9. Which product does each model represent? Write a multiplication equation for each model.
   a) Deposit 5 sets of 2 red tiles.
   b) Deposit 5 sets of 2 yellow tiles.
   c) Withdraw 7 sets of 3 red tiles.
   d) Withdraw 9 sets of 4 yellow tiles.
   e) Deposit 11 sets of 3 yellow tiles.
   f) Withdraw 10 sets of 5 red tiles.

Apply

10. Use a circle and coloured tiles. Find each product. Sketch the tiles you used.
   a) (+1) × (+5)   b) (+8) × (+3)
   c) (+7) × (−2)   d) (+8) × (−3)
   e) (−5) × (+6)   f) (−4) × (−8)

11. Use coloured tiles or a number line. Find each product.
   a) (+4) × (+2)   b) (−4) × (−2)
   c) (+2) × (+8)   d) (+5) × (−6)
   e) (−4) × (+6)   f) (−7) × (−3)

12. The temperature rose 2°C each hour for 9 h. Use integers to find the total change in temperature.

13. Donovan was draining an above-ground swimming pool. The water level dropped 3 cm each hour for 11 h. Use integers to find the change in the water level after 11 h.

14. Lissa used the expression $(+8) \times (-6)$ to solve a word problem.
   a) What might the word problem have been? Solve the problem.
   b) Compare your word problem with those of your classmates. Which is your favourite problem?

15. Assessment Focus How many different ways can you model the product $(-7) \times (-8)$?
   Show each strategy. Which strategy do you prefer? Explain your choice.

16. Rema was playing a board game. She moved back 4 spaces on each of 4 consecutive turns. Use integers to find her total change in position.

17. Ellen's dad spends $5 a week on newspapers.
   a) How much less newspaper money will he have 8 weeks from now?
   b) How much more newspaper money did he have 2 weeks ago?
   Draw a model to represent each answer. Write the equation that each model represents.

18. A toy car travels along a number line marked in centimetres. A distance of 1 cm to the right is represented by +1. A distance of 1 cm to the left is represented by −1. The car moves 4 cm to the left each second.
   a) The car is at 0 now. Where will the car be 10 s from now?
   b) Where was the car 3 s ago?
   c) How can you use an operation with integers to answer parts a and b?

19. Hugh used the expression $(-7) \times (+6)$ to solve a word problem. What might the word problem have been? Solve the problem.

20. Take It Further Use coloured tiles or a number line. Find each product.
   a) $(+3) \times (-2) \times (+4)$
   b) $(-5) \times (-1) \times (+3)$
   c) $(-5) \times (-2) \times (-3)$
   d) $(+2) \times (-3) \times (-6)$

Reflect

How did your knowledge of adding integers help you with this lesson?
How did your knowledge of opposites change with this lesson?

2.2 Developing Rules to Multiply Integers

Focus Use patterns to develop rules for multiplying integers.

We can write the number of tiles in this array in two ways.
As a sum:
(+5) + (+5) + (+5) = +15

As a product:
(+3) × (+5) = +15

How can you use integers to write the number of tiles in this array in two ways?

Investigate

Work with a partner.

Your teacher will give you a large copy of this multiplication table.

Fill in the products that you know best. Use any patterns you see to help you complete the table.

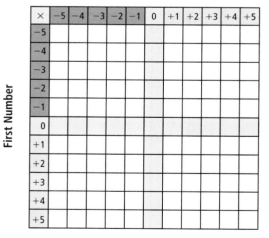

Compare your completed table with that of another pair of classmates.
Explain a strategy you could use to multiply any negative integer by any positive integer.
Explain a strategy you could use to multiply any two negative integers.

Connect

These properties of whole numbers are also properties of integers.

Multiplying by 0 (Zero property)
$3 \times 0 = 0$ and $0 \times 3 = 0$
So, $(-3) \times 0 = 0$ and $0 \times (-3) = 0$

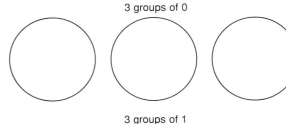
3 groups of 0

Multiplying by 1 (Multiplicative Identity)
$3 \times 1 = 3$ and $1 \times 3 = 3$
So, $(-3) \times (+1) = -3$ and $(+1) \times (-3) = -3$

3 groups of 1

Since multiplying by 1 does not change the identity of a number, we call 1 the *multiplicative identity*.

1 group of 3

Commutative Property
$3 \times 4 = 12$ and $4 \times 3 = 12$
So, $(-3) \times (+4) = -12$ and $(+4) \times (-3) = -12$

Distributive Property
$3 \times (4 + 5) = 3 \times 4 + 3 \times 5$
$ = 12 + 15$
$ = 27$

So, $(+3) \times [(-4) + (-5)] = [(+3) \times (-4)] + [(+3) \times (-5)]$
$ = (-12) + (-15)$
$ = -27$

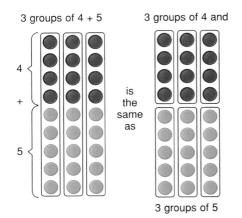

3 groups of 4 + 5 is the same as 3 groups of 4 and 3 groups of 5

2.2 Developing Rules to Multiply Integers

Example 1

Find each product.

a) $(-9) \times (+4)$ b) $(-4) \times (-9)$ c) $(+4) \times (+9)$

▶ **A Solution**

a) Multiply the numbers as if they were positive.
$9 \times 4 = 36$
The integers have opposite signs, so the product is negative.
So, $(-9) \times (+4) = -36$

b) The integers have the same sign, so the product is positive.
So, $(-4) \times (-9) = +36$

c) The integers have the same sign, so the product is positive.
So, $(+4) \times (+9) = +36$

Example 2

Find the product: $(+20) \times (-36)$

▶ **A Solution**

$(+20) \times (-36) = (+20) \times [(-30) + (-6)]$
$= [(+20) \times (-30)] + [(+20) \times (-6)]$
$= (-600) + (-120)$
$= -720$
So, $(+20) \times (-36) = -720$

Write –36 in expanded form. Use the distributive property.

Example 3

Find the product: $(-25) \times (-48)$

▶ **A Solution**

Multiply the numbers as if they were positive: 25×48
Use a rectangle model.
$(25) \times (48) = (20 \times 40) + (5 \times 40) + (20 \times 8) + (5 \times 8)$
$= 800 + 200 + 160 + 40$
$= 1200$
The integers have the same sign, so the product is positive.
So, $(-25) \times (-48) = +1200$

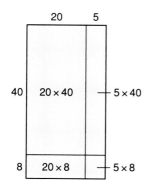

When we write the product of integers, we do not need to write the multiplication sign. That is, we may write $(-8) \times (-9)$ as $(-8)(-9)$.

Discuss the ideas

1. Think about the work of this lesson and the previous lesson. What is the sign of the product when you multiply 2 integers:
   - if both integers are positive?
   - if one integer is positive and the other integer is negative?
   - if both integers are negative?
2. Explain your strategy to multiply 2 integers.

Practice

Check

3. Will each product be positive or negative? How do you know?
   a) $(-6) \times (+2)$ b) $(+6) \times (+4)$
   c) $(+4) \times (-2)$ d) $(-7) \times (-3)$

4. Find each product.
   a) $(+8)(-3)$ b) $(-5)(-4)$
   c) $(-3)(+9)$ d) $(+7)(-6)$
   e) $(+10)(-3)$ f) $(-7)(-6)$
   g) $(0)(-8)$ h) $(+10)(-1)$
   i) $(-7)(-8)$ j) $(+9)(-9)$

5. a) Find the product of each pair of integers.
      i) $(+3)(-7)$ and $(-7)(+3)$
      ii) $(+4)(+8)$ and $(+8)(+4)$
      iii) $(-5)(-9)$ and $(-9)(-5)$
      iv) $(-6)(+10)$ and $(+10)(-6)$
   b) Use the results of part a. Does the order in which integers are multiplied affect the product? Explain.

6. Find each product.
   a) $(+20) \times (+15)$ b) $(-30) \times (-26)$
   c) $(+50) \times (-32)$ d) $(-40) \times (+21)$
   e) $(-60) \times (+13)$ f) $(+80) \times (-33)$
   g) $(+70) \times (+47)$ h) $(-90) \times (-52)$

Apply

7. Find each product.
   a) $(+25) \times (-12)$ b) $(-45) \times (+21)$
   c) $(-34) \times (-16)$ d) $(-37) \times (+18)$
   e) $(+17)(+13)$ f) $(+84)(-36)$
   g) $(-51)(-25)$ h) $(+29)(+23)$

8. Copy each equation. Replace ☐ with an integer to make the equation true.
   a) $(+5) \times ☐ = +20$
   b) $☐ \times (-9) = +27$
   c) $(-9) \times ☐ = -54$
   d) $☐ \times (-3) = +18$
   e) $☐ \times (+5) = -20$
   f) $☐ \times (-12) = +144$
   g) $☐ \times (-6) = +180$
   h) $☐ \times (-4) = +24$

2.2 Developing Rules to Multiply Integers

9. Write the next 3 terms in each pattern. Then write the pattern rule.
   a) +1, +2, +4, +8, …
   b) +1, −6, +36, −216, …
   c) −1, +3, −9, +27, …
   d) −4, +4, −4, +4, …

10. Gaston withdrew $26 from his bank account each week for 17 weeks. Use integers to find the total amount Gaston withdrew over the 17 weeks.
    Show your work.

11. **Assessment Focus** Use the integers: −5, +9, −8, +4, −2
    a) Which two integers have the greatest product?
    b) Which two integers have the least product?
    c) Provide a convincing argument that your answers to parts a and b are correct.

12. a) Find each product. Then use a calculator to extend the pattern 4 more rows.
    i) (−2)(−3)
    ii) (−2)(−3)(−4)
    iii) (−2)(−3)(−4)(−5)
    iv) (−2)(−3)(−4)(−5)(−6)
    b) Use the results in part a.
       i) What is the sign of a product when it has an even number of negative factors? Explain.
       ii) What is the sign of a product when it has an odd number of negative factors? Explain.
    c) Investigate what happens when a product has positive and negative factors. Do the rules in part b still apply? Explain.

13. Amelie was doing a math question. The answer she got did not match the answer in the answer key. So, she asked a friend to look at her work.

    (+60) × (−18)
     = (+60) × [(−20) + (+2)]
     = [(+60) × (−20)] + [(+60) × (+2)]
     = (+1200) + (+120)
     = +1320

    a) What was Amelie's error?
    b) Correct Amelie's error.
       What is the correct answer?

14. Gavin used the expression (+15) × (−8) to solve a word problem. What might the word problem have been? Solve the problem.

15. Explain why an integer multiplied by itself can never result in a negative product.

16. Bridget used the expression (−12) × (+7) to solve a word problem. What might the word problem have been? Solve the problem.

17. Write −36 as the product of two or more integer factors. Do this as many different ways as you can. Show your work.

18. The product of two integers is −144. The sum of the integers is −7. What are the two integers?

19. **Take It Further** When you multiply two natural numbers, the product is never less than either of the two numbers. Is the same statement true for the product of any two integers? Investigate, then write what you find out.

> Natural numbers are 1, 2, 3, 4, …, and so on.

20. **Take It Further** The product of two integers is between +160 and +200. One integer is between −20 and −40.
  a) What is the greatest possible value of the other integer?
  b) What is the least possible value of the other integer?

21. **Take It Further** How can you use the rules for multiplying two integers to multiply more than two integers? Use an example to illustrate your strategy.

Reflect

Suppose your friend missed this lesson. Explain to her how to multiply two integers. Use examples in your explanation.

Game

What's My Product?

In this game, a black card represents a positive integer; for example, the 5 of spades is +5.
A red card represents a negative integer; for example, the 6 of hearts is –6.

HOW TO PLAY

1. Remove the face cards (Jacks, Queens, Kings) from the deck. Use the remaining cards. An ace is 1.

2. Shuffle the cards. Place them face down in a pile. Players take turns to turn over the top 2 cards, then find the product. The person with the greater product keeps all 4 cards.

3. If there is a tie, each player turns over 2 more cards and the player with the greater product keeps all 8 cards.

4. Play continues until all cards have been used. The winner is the player with more cards.

YOU WILL NEED
A standard deck of playing cards

NUMBER OF PLAYERS
2

GOAL OF THE GAME
To have more cards

2.3 Using Models to Divide Integers

Focus Use a model to divide integers.

We can think of division as the opposite of multiplication.
12 ÷ 4 = ?
This can mean how many sets of 4 will give a product of 12:
? × 4 = 12

There are 3 sets of 4.
So, 3 × 4 = 12

We can use a "bank" model to multiply 2 integers.
- A circle represents the "bank."
  We start with the bank having zero value.
- The first integer tells us to deposit (put in) or to withdraw (take out).
- The second integer tells us what to put in or take out.
- We can use this model to multiply 3 × 4.

How can we use this model to find 12 ÷ 4?

Make 3 deposits of 4 yellow tiles.
There are 12 yellow tiles.
So, 3 × 4 = 12

Investigate

Work with a partner.
You will need coloured tiles.
Choose 2 positive integers between 1 and 20
whose quotient is an integer.
For example: 12 and 3, since 12 ÷ 3 = 4
Use these integers and their opposites.
Write all possible division expressions.

Use the tiles and a "bank" model to divide.
Sketch the tiles you used in each case.
Write a division equation each time.

A quotient is the number
that results from the division
of one number by another.

Share your results with the class.
What patterns do you notice?
How can you predict the quotient of two integers?

Connect

We can extend the use of a number line to model the division of two integers.

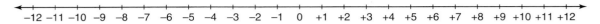

Visualize walking the line to divide integers.
This time, the direction you end up facing determines the sign of the quotient.

▶ Divide: (+9) ÷ (+3)
We need to find how many steps of +3 make +9.
The step size, +3, is positive; so, walk forward.
Start at 0. Take steps forward to end up at +9.

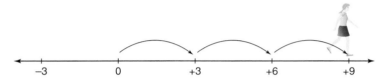

We took 3 steps. We are facing the positive end of the line.
So, (+9) ÷ (+3) = +3

▶ Divide: (−9) ÷ (−3)
We need to find how many steps of −3 make −9.
The step size, −3, is negative; so, walk backward.
Start at 0. Take steps backward to end up at −9.

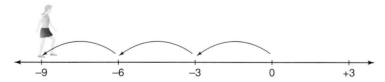

We took 3 steps. We are facing the positive end of the line.
So, (−9) ÷ (−3) = +3

▶ Divide: (−9) ÷ (+3)
We need to find how many steps of +3 take us to −9.
The step size, +3, is positive; so, walk forward.
Start at 0. To end up at −9, we took 3 steps forward.

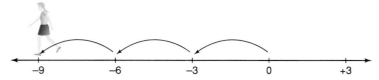

We are facing the negative end of the line.
So, $(-9) \div (+3) = -3$

▶ Divide: $(+9) \div (-3)$

We need to find how many steps of -3 take us to $+9$.
The step size, -3, is negative; so, walk backward.
Start at 0. To end up at $+9$, we took 3 steps backward.

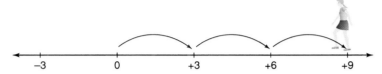

We are facing the negative end of the line.
So, $(+9) \div (-3) = -3$

Example 1

The 1850-km Iditarod dogsled race lasts from 10 to 17 days.
One night, the temperature fell 2°C each hour for a total change of −12°C.
Use integers to find how many hours this change in temperature took.

▶ ### A Solution

−2 represents a fall of 2°C.
−12 represents a change of −12°C.
Using integers, we need to find how many −2s take us to −12;
that is, $(-12) \div (-2)$.

Start at 0.
Move 2 units left.
Continue to move 2 units left until you reach −12.

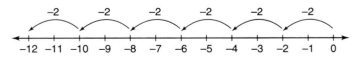

Six moves of 2 units left were made.
So, $(-12) \div (-2) = +6$

2.3 Using Models to Divide Integers

Example 2

Use a model to find the quotient: (−12) ÷ (+4)

▶ **A Solution**

Divide: (−12) ÷ (+4)
How many groups of +4 will make −12?
Use coloured tiles and the "bank" model.

Start with a value of 0 in the circle.
To get a product of −12, 12 red tiles must be left in the circle.
So, model 0 with 12 zero pairs.
+4 is modelled with 4 yellow tiles.
Take out sets of 4 yellow tiles.

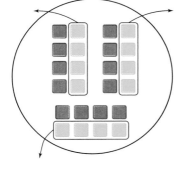

3 sets were removed.
So, (−12) ÷ (+4) = −3

The negative sign in the quotient indicates removal.

Discuss the ideas

1. Which model do you prefer to use to divide integers?
2. How can you use the inverse operation to check your answers?

Practice

Check

3. Write a related multiplication equation for each division equation.
   a) (+25) ÷ (+5) = +5
   b) (+24) ÷ (−2) = −12
   c) (−14) ÷ (−7) = +2
   d) (−18) ÷ (+6) = −3

4. Which integer division does each number line represent? Find each quotient.

a)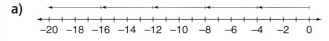

b)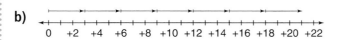

c)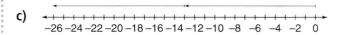

5. Enrico walked a number line to model a division. He started at 0. Enrico took steps forward of size 4. He ended up at −24. Which division did Enrico model? How did you find out?

6. Use a number line. Find each quotient.
   a) (+8) ÷ (+1)   **b)** (−6) ÷ (−2)
   c) (−16) ÷ (+8)   **d)** (−3) ÷ (−1)
   e) (+15) ÷ (−3)   **f)** (−20) ÷ (+2)

7. a) How many sets?
   i) 12 yellow tiles grouped in sets of 6
   ii) 15 red tiles grouped in sets of 3
   b) How many in each set?
   i) 8 yellow tiles shared among 2 sets
   iii) 21 red tiles shared among 7 sets

Apply

8. Use coloured tiles to represent each division. Find each quotient. Sketch the tiles you used.
   a) (+18) ÷ (+6)   **b)** (−18) ÷ (+9)
   c) (−16) ÷ (−4)   **d)** (+21) ÷ (−7)
   e) (+15) ÷ (−5)   **f)** (−16) ÷ (−8)

9. Use coloured tiles, a number line, or another model to show your thinking clearly. Find each quotient.
   a) (+8) ÷ (+4)   **b)** (−8) ÷ (−4)
   c) (+8) ÷ (−4)   **d)** (−8) ÷ (+4)
   Compare the quotients. What do you notice?

10. Use coloured tiles, a number line, or another model to show your thinking clearly. Find each quotient.
   a) (+24) ÷ (+8)   **b)** (−20) ÷ (−5)
   c) (+28) ÷ (−7)   **d)** (−25) ÷ (+5)
   e) (−14) ÷ (+2)   **f)** (+18) ÷ (−9)

11. The temperature rose 3°C each hour for a total change of +12°C. Use integers to find the number of hours the change in temperature took.

12. The temperature fell 4°C each hour for a total change of −20°C. Use integers to find the number of hours the change in temperature took.

13. A submarine was at the surface of the ocean. It made 4 identical plunges in a row. Its final depth was 148 m below sea level. What was the depth of each plunge?

14. Assessment Focus Heather used the expression (+45) ÷ (−5) to solve a word problem. What might the word problem have been? Show as many different ways as you can to solve the problem.

15. Maddie used the expression (−12) ÷ (+6) to solve a word problem. What might the word problem have been? Solve the problem.

2.3 Using Models to Divide Integers

16. A snail travels along a number line marked in centimetres. A distance of 1 cm to the right is represented by +1. A distance of 1 cm to the left is represented by −1. The snail moves 6 cm to the left each minute.

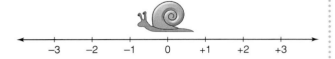

a) The snail is at 0 now. After how many minutes will the snail be at −36 on the number line?
b) When was the snail at +18 on the number line?
Draw a model to represent each answer. Write a division equation for each model.

17. Take It Further Abraham used the Internet to find the low temperature in six Western Canadian cities on a particular day in January. He recorded the temperatures in a table.

City	Low Temperature (°C)
Whitehorse	−10
Iqaluit	−7
Vancouver	+9
Edmonton	−1
Winnipeg	−9
Saskatoon	−6

a) Find the mean low temperature for these cities on that day.
b) The low temperature in Regina for the same day was added to the table. The mean low temperature for the seven cities was −3°C. What was the temperature in Regina?
Communicate your thinking clearly.

18. Take It Further Reena deposits $4 into her savings account each week. Today, Reena's account has a balance of $16.
a) How many weeks from now will Reena's account have a balance of $40?
b) What was the balance in Reena's account 2 weeks ago?
Explain how you can use integers to model each situation.

Reflect

How is the division of integers similar to the division of whole numbers? How is it different? Use examples to explain.

Mid-Unit Review

LESSON

2.1

1. Use a model to represent each product. Draw the model you used each time.
   a) $(-9) \times (+4)$ b) $(-7) \times (-5)$
   c) $(+4) \times (+8)$ d) $(+3) \times (-5)$

2. A glacier retreated about 2 m per day for 7 days. Use integers to find the total change in the length of the glacier.

3. The temperature rose 4°C each hour for 5 h. Use integers to find the total change in temperature.

2.2

4. Will each product be positive or negative? How do you know?
   a) $(-8) \times (+5)$ b) $(-5) \times (-3)$
   c) $(+12) \times (-4)$ d) $(+8) \times (+9)$

5. Find each product in question 4.

6. Find each product.
   a) $(-20)(+14)$
   b) $(-19)(-24)$
   c) $(+40)(+27)$
   d) $(+13)(-31)$

7. A swimming pool drains 35 L of water in 1 min. Find how much water drained out of the pool in 30 min. How can you model this situation with integers?

8. Copy each equation. Replace ☐ with an integer to make the equation true.
   a) $(+4) \times \square = -32$
   b) $\square \times (-6) = +54$
   c) $(-8) \times \square = -56$
   d) $\square \times (-1) = +12$

2.3

9. Write 2 related multiplication equations for each division equation.
   a) $(+27) \div (+3) = +9$
   b) $(+14) \div (-7) = -2$
   c) $(-21) \div (-3) = +7$
   d) $(-26) \div (+2) = -13$

10. Use coloured tiles, a number line, or another model. Find each quotient.
    a) $(+20) \div (+4)$ b) $(-24) \div (-6)$
    c) $(+32) \div (-8)$ d) $(-36) \div (+4)$

11. The water level in a well dropped 5 cm each hour. The total drop in the water level was 30 cm. Use integers to find how long it took for the water level to change.

12. Maurice used the expression $(-18) \div (+3)$ to solve a word problem. What might the word problem have been? Solve the problem.

13. Explain how you can use a number line to model the quotient of $(+64) \div (-8)$.

2.4 Developing Rules to Divide Integers

Focus Use patterns to develop the rules for dividing integers.

For any multiplication fact with two different whole number factors, you can write two related division facts.
For example: $9 \times 7 = 63$
What are the related division facts? What strategies did you use?

We can apply the same strategies to multiplication facts with two different integer factors.

Investigate

Work in groups of 4.
Choose 2 positive integers and 2 negative integers.
Express each integer as a product of integers.
Do this in two ways.
For each product, write two related division facts.

Compare your division facts with those of another group of classmates. Generalize a rule you can use to determine the sign of the quotient when you divide:
- a positive integer by a positive integer
- a positive integer by a negative integer
- a negative integer by a positive integer
- a negative integer by a negative integer

For each rule, create two examples of your choice.

Connect

To divide integers, we can use the fact that division is the inverse of multiplication.

▶ We know that:
$(+5) \times (+3) = +15$

So, $(+15) \div (+5) = +3$ and $(+15) \div (+3) = +5$

dividend divisor quotient

When the dividend and the divisor are positive, the quotient is positive.

▶ We know that:
$(-5) \times (+3) = -15$

So, $(-15) \div (+3) = -5$ and $(-15) \div (-5) = +3$

The dividend is negative and the divisor is positive. The quotient is negative.

Both the dividend and the divisor are negative. The quotient is positive.

▶ We know that:
$(-5) \times (-3) = +15$

So, $(+15) \div (-5) = -3$ and $(+15) \div (-3) = -5$

When the dividend is positive and the divisor is negative, the quotient is negative.

A division expression can be written with a division sign: $(-48) \div (-6)$; or, as a fraction: $\frac{-48}{-6}$

When the expression is written as a fraction, we do not need to use brackets. The fraction bar acts as a **grouping symbol**.
A grouping symbol keeps terms together, just like brackets.

Example 1

Divide.
a) $(-8) \div (-4)$ b) $\frac{-9}{+3}$

▶ **A Solution**

a) Divide the numbers as if they were positive.
$8 \div 4 = 2$
The integers have the same sign; so, the quotient is positive.
$(-8) \div (-4) = +2$

b) The integers have opposite signs; so, the quotient is negative.
$\frac{-9}{+3} = -3$

Example 2

Divide: $\frac{+96}{-6}$

▶ **A Solution**

Divide as you would whole numbers.

$$6\overline{)9^{3}6}^{16}$$

The integers have opposite signs; so, the quotient is negative.
So, $\frac{+96}{-6} = -16$

Example 3

Shannon made withdrawals of $14 from her bank account.
She withdrew a total of $98.
Use integers to find how many withdrawals Shannon made.

▶ **A Solution**

−14 represents a withdrawal of $14.
−98 represents a total withdrawal of $98.
Divide to find the number of withdrawals.
Since each integer has 2 digits, use a calculator.
$(-98) \div (-14) = +7$
Shannon made 7 withdrawals of $14.

Discuss the ideas

1. Think about the work of this lesson and the previous lesson. What is the sign of the quotient when you divide 2 integers:
   - if both integers are positive?
   - if one integer is positive and the other integer is negative?
   - if both integers are negative?
2. Explain your strategy to divide 2 integers.
3. How do these rules compare to the rules for multiplying integers?

Practice

Check

4. Will each quotient be positive or negative? How do you know?
   a) $(-45) \div (+5)$
   b) $(+16) \div (+8)$
   c) $(+24) \div (-2)$
   d) $(-30) \div (-6)$

5. Find each quotient.
   a) $(+12) \div (+4)$
   b) $(-15) \div (-3)$
   c) $(-18) \div (+9)$
   d) $(+81) \div (-9)$
   e) $(+72) \div (-8)$
   f) $(-64) \div (-8)$
   g) $(-14) \div (+1)$
   h) $(+54) \div (-6)$
   i) $(-27) \div (-3)$
   j) $(+32) \div (+4)$

6. Copy and continue each pattern until you have 8 rows. What does each pattern illustrate?
   a) $(-12) \div (+3) = -4$
   $(-9) \div (+3) = -3$
   $(-6) \div (+3) = -2$
   $(-3) \div (+3) = -1$
   b) $(+25) \div (-5) = -5$
   $(+15) \div (-3) = -5$
   $(+5) \div (-1) = -5$
   $(-5) \div (+1) = -5$

   c) $(+8) \div (+2) = +4$
   $(+6) \div (+2) = +3$
   $(+4) \div (+2) = +2$
   $(+2) \div (+2) = +1$
   d) $(+14) \div (+7) = +2$
   $(+10) \div (+5) = +2$
   $(+6) \div (+3) = +2$
   $(+2) \div (+1) = +2$
   e) $(-14) \div (+7) = -2$
   $(-10) \div (+5) = -2$
   $(-6) \div (+3) = -2$
   $(-2) \div (+1) = -2$
   f) $(-10) \div (-5) = +2$
   $(-5) \div (-5) = +1$
   $(0) \div (-5) = 0$
   $(+5) \div (-5) = -1$

7. a) Use each multiplication fact to find a related quotient.
   i) Given $(+8) \times (+3) = +24$, find $(+24) \div (+3) = \square$.
   ii) Given $(-5) \times (-9) = +45$, find $(+45) \div (-9) = \square$.
   iii) Given $(-7) \times (+4) = -28$, find $(-28) \div (+4) = \square$.
   b) For each division fact in part a, write a related division fact.

Apply

8. Write 2 related division facts for each multiplication fact.
   a) $(-6) \times (+5) = -30$
   b) $(+7) \times (+6) = +42$
   c) $(+9) \times (-4) = -36$
   d) $(-4) \times (-8) = +32$

9. Divide.
   a) $\frac{-20}{-5}$ b) $\frac{+21}{-7}$
   c) $\frac{-36}{+4}$ d) $\frac{0}{-8}$

10. Copy each equation. Replace ☐ with an integer to make the equation true.
    a) $(+25) \div ☐ = +5$
    b) $☐ \div (-9) = +10$
    c) $(-63) \div ☐ = -7$
    d) $☐ \div (-3) = +7$
    e) $☐ \div (+5) = -12$
    f) $☐ \div (-7) = -7$
    g) $☐ \div (-6) = +8$
    h) $☐ \div (-4) = -11$

11. Nirmala borrowed $7 every day. She now owes $56. For how many days did Nirmala borrow money?
    a) Write this problem as a division expression using integers.
    b) Solve the problem.

12. The temperature dropped a total of 15°C over a 5-h period. The temperature dropped by the same amount each hour. Find the hourly drop in temperature.
    a) Write this problem as a division expression using integers.
    b) Solve the problem.

13. Winnie used the money in her savings account to pay back a loan from her mother. Winnie paid back her mother in 12 equal weekly payments. Over the 12 weeks, the balance in Winnie's savings account decreased by $132.
    By how much did her balance change each week?

14. An equestrian was penalized a total of 24 points over a number of performances. The mean number of points lost per performance was –6. How many performances did the equestrian make?
    a) Write this problem as a division expression using integers.
    b) Solve the problem.

15. Write the next three terms in each pattern. What is each pattern rule?
    a) $+1, -3, +9, -27, \ldots$
    b) $+6, -12, +18, -24, \ldots$
    c) $+5, +20, -10, -40, +20, +80, \ldots$
    d) $+128, -64, +32, -16, \ldots$
    e) $-1\,000\,000, +100\,000, -10\,000, +1000, \ldots$

16. **Assessment Focus** Suppose you divide two integers. The quotient is an integer. When is the quotient:
    a) less than both integers?
    b) greater than both integers?
    c) between the two integers?
    d) equal to +1?
    e) equal to −1?
    f) equal to 0?
    Use examples to illustrate your answers. Show your work.

17. Find all the divisors of −32. Write a division equation each time.

 18. Divide.
    a) $(+60) \div (-12)$ b) $(-90) \div (-15)$
    c) $(-77) \div (+11)$ d) $(-80) \div (-20)$
    e) $(+56) \div (+14)$ f) $(+90) \div (-18)$

19. Calculate the mean of these bank deposits and withdrawals. Show your work.
    +$36, −$20, −$18, +$45, +$27, −$16

 20. Bea used the expression $(+78) \div (-13)$ to solve a word problem. What might the word problem have been? Solve the problem.

21. **Take It Further** A warm front caused the temperature to rise 20°C over a 10-h period.
    a) What was the average rise in temperature per hour?
    b) After the rise in temperature, the precipitation that falls is snow. What could the temperature have been 10 h ago?
    How many different answers can you find? Which answers are most reasonable? Explain.

22. **Take It Further** Find as many examples as you can of three different 1-digit numbers that are all divisible by +2 and have a sum of +4.

23. **Take It Further** Find all the divisors of −36. Write a division equation each time. Do you think −36 is a square number? Justify your answer.

24. **Take It Further** The mean daily high temperature in Rankin Inlet, Nunavut, during one week in January was −20°C. What might the temperatures have been on each day of the week? How many different possible answers can you find? Explain.

Reflect

You have learned different strategies for dividing integers. Which strategy do you prefer to use? Justify your choice. Listen to a classmate who has a different choice.

2.5 Order of Operations with Integers

Focus Apply the order of operations with integers.

How many different ways can you evaluate this expression?

$9 \times 6 + 36 \div 4 - 1$

The expression can also be written as: $9(6) + \frac{36}{4} - 1$

To ensure everyone gets the same value, use the order of operations.

Recall the order of operations with whole numbers.
- Do the operations in brackets first.
- Multiply and divide, in order, from left to right.
- Add and subtract, in order, from left to right.

The same order of operations applies to all integers.

Investigate

Work in groups of 4.
Choose 5 different integers between −10 and +10.
Use any operations or brackets.
Find the expression that has the greatest integer value.
Find the expression that has the least integer value.

Trade expressions with another group of classmates.
Find the values of your classmates' expressions.
Check that you and your classmates get the same answers.

Connect

Since we use curved brackets to show an integer; for example, (−2), we use square brackets to group terms. For example, $[(+9) - (-2)] \times (-3)$

When an expression is written as a fraction, the fraction bar indicates division. The fraction bar also acts like a grouping symbol. That is, the operations in the numerator and the denominator must be done first before dividing the numerator by the denominator.

Example 1

Evaluate: $[(-6) + (-2)] \div (-4) + (-5)$

▶ **A Solution**

$[(-6) + (-2)] \div (-4) + (-5)$ Do the operation in square brackets first.

$= (-8) \div (-4) + (-5)$ Divide.
$= (+2) + (-5)$ Add.
$= -3$

Example 2

Evaluate: $\dfrac{2 + 4 \times (-8)}{-6}$

▶ **A Solution**

$\dfrac{2 + 4 \times (-8)}{-6}$ Evaluate the numerator.
             Multiply.

$= \dfrac{2 + (-32)}{-6}$ Add.

$= \dfrac{-30}{-6}$ Divide.

$= 5$

If an integer does not have a sign, it is assumed to be positive; for example, 2 = +2. Then we do not need to put the number in brackets.

Example 3

Evaluate: $\dfrac{[18 - (-6)] \times (-2)}{3(-4)}$

▶ **A Solution**

$\dfrac{[18 - (-6)] \times (-2)}{3(-4)}$ Evaluate the numerator and denominator separately. Do the square brackets first.

$= \dfrac{24 \times (-2)}{3(-4)}$ Multiply.

$= \dfrac{-48}{-12}$ Divide.

$= 4$

Discuss the ideas

1. Why are the square brackets unnecessary in this expression?
   $(-3) + [12 \div (-4)]$
2. In *Example 3*, why were the numerator and denominator evaluated separately?

Practice

Check

3. State which operation you do first.
   a) $7 + (-1) \times (-3)$
   b) $(-18) \div (-6) - (-4)$
   c) $6 + (-4) - (-2)$
   d) $(-2)[7 + (-5)]$
   e) $(-3) \times (-4) \div (-1)$
   f) $8 - 3 + (-4) \div (-1)$

4. Evaluate each expression in question 3. Show all steps.

5. Elijah evaluated this expression as shown.

   $3 - (-5) + 8(-4) = 3 - (-5) + (-32)$
   $= 3 - (-37)$
   $= 40$

   Is Elijah's solution correct? If your answer is yes, explain the steps Elijah took. If your answer is no, what error did Elijah make? What is the correct answer? Show your work.

6. a) Evaluate.
      i) $12 \div (2 \times 3) - 2$
      ii) $12 \div 2 \times (3 - 1)$
   b) Why are the answers different? Explain.

Apply

7. Evaluate. State which operation you do first.
   a) $7(4) - 5$
   b) $6[2 + (-5)]$
   c) $(-3) + 4(7)$
   d) $(-6) + 4(-2)$
   e) $15 \div [10 \div (-2)]$
   f) $18 \div 2(-6)$

8. Evaluate. Show all steps.
   a) $6(5 - 7) - 3$
   b) $4 - [5 + (-11)]$
   c) $[4 - (-8)] \div 6$
   d) $8 - 66 \div (-11)$
   e) $(-24) \div 12 + (-3)(-4)$
   f) $6(-3) + (-8)(-4)$

9. Evaluate. Show all steps.
   a) $\frac{(-7) \times 4 + 8}{4}$
   b) $\frac{4 + (-36) \div 4}{-5}$
   c) $\frac{-32}{(-6)(-2) - (-4)}$
   d) $\frac{9}{(-3)+(-18) \div 3}$

10. Evaluate. Show all steps.
    a) $\frac{4(-3) + 7(-4)}{5(-1)}$
    b) $\frac{[19-(-5)] \div (-3)}{2(-2)}$
    c) $\frac{32 \div 4 - (-28) \div 7}{12 \div (-4)}$
    d) $\frac{12 - 4(-6)}{[3 - (-3)] \times (-3)}$

11. **Assessment Focus** Robert, Brenna, and Christian got different answers for this problem: $(-40) - 2[(-8) \div 2]$
Robert's answer was −32, Christian's answer was −48, and Brenna's answer was 168.
   a) Which student had the correct answer?
   b) Show and explain how the other two students got their answers. What errors did they make?

12. Evaluate each expression. Then insert one pair of square brackets in each expression so it evaluates to −5.
   a) $(-20) \div 2 - (-2)$
   b) $(-21) + 6 \div 3$
   c) $10 + 3 \times 2 - 7$

13. Keisha had $405 in her bank account. In one month, she made 4 withdrawals of $45 each. What is the balance in her account? Write an integer expression to represent this problem. Solve the problem. How did you decide which operations to use?

14. Use three −4s and any operations or brackets. Write an expression with a value of:
   a) −12   b) −4   c) 0
   d) −3   e) 5   f) 2

15. **Take It Further** The daily highest temperatures for one week in February were: −2°C, +5°C, −8°C, −4°C, −11°C, −10°C, −5°C
Find the mean highest temperature. How did you decide which operations to use?

16. **Take It Further** Write an expression for each statement. Evaluate each expression.
   a) Divide the sum of −24 and 4 by −5.
   b) Multiply the sum of −4 and 10 by −2.
   c) Subtract 4 from −10, then divide by −2.

17. **Take It Further** Copy each equation. Replace each ☐ with the correct sign $(+, -, \times, \div)$ to make each equation true.
   a) $(-10) \,\square\, (-2) \,\square\, 1 = 21$
   b) $(-5) \,\square\, (-2) \,\square\, 4 = 1$
   c) $6 \,\square\, (-7) \,\square\, 2 = -44$
   d) $(-2)(-2) \,\square\, 8 = -4$

Reflect

Suppose you evaluate an expression that has different operations. How do you know where to begin? How do you know what to do next? Make up an integer expression that has three operations. Explain how you evaluate it.

Understanding the Problem

Have you ever tried to solve
a math problem you didn't understand?
The first step in solving a problem is understanding it.

Consider this problem:
Work with a partner.
Use these integers: –9, –5, –2, 0, 1, 3, 5
Replace each * in the expression below
with an integer to get the greatest value.
Each integer can only be used once.

$(*)(*) + (*) \div (*) - (*)$

Explain why you placed the negative integers
where you did.
Why did this help produce the greatest value?

Suppose you don't understand a problem.
Here are some strategies you can use to help.

- Explain the problem to someone else in your own words.
- Break the problem down into parts.
  Read each part on its own.
  Think about what each part means.
- Highlight the important words.
- Decide what your answer will look like.
  Will your answer include:
  – a graph?
  – a number?
  – a table?
  – a diagram?
  – a written explanation?
- Think about how many parts your answer needs.

Here is one way to think about the problem:
Work with a partner.

Use these integers: −9, −5, −2, 0, 1, 3, 5
Replace each * in the expression below with an integer
to get the greatest value.

> These words help me make sense of what to do.

Each integer can only be used once.

(*) (*) + (*) ÷ (*) − (*)

Explain why you placed the negative integers
where you did.
Why did this help produce the greatest value?

> I should write a two-part answer.

Use at least one of the strategies on page 94 to help you understand
each of these problems. Solve each problem.

1. Replace each * with a positive integer to make a true statement.
Find at least 4 different ways to do this.
(*) − (*) = −4

2. Insert one pair of square brackets in this expression so it has
the least value. Explain your strategy.
(−5) + (+4) − (−2) ÷ (−1) × (−3)

3. Robbie collected these daily high temperatures for one week in January:
−10°C, −5°C, −1°C, 0°C, 1°C, 5°C
He forgot to record the temperature on the seventh day.
The newspaper reported that the mean high
temperature for the week was 0°C.
What was the high temperature on the seventh day?
How do you know?

Strategies for Success: Understanding the Problem

Unit Review

What Do I Need to Know?

✓ **Multiplying Integers**

The product of two integers with the same sign is a positive integer.
$(+6) \times (+4) = +24$; $(-18) \times (-3) = +54$

The product of two integers with different signs is a negative integer.
$(-8) \times (+5) = -40$; $(+9) \times (-6) = -54$

The sign of a product with an even number of negative factors is positive.
$(-2) \times (-2) \times (-2) \times (-2) = +16$

The sign of a product with an odd number of negative factors is negative.
$(-2) \times (-2) \times (-2) \times (-2) \times (-2) = -32$

✓ **Dividing Integers**

The quotient of two integers with the same sign is a positive integer.
$(+56) \div (+8) = \frac{+56}{+8} = +7$; $(-24) \div (-6) = \frac{-24}{-6} = +4$

The quotient of two integers with different signs is a negative integer.
$(-30) \div (+6) = \frac{-30}{+6} = -5$; $(+56) \div (-7) = \frac{+56}{-7} = -8$

✓ **Order of Operations**

- Do the operations in brackets first.
- Multiply and divide, in order, from left to right.
- Add and subtract, in order, from left to right.

When the expression is written as a fraction:
- Evaluate the numerator and denominator separately.
- Then divide the numerator by the denominator.

What Should I Be Able to Do?

LESSON

2.1

1. Write each multiplication as a repeated addition. Then use coloured tiles to find each sum.
   a) $(+2) \times (-1)$ b) $(+2) \times (+9)$
   c) $(+3) \times (-3)$ d) $(+3) \times (+7)$

2. Use a model to find each product.
   a) $(-7) \times (-5)$ b) $(+10) \times (-6)$
   c) $(-4) \times (+4)$ d) $(+6) \times (+8)$

3. The temperature change in a chemistry experiment was $-2°C$ every 30 min. The initial temperature was $6°C$. What was the temperature after 4 h?

4. Will each product be positive or negative? How do you know?
   a) $(+25) \times (-31)$ b) $(-13) \times (-15)$
   c) $(-11) \times (+12)$ d) $(+9) \times (+13)$

2.2

5. Find each product.
   a) $(+9) \times (-7)$ b) $(+4) \times (+7)$
   c) $(-11) \times (+13)$ d) $(-40) \times (-22)$
   e) $(-1) \times (+17)$ f) $(-37) \times 0$

6. Copy each equation. Replace ☐ with an integer to make the equation true.
   a) $(-12) \times ☐ = +72$
   b) $☐ \times (+8) = +80$
   c) $(+7) \times ☐ = 0$
   d) $☐ \times (-4) = -60$

7. An old bucket has a small leak. Fifty-five millilitres of water leak out in 1 h. Use integers to find how much water leaks out in 6 h.

8. Write a word problem that could be solved using the expression $(+5) \times (-7)$. Solve the problem.

2.3

9. Use coloured tiles. Find each quotient. Sketch the tiles you used.
   a) $(+15) \div (+3)$ b) $(+36) \div (-9)$
   c) $(-21) \div (+7)$ d) $(-27) \div (-3)$

10. Use a model to find each quotient.
    a) $(+18) \div (-3)$ b) $(+14) \div (+2)$
    c) $(-28) \div (+4)$ d) $(-30) \div (-6)$

11. Tyler decides that, starting this week, he will withdraw $5 from his savings account each week.
    a) How many weeks from now will Tyler have withdrawn $65?
    b) Explain how you can use integers to model this situation.
    c) What assumptions do you make?

2.4

12. Will each quotient be positive or negative? How do you know?
    a) $(+26) \div (-2)$ b) $(-32) \div (-8)$
    c) $(-1) \div (+1)$ d) $(+42) \div (+7)$

LESSON

13. Divide.
   a) $(-56) \div (-7)$ b) $(+40) \div (-5)$
   c) $(-88) \div (+8)$ d) $(+28) \div (+2)$

14. Divide.
   a) $\frac{-18}{-2}$ b) $\frac{+16}{-4}$
   c) $\frac{-18}{+6}$ d) $\frac{0}{-9}$

15. Divide.
   a) $(+24) \div (-12)$ b) $(-63) \div (+21)$
   c) $(+75) \div (+15)$ d) $(-99) \div (-11)$

16. Moira removed 3 candies from the jar every day. She now has removed 63 candies. For how many days did Moira remove candies?
   a) Write this problem as a division expression using integers.
   b) Solve the problem.
   c) What assumptions do you make?

17. Write a word problem that could be solved using the expression $(+72) \div (-9)$. Solve the problem.

18. Find all the divisors of -21. Write a division equation each time.

2.5
19. State which operation you do first.
   a) $4 - 6(-2)$
   b) $(-18) \div (-9) - 3$
   c) $[7 - (-3)] \div 5$
   d) $4(-6) \div (-2)$

20. Evaluate each expression in question 19. Show all steps.

21. Evaluate.
   a) $(-8) \div (-4) + 6(-3)$
   b) $(-5) + (-12) \div (-3)$
   c) $18 + 3[10 \div (-5)]$
   d) $(-16) \div 8[7 - (-2)]$

22. Evaluate. Show all steps.
   a) $\frac{3(-6) - 3}{-7}$
   b) $\frac{(-4) + [(-7) - (-2)]}{3}$
   c) $\frac{20}{(-3) + (-14) \div 7}$

23. Evaluate. Show all steps.
   a) $\frac{[18 - (-4)] \div (-11)}{(-4) + 2}$
   b) $\frac{5(-2) + (-12) \div 3}{28 \div (-4)}$
   c) $\frac{(-8)(-3)}{(-16) \div [(-13) - (-9)]}$

24. In a darts game, Suzanne and Corey each threw the darts 10 times. Corey had three $(+2)$ scores, three (-3) scores, and four $(+1)$ scores. Suzanne had four $(+2)$ scores, four (-3) scores, and two $(+1)$ scores.

   a) What was each person's final score? How did you decide which operations to use?
   b) The winner had the greater score. Who won the game? Explain.

Practice Test

1. Evaluate.
   a) $(+9) \times (+10)$
   b) $(+6) \times (-11)$
   c) $(+96) \div (-16)$
   d) $(+39) \div (+3)$
   e) $(-8) \times (+6)$
   f) $(-36) \div (+9)$
   g) $(-44) \div (-4)$
   h) $(-5) \times (-1)$

2. Evaluate.
   a) $(-20)(-5) + 16 \div (-8)$
   b) $\dfrac{14 - 10 \div 2}{-3}$
   c) $\dfrac{[(-9) - (-2)] \times [8 + (-4)]}{(-14) \div (-2)}$
   d) $[7 - (-2)] \times 2 + (-12) \div (-4)$

3. The temperature on Sunday was 4°C. The temperature dropped 8°C on Monday and dropped twice as much on Tuesday. What was the temperature on Tuesday? How did you decide which operations to use?

4. Suppose you own a store. Use integers to model each situation. For each situation, calculate the money you receive or spend.
   a) Six people come into your store. Each person buys items worth $15.
   b) You pay 3 bills. Each bill is for $35.
   c) A supplier gives you $7 for each case of his product that you sell. You sold 9 cases this month.

5. Use the integers below.
   $0, -2, +3, -1, +1, +2, +4$
   a) Find two pairs of integers that have a quotient of -2.
   b) Which two integers have the greatest product?
   c) Which two integers have the least sum?
   d) Which two integers have a quotient less than -3?
   e) Write your own problem using two of the integers. Solve your problem. Show your work.

Unit Problem Charity Golf Tournament

A Grade 8 class and a local bank sponsor a golf tournament to raise money for local charities.
The bank provides these prizes:

> 1st place—$5000 to a charity of the player's choice
> 2nd and 3rd places—$1000 to a charity of the player's choice

Golf Terms

- "Par" is the number of strokes it should take for a player to reach the hole.
  If par is 3 and you take 5 strokes, then your score in relation to par is +2, or 2 over.
  If par is 3 and you take 2 strokes, then your score in relation to par is −1, or 1 under.
- A *bogey* is 1 stroke more than par, or 1 over par.
- A *double bogey* is 2 strokes more than par, or 2 over par.
- A *birdie* is 1 stroke less than par, or 1 under par.
- An *eagle* is 2 strokes less than par, or 2 under par.

Here are the top 6 golfers:
Chai Kim, Delaney, Hamid, Hanna, Kyle, and Weng Kwong

1. The golf course has 9 holes. Here is one person's results: par on 3 holes, a bogey on 2 holes, a birdie on 1 hole, an eagle on 2 holes, and a double bogey on 1 hole
   a) Write an integer expression to represent these results. How did you decide which operations to use?
   b) Evaluate the expression in part a to calculate the score in relation to par.
2. Chai Kim wrote his results in a table like this.

Hole	1	2	3	4	5	6	7	8	9
Par	3	4	3	3	5	4	4	3	3
Under/Over Par	0	−1	+2	0	−1	0	0	−1	0
Score	3	3	5	3					

a) Copy and complete the table. Use the following information:
   Hole 1, 4, 6, 7, 9   Par
   Hole 2, 5, 8   Birdie
   Hole 3   Double bogey
b) What was Chai Kim's final score?
c) What was his final score in relation to par?

3. For each person below, make a table similar to the table in question 2. Use the information below. What is each golfer's final score?

a) Kyle: Bogey holes   1, 3, 5, 9
            Birdie hole   6
            Par holes   2, 4, 7, 8
b) Delaney: Bogey holes   3, 4, 6
            Birdie holes   1, 2, 7, 8, 9
            Eagle hole   5
c) Hamid: Birdie every hole except hole 8
            Double bogey hole 8

4. a) Hanna had a score of −5 in relation to par.
   Weng Kwong had a score of +3 in relation to par.
   Use the information in questions 2 and 3.
   Rank the players in order from least to greatest score.
b) Who won the tournament and the $5000 prize?
   What was the score in relation to par?
c) Who won the $1000 prizes?
   What were the scores in relation to par?

5. Use a table similar to that in question 2.
Complete the table with scores of your choice.
Calculate the final score, and the final score in relation to par.

Check List

Your work should show:
- ✓ how you used integers to solve the problems
- ✓ accurate calculations and ordering of integers
- ✓ tables you constructed to display the scores
- ✓ clear explanations, with correct use of mathematical language

Reflect on Your Learning

What did you find easy about working with integers? What was difficult for you? Give examples to illustrate your answers.

UNIT 3
Operations with Fractions

Look at the shapes at the right. How is each shape made from the previous shape? The equilateral triangle in *Step 1* has side length 1 unit. What is the perimeter of each shape?

What patterns do you see? Suppose the patterns continue. What will the perimeter of the shape in *Step 4* be?

For any shape after *Step 1*, how is its perimeter related to the perimeter of the preceding shape?

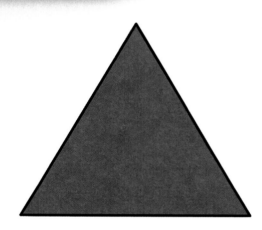

Step 1

What You'll Learn

- Estimate the products and quotients of fractions and mixed numbers.
- Multiply a fraction by a whole number and by a fraction.
- Divide a fraction by a whole number and by a fraction.
- Multiply and divide mixed numbers.
- Use the order of operations with fractions.

Why It's Important

You use fractions when you shop, measure, and work with percents and decimals. You also use fractions in sports, cooking, and business.

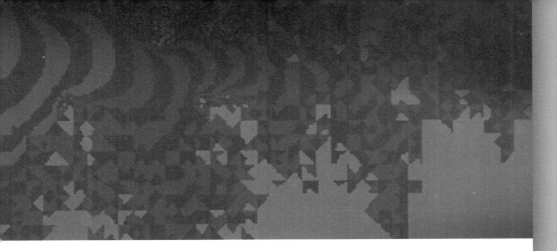

Step 2   Step 3

Key Words

- product
- factor
- proper fraction
- equivalent fraction
- reciprocal
- mixed number
- improper fraction
- quotient
- divisor
- dividend

3.1 Using Models to Multiply Fractions and Whole Numbers

Focus Use models to multiply a fraction by a whole number.

How many ways can you find this sum?
$2 + 2 + 2 + 2 + 2 + 2 + 2 + 2 + 2$

Investigate

Work with a partner.
Use any models to help.

Each of 4 students needs $\frac{5}{6}$ of a bag of oranges to make a pitcher of freshly squeezed orange juice.
Each bag of oranges contains 12 oranges.
How many bags of oranges are used?

Compare your strategy for solving the problem with that of another pair of classmates.
Did you use addition to solve the problem?
If so, which addition expression did you use?
How could you represent the problem with a multiplication expression?

104   UNIT 3: Operations with Fractions

Connect

We can find a meaning for: $\frac{1}{5} + \frac{1}{5} + \frac{1}{5} + \frac{1}{5} = \frac{4}{5}$

➤ Repeated addition can be written as multiplication.
$\frac{1}{5}$ is added 4 times.
$\frac{1}{5} + \frac{1}{5} + \frac{1}{5} + \frac{1}{5} = 4 \times \frac{1}{5} = \frac{4}{5}$
$\frac{1}{5} + \frac{1}{5} + \frac{1}{5} + \frac{1}{5} = \frac{1}{5} \times 4 = \frac{4}{5}$
We can show this as a picture.

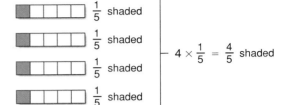

Similarly: $\frac{3}{4} + \frac{3}{4} + \frac{3}{4} + \frac{3}{4} + \frac{3}{4} + \frac{3}{4} + \frac{3}{4} = \frac{21}{4}$

But, $\frac{3}{4} + \frac{3}{4} + \frac{3}{4} + \frac{3}{4} + \frac{3}{4} + \frac{3}{4} + \frac{3}{4} = 7 \times \frac{3}{4} = \frac{21}{4}$

So, $7 \times \frac{3}{4} = \frac{21}{4}$

Also, $\frac{3}{4} \times 7 = \frac{21}{4}$

➤ We can use a number line divided into fourths to show that $7 \times \frac{3}{4} = \frac{21}{4}$.

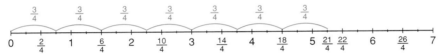

We can use a different number line to show that $\frac{3}{4} \times 7 = \frac{21}{4}$.

For $\frac{3}{4} \times 7$, we can think: $\frac{3}{4}$ of 7

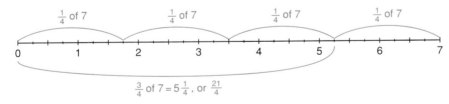

➤ Another way to multiply $7 \times \frac{3}{4}$:
Sketch a rectangle with base 7 units and height 1 unit.
Divide the height into fourths.
Shade the rectangle with base 7 and height $\frac{3}{4}$.

The area of the shaded rectangle is: base × height = $7 \times \frac{3}{4}$

Each small rectangle has area: $1 \times \frac{1}{4} = \frac{1}{4}$

So, the area of the shaded rectangle is: $21 \times \frac{1}{4} = \frac{21}{4}$

Then, $7 \times \frac{3}{4} = \frac{21}{4}$ **This is a multiplication equation.**

Example 1

New flooring has been installed in two-thirds of the classrooms in the school.
There are 21 classrooms in the school.
How many classrooms have new flooring?

▶ **A Solution**

Multiply: $21 \times \frac{2}{3}$
Use a number line divided into thirds.

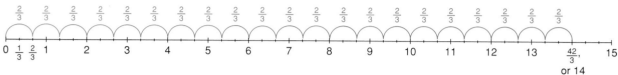

So, $21 \times \frac{2}{3} = \frac{42}{3}$, or 14
Fourteen classrooms have new flooring.

▶ **Example 1**
Another Solution

Find: $\frac{2}{3}$ of 21
Use counters.
Model 21 with counters.

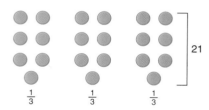

Find thirds by dividing the counters into 3 equal groups.
Each group contains 7 counters.
$\frac{1}{3}$ of $21 = 7$
So, $\frac{2}{3}$ of $21 = 14$
Fourteen classrooms have new flooring.

Example 2

An office building with four floors has rented out $\frac{3}{5}$ of each floor.
How many floors of the building have been rented?

▶ **A Solution**

Multiply: $4 \times \frac{3}{5}$

$4 \times \frac{3}{5} = \frac{3}{5} + \frac{3}{5} + \frac{3}{5} + \frac{3}{5}$

Model the expression $\frac{3}{5} + \frac{3}{5} + \frac{3}{5} + \frac{3}{5}$ with fraction circles.

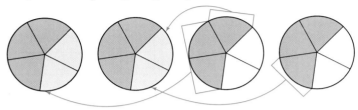

Put the fifths together to make wholes.
2 wholes and two fifths equal $2\frac{2}{5}$.

So, $4 \times \frac{3}{5} = 2\frac{2}{5}$

$2\frac{2}{5}$ floors of the office building have been rented.

▶ **Example 2**
Another Solution

Multiply: $4 \times \frac{3}{5}$

Sketch a rectangle with base 4 units and height 1 unit.
Divide the height into fifths.
Shade the rectangle with base 4 and height $\frac{3}{5}$.
The area of the shaded rectangle is:
base × height = $4 \times \frac{3}{5}$

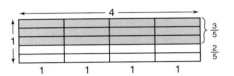

Each small rectangle has area: $1 \times \frac{1}{5} = \frac{1}{5}$

So, the shaded area is: $12 \times \frac{1}{5} = \frac{12}{5}$, or $2\frac{2}{5}$

So, $4 \times \frac{3}{5} = 2\frac{2}{5}$

$2\frac{2}{5}$ floors of the office building have been rented.

Discuss the ideas

1. Why can a product be written as repeated addition?
2. When might you not want to use repeated addition to find a product?
3. How could you use a rectangle model to solve the problem in *Example 1*?
4. How could you use a number line to solve the problem in *Example 2*?

Practice

Check

5. Write each statement as a multiplication statement in two ways.
  a) $\frac{5}{9}$ of 45   b) $\frac{3}{8}$ of 32
  c) $\frac{1}{12}$ of 36   d) $\frac{4}{5}$ of 25

6. Write each repeated addition as a multiplication statement in two ways.
  a) $\frac{1}{4} + \frac{1}{4} + \frac{1}{4}$
  b) $\frac{2}{5} + \frac{2}{5} + \frac{2}{5} + \frac{2}{5} + \frac{2}{5} + \frac{2}{5} + \frac{2}{5}$
  c) $\frac{3}{10} + \frac{3}{10} + \frac{3}{10} + \frac{3}{10}$

7. Use fraction circles to find: $\frac{2}{3} \times 6$
  a) Write the multiplication as repeated addition.
  b) Use fraction circles to find the sum.
  c) Sketch the fraction circles.
  d) Write the multiplication equation the fraction circles represent.

8. Write the two multiplication equations each number line represents.
  a)
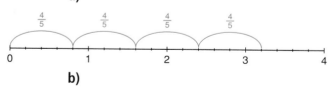
  b)
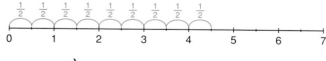
  c)
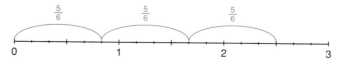

Apply

9. For each diagram below, state the product the shaded area represents.
  a)

  b)

10. Write the two multiplication statements each set of fraction circles represents. Then find each product.
  a)
  b)

11. Use fraction circles to find each product. Sketch the fraction circles. Write a multiplication equation each time.
  a) $5 \times \frac{1}{8}$ b) $\frac{2}{5} \times 3$ c) $4 \times \frac{5}{12}$

12. Use counters to help you find each product.
  a) $\frac{1}{2} \times 24$ b) $\frac{1}{3} \times 24$ c) $\frac{1}{4} \times 24$
  d) $\frac{1}{6} \times 24$ e) $\frac{1}{8} \times 24$ f) $\frac{1}{12} \times 24$

13. Use the results in question 12 to find each product.
  a) $\frac{2}{2} \times 24$ b) $\frac{2}{3} \times 24$ c) $\frac{3}{4} \times 24$
  d) $\frac{5}{6} \times 24$ e) $\frac{3}{8} \times 24$ f) $\frac{5}{12} \times 24$

14. Multiply. Draw a picture or number line to show each product.
   a) $3 \times \frac{4}{7}$
   b) $\frac{2}{15} \times 10$
   c) $4 \times \frac{9}{4}$
   d) $\frac{2}{5} \times 7$

15. Draw and shade rectangles to find each product.
   a) $\frac{1}{3} \times 12$
   b) $\frac{1}{5} \times 15$
   c) $\frac{3}{5} \times 15$
   d) $\frac{3}{8} \times 16$

16. Multiply.
   a) $3 \times \frac{4}{5}$
   b) $5 \times \frac{7}{9}$
   c) $\frac{5}{3} \times 6$
   d) $\frac{1}{2} \times 5$
   e) $12 \times \frac{7}{8}$
   f) $\frac{2}{4} \times 9$

17. It takes $\frac{2}{3}$ h to pick all the apples on one tree at Springwater Farms. There are 24 trees. How long will it take to pick all the apples? Show your work.

18. Assessment Focus
   a) Describe a situation that could be represented by $5 \times \frac{3}{8}$.
   b) Draw a picture to show $5 \times \frac{3}{8}$.
   c) What meaning can you give to $\frac{3}{8} \times 5$? Draw a picture to show your thinking.

19. Parri used the expression $\frac{5}{8} \times 16$ to solve a word problem. What might the word problem be? Solve the problem.

20. Naruko went to the West Edmonton Mall. She took $28 with her. She spent $\frac{4}{7}$ of her money on rides. How much money did Naruko spend on rides? Use a model to show your answer.

21. Take It Further
   a) Use models. Multiply.
      i) $2 \times \frac{1}{2}$
      ii) $3 \times \frac{1}{3}$
      iii) $4 \times \frac{1}{4}$
      iv) $5 \times \frac{1}{5}$
   b) Look at your answers to part a. What do you notice? How can you explain your findings?
   c) Write two different multiplication statements with the same product as in part a.

22. Take It Further Jacques takes $\frac{3}{4}$ h to fill one shelf at the supermarket. Henri can fill the shelves in two-thirds Jacques' time. There are 15 shelves. Henri and Jacques work together. How long will it take to fill the shelves? Justify your answer.

Reflect

Explain how your knowledge of adding fractions helped you in this lesson. Include an example.

3.2 Using Models to Multiply Fractions

Focus Use models to multiply fractions.

Describe a picture that shows 38 × 25.

Investigate

Work with a partner.
Model this problem with concrete materials.

> One-quarter of a cherry pie was left over after dinner.
> Graham ate one-half of the leftover pie for lunch the next day.
> What fraction of the whole pie did he have for lunch?
> What if Graham had eaten only one-quarter of the leftover pie.
> What fraction of the whole pie would he have eaten?

How did you solve the problem?
Compare your solutions and strategies with those of another pair of classmates.
Was one strategy more efficient than another? Explain.

Connect

We can use different models to find the product of two fractions.
The following *Examples* show how we can use Pattern Blocks, counters, and a rectangle model.

Example 1

Sandi cut $\frac{2}{3}$ of the grass on a lawn.
Akiva cut $\frac{1}{2}$ of the remaining grass.
What fraction of the lawn did Akiva cut?

▶ **A Solution**

Use Pattern Blocks.
Let the yellow hexagon represent the lawn.
6 green triangles cover the yellow hexagon.
So, 6 green triangles also represent the lawn.
4 green triangles represent the grass cut by Sandi.
2 green triangles represent the remaining grass.

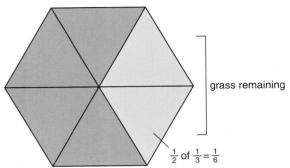

Akiva cut $\frac{1}{2}$ of the remaining grass.
One-half of the remaining grass is 1 green triangle.
One green triangle represents $\frac{1}{6}$.
So, $\frac{1}{2} \times \frac{1}{3} = \frac{1}{6}$
Akiva cut $\frac{1}{6}$ of the lawn.

Example 2

Multiply: $\frac{2}{3} \times \frac{6}{8}$

▶ **A Solution**

Use counters.

Think: We want $\frac{2}{3}$ of $\frac{6}{8}$ of one whole set of counters.

Model one whole set of eighths with eight counters.
There are 6 counters in $\frac{6}{8}$.
To model thirds, arrange the 6 counters into 3 equal groups.
Each group of 2 counters represents $\frac{1}{3}$.

So, $\frac{2}{3}$ of 6 counters is 4 counters.
But, 4 counters are part of a whole set of 8 counters.
So, 4 counters represent $\frac{4}{8}$, or $\frac{1}{2}$ of the original whole set.
So, $\frac{2}{3} \times \frac{6}{8} = \frac{1}{2}$

Example 3

One-half of the Grade 8 students tried out for the school's lacrosse team. Three-quarters of these students were successful. What fraction of the Grade 8 students are on the team?

▶ **A Solution**

Three-quarters of one-half of the Grade 8 students are on the team.

Draw a rectangle.
Show $\frac{1}{2}$ of the rectangle.

Divide $\frac{1}{2}$ of the rectangle into quarters.
Shade $\frac{3}{4}$.

Use broken lines to divide the whole rectangle into equal parts.
There are 8 equal parts.
Three parts are shaded.

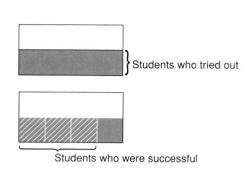

$\frac{3}{4} \times \frac{1}{2} = \frac{3}{8}$

So, $\frac{3}{8}$ of the Grade 8 students are on the team.

 **Discuss the ideas**

1. In *Example 1*, suppose Akiva cut $\frac{1}{4}$ of the remaining grass. Could we use Pattern Blocks to model the problem? Justify your answer.
2. In *Example 2*, why did we start with 8 counters? Could we have started with a different number of counters? Justify your answer.
3. Look at the area model in *Example 3*. The factors are fractions. How do the numbers of rows and columns in the product relate to the factors? How does the number of parts in the whole relate to the factors?
4. In *Example 3*, could you have divided the rectangle in half a different way? Would you get the same answer? Explain.

Practice

Check

5. Draw a rectangle to multiply $\frac{3}{5} \times \frac{1}{4}$.
  a) Divide the rectangle into fourths vertically. How many fourths will you shade?
  b) Divide the rectangle into fifths horizontally. How many fifths will you shade?
  c) How many equal parts does the rectangle have?
  d) How many of these parts have been shaded twice?
  e) What is the product of $\frac{3}{5} \times \frac{1}{4}$?

6. Copy each rectangle onto grid paper. Shade the rectangle to find each product.
  a) $\frac{1}{2} \times \frac{3}{4}$
  b) $\frac{3}{4} \times \frac{2}{3}$
  c) $\frac{2}{5} \times \frac{1}{2}$
  d) $\frac{5}{6} \times \frac{1}{2}$
  e) $\frac{3}{5} \times \frac{7}{8}$
  f) $\frac{4}{5} \times \frac{3}{4}$

7. Use counters to find each product. Draw a diagram to record your work.
  a) $\frac{3}{4} \times \frac{12}{15}$
  b) $\frac{4}{5} \times \frac{10}{18}$
  c) $\frac{1}{2} \times \frac{4}{12}$
  d) $\frac{1}{4} \times \frac{8}{9}$
  e) $\frac{5}{9} \times \frac{18}{24}$
  f) $\frac{2}{3} \times \frac{15}{20}$

Apply

8. Find each product.
  a) $\frac{3}{4} \times \frac{5}{8}$
  b) $\frac{4}{9} \times \frac{2}{5}$
  c) $\frac{1}{4} \times \frac{2}{3}$
  d) $\frac{6}{7} \times \frac{2}{3}$
  e) $\frac{2}{3} \times \frac{1}{3}$
  f) $\frac{4}{5} \times \frac{4}{5}$

9. Write 3 multiplication statements using proper fractions. Make sure each statement is different from any statements you have worked with so far. Use a model to illustrate each product.

> Recall that, in a proper fraction, the numerator is less than the denominator.

10. Write the multiplication equation represented by each diagram below.
  a)
  b)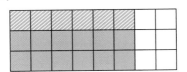
  c)
  d)

11. Barry used $\frac{5}{8}$ of the money he had saved to buy a DVD player and 4 DVDs. The cost of the DVD player was $\frac{2}{5}$ of the amount he spent. What fraction of his savings did Barry spend on the DVD player?

3.2 Using Models to Multiply Fractions

12. **Assessment Focus**
    a) Find each product.
        i) $\frac{3}{4} \times \frac{2}{5}$ ii) $\frac{2}{4} \times \frac{3}{5}$
        iii) $\frac{1}{4} \times \frac{3}{8}$ iv) $\frac{3}{4} \times \frac{1}{8}$
        v) $\frac{3}{5} \times \frac{4}{6}$ vi) $\frac{3}{6} \times \frac{4}{5}$
    b) What patterns do you see in the answers for part a? Use a model to illustrate the patterns.
    c) Write some other products similar to those in part a.
    Show your work.

13. Use a model to answer each question.
    a) One-third of the students in a class wear glasses. One-half of the students who wear glasses are girls. What fraction of the class is girls who wear glasses?
    b) John has $\frac{2}{3}$ of a tank of gas. He uses $\frac{3}{4}$ of the gas to get home. What fraction of a tank of gas does John use to get home? What fraction of the tank of gas is left?
    c) Justin ate $\frac{3}{5}$ of a box of raisins. His sister then ate $\frac{1}{4}$ of the raisins left in the box. What fraction of the box of raisins did Justin's sister eat? What fraction of the box of raisins remained?

14. Gwen used the expression $\frac{4}{9} \times \frac{1}{5}$ to solve a word problem. What might the word problem be? Solve the problem.

15. **Take It Further** One-eighth of the seats in the movie theatre were empty. $\frac{2}{7}$ of the seats that were full were filled with teenagers. What fraction of the theatre was filled with teenagers?

16. **Take It Further** Why is $\frac{5}{8}$ of $\frac{3}{12}$ equal to $\frac{3}{8}$ of $\frac{5}{12}$? Use a model to explain your answer.

17. **Take It Further**
    a) Look at the diagram. How does it show $\frac{3}{5}$ of one whole?

    b) Use the same diagram. Explain how it shows that one whole is $\frac{5}{3}$ of $\frac{3}{5}$.

Reflect

When you use an area model to multiply two fractions, how do you decide how to draw the rectangle?
Include an example in your explanation.

3.3 Multiplying Fractions

Focus Develop an algorithm to multiply fractions.

Which multiplication equation does this diagram represent?
How do you know?

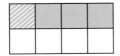

Investigate

Work with a partner.
Use an area model to find each product.

- $\frac{2}{3} \times \frac{4}{5}$
- $\frac{1}{2} \times \frac{3}{8}$
- $\frac{3}{5} \times \frac{4}{7}$
- $\frac{2}{5} \times \frac{3}{8}$

Write the multiplication equations in a table.
Look at the table.
What patterns do you notice?
How can you use patterns to multiply $\frac{2}{3} \times \frac{4}{5}$?
Use your patterns to calculate $\frac{7}{8} \times \frac{3}{10}$.
Use an area model to check your product.

Compare your strategies with those of another pair of classmates.
How does your strategy work?
Does your strategy work with $\frac{2}{5} \times \frac{3}{4}$?
Do you think your strategy will work with all fractions? Explain.

Connect

Here is an area model to show: $\frac{4}{7} \times \frac{2}{5} = \frac{8}{35}$

The product of the numerators is:
$4 \times 2 = 8$
The product of the denominators is:
$7 \times 5 = 35$
That is, $\frac{4}{7} \times \frac{2}{5} = \frac{4 \times 2}{7 \times 5}$
$= \frac{8}{35}$

Check if there are common factors in the numerator and denominator.

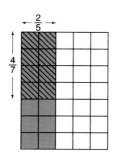

So, to multiply two fractions, multiply the numerators and multiply the denominators.

We can use this method to multiply proper fractions and improper fractions.

Example 1

Multiply. Estimate to check the product is reasonable.
$\frac{7}{5} \times \frac{8}{3}$

▶ **A Solution**

$\frac{7}{5} \times \frac{8}{3}$

There are no common factors in the numerators and denominators.
So, $\frac{7}{5} \times \frac{8}{3} = \frac{7 \times 8}{5 \times 3}$
$= \frac{56}{15}$
$= \frac{45}{15} + \frac{11}{15}$
$= 3 + \frac{11}{15}$, or $3\frac{11}{15}$

Recall that factors are the numbers that are multiplied to get a product; for example, 2 and 5 are factors of 10 because $2 \times 5 = 10$.

45 is the multiple of 15 that is closest to 56, and less than 56.

Estimate to check.
$\frac{7}{5}$ is between 1 and 2, but closer to 1.
$\frac{8}{3}$ is between 2 and 3, but closer to 3.

So, the product is about $1 \times 3 = 3$.

Since $3\frac{11}{15}$ is close to 3, the product is reasonable.

116 UNIT 3: Operations with Fractions

Example 2

Three-eighths of the animals in a pet store are fish.
Two-fifteenths of the fish are tropical fish.
What fraction of the animals in the pet store are tropical fish?
Use benchmarks to check the solution is reasonable.

▶ **A Solution**

Since $\frac{3}{8}$ of the animals are fish and $\frac{2}{15}$ of the fish are tropical fish,
then the fraction of animals that are tropical fish is $\frac{2}{15}$ of $\frac{3}{8}$, or $\frac{2}{15} \times \frac{3}{8}$.

$\frac{2}{15} \times \frac{3}{8} = \frac{2 \times 3}{15 \times 8}$ Multiply the numerators and multiply the denominators.

$\phantom{\frac{2}{15} \times \frac{3}{8}} = \frac{6}{120}$ Simplify. Divide by the common factor, 6.

$\phantom{\frac{2}{15} \times \frac{3}{8}} = \frac{6 \div 6}{120 \div 6}$

$\phantom{\frac{2}{15} \times \frac{3}{8}} = \frac{1}{20}$

Estimate to check.
$\frac{2}{15}$ is close to 0.
$\frac{3}{8}$ is about $\frac{1}{2}$.
So, $\frac{2}{15} \times \frac{3}{8}$ is close to 0.

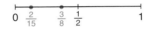

Since $\frac{1}{20}$ is close to 0, the product is reasonable.
One-twentieth of the animals in the pet store are tropical fish.

▶ **Example 2
Another Solution**

Here is another way to calculate.
$\frac{2}{15} \times \frac{3}{8} = \frac{2 \times 3}{15 \times 8}$
Notice that the numerator and denominator have common
factors 2 and 3.
To simplify first, divide the numerator
and denominator by these factors.

$\frac{2}{15} \times \frac{3}{8} = \frac{\cancel{2}^1 \times \cancel{3}^1}{\cancel{15}_5 \times \cancel{8}_4}$

$\phantom{\frac{2}{15} \times \frac{3}{8}} = \frac{1 \times 1}{5 \times 4}$ $2 \div 2 = 1$ $3 \div 3 = 1$
$\phantom{\frac{2}{15} \times \frac{3}{8} = } \phantom{\frac{1 \times 1}{5 \times 4}}$ $15 \div 3 = 5$ $8 \div 2 = 4$

$\phantom{\frac{2}{15} \times \frac{3}{8}} = \frac{1}{20}$

> Dividing a fraction by a common factor of the numerator and denominator produces an equivalent fraction.

One-twentieth of the animals in the pet store are tropical fish.

▶ **Example 2**
Another Solution

$\frac{2}{15} \times \frac{3}{8} = \frac{2 \times 3}{15 \times 8}$

The numerator and denominator have common factors 2 and 3.
Write the denominator to show the common factors.

$\frac{2}{15} \times \frac{3}{8} = \frac{2 \times 3}{3 \times 5 \times 2 \times 4}$ Rewrite making fractions that equal 1.

$= \frac{2}{2} \times \frac{3}{3} \times \frac{1}{5 \times 4}$

$= 1 \times 1 \times \frac{1}{20}$

When multiplying by 1, the value of the fraction does not change.

$= \frac{1}{20}$

One-twentieth of the animals in the pet store are tropical fish.

Discuss the ideas

1. Why is it important to estimate to check the product?
2. Look at the different solutions to *Example 2*. Why is it often helpful to simplify the fractions before multiplying?
3. How do you recognize when fractions can be simplified before you multiply them?

Practice

Check

4. Find the common factors of each pair of numbers.
   a) 4, 12　　b) 14, 21　　c) 8, 16
   d) 6, 9　　 e) 10, 15　　f) 18, 24

5. Multiply: $\frac{5}{6} \times \frac{3}{20}$
   a) Multiply. Simplify first.
   b) Use benchmarks to estimate the product.
   c) Is the product reasonable? How do you know?

6. In a First Nations school, five-eighths of the Grade 8 students play the drums. Of these students, three-tenths also play the native flute. What fraction of the Grade 8 students play both the drums and the native flute? Estimate to check the solution is reasonable.

118   UNIT 3: Operations with Fractions

Apply

7. Multiply. Simplify before multiplying. Use benchmarks to estimate to check the product is reasonable.
   a) $\frac{3}{4} \times \frac{8}{5}$ b) $\frac{1}{3} \times \frac{9}{10}$ c) $\frac{7}{5} \times \frac{15}{21}$
   d) $\frac{5}{9} \times \frac{3}{5}$ e) $\frac{2}{9} \times \frac{15}{4}$ f) $\frac{7}{3} \times \frac{9}{14}$

8. Multiply. Use benchmarks to estimate to check the product is reasonable.
   a) $\frac{3}{5} \times \frac{2}{3}$ b) $\frac{1}{2} \times \frac{5}{10}$ c) $\frac{1}{6} \times \frac{1}{4}$
   d) $\frac{13}{8} \times \frac{3}{2}$ e) $\frac{5}{4} \times \frac{11}{10}$ f) $\frac{7}{3} \times \frac{7}{8}$
   Which of these questions could have been solved using mental math? Justify your choice.

9. Solve each problem. Estimate to check the solution is reasonable.
   a) Josten took $\frac{3}{8}$ of his savings on a shopping trip. He used $\frac{1}{4}$ of the money to buy a new coat. What fraction of his savings did Josten spend on the coat?
   b) Gervais ate $\frac{1}{3}$ of a baguette with his dinner. Chantel ate $\frac{1}{4}$ of the leftover baguette as an evening snack. What fraction of the baguette did Chantel eat as a snack?

10. Write a story problem that can be represented by the expression $\frac{7}{8} \times \frac{1}{2}$. Solve your problem. Trade problems with a classmate. Solve your classmate's problem. Check to see that your solutions are the same.

11. Eeva spent $\frac{5}{6}$ of $\frac{3}{4}$ of her total allowance on a hair crimper. What fraction of her total allowance did Eeva have left?

12. a) Find each product.
    i) $\frac{3}{4} \times \frac{4}{3}$ ii) $\frac{1}{5} \times \frac{5}{1}$
    iii) $\frac{7}{2} \times \frac{2}{7}$ iv) $\frac{5}{6} \times \frac{6}{5}$
    b) What do you notice about the products in part a? Write 3 more pairs of fractions that have the same product. What can you say about the product of a fraction and its reciprocal?

    $\frac{11}{12}$ and $\frac{12}{11}$ are reciprocals.

13. **Assessment Focus** In question 12, each product is 1.
    a) Write a pair of fractions that have each product.
       i) 2   ii) 3   iii) 4   iv) 5
    b) Write a pair of fractions that have the product 1. Change only one numerator or denominator each time to write a pair of fractions that have each product.
       i) 2   ii) 3   iii) 4   iv) 5
    c) How can you write a pair of fractions that have the product 10? Show your work.

14. The sum of two fractions is $\frac{7}{12}$. The product of the same two fractions is $\frac{1}{12}$. What are the two fractions? Describe the strategy you used.

15. Multiply. Estimate to check the product is reasonable.
   a) $\frac{33}{40} \times \frac{15}{55}$
   b) $\frac{26}{39} \times \frac{9}{13}$
   c) $\frac{51}{64} \times \frac{8}{17}$
   d) $\frac{76}{91} \times \frac{7}{19}$

16. a) Multiply $\frac{24}{25} \times \frac{85}{96}$ using each strategy below.
   i) Simplify before multiplying.
   ii) Multiply first, then simplify.
   b) Which strategy in part a did you find easier? Justify your choice.

17. The product of 2 fractions is $\frac{3}{4}$. What might the fractions be? How many pairs of fractions could have a product of $\frac{3}{4}$? How do you know?

18. Take It Further Keydon baked a wild blueberry upside-down cobbler.
Shawnie ate $\frac{1}{6}$ of the cobbler.
Iris ate $\frac{1}{5}$ of what was left.
Chan ate $\frac{1}{4}$ of what was left after that.
Cami ate $\frac{1}{3}$ of what was left after that.
Demi ate $\frac{1}{2}$ of what was left after that.
How much of the original cobbler remained?

19. Take It Further The product of two fractions is $\frac{2}{3}$. One fraction is $\frac{3}{5}$. What is the other fraction? How do you know?

20. Take It Further Eddie used the expression $\frac{4}{7} \times \frac{3}{5}$ to solve a word problem. Which of these word problems better fits the expression? How do you know? Solve the problem.
   a) $\frac{4}{7}$ of the Grade 8 students voted to have Spirit Day. $\frac{3}{5}$ of those students wanted Spirit Day to be on the first day of classes. What fraction of the Grade 8 students wanted Spirit Day to be on the first day of classes?
   b) $\frac{3}{5}$ of the Grade 7 students voted to have a school dance. $\frac{4}{7}$ of those students wanted the dance to be on the day before Spring Break. What fraction of the Grade 7 students wanted the dance to be on the day before Spring Break?

21. Take It Further Find each square root. Explain the strategy you used.
   a) $\sqrt{\frac{4}{9}}$
   b) $\sqrt{\frac{16}{25}}$
   c) $\sqrt{\frac{36}{81}}$
   d) $\sqrt{\frac{49}{169}}$

Reflect

When we multiply 2 whole numbers, the product is always greater than either factor.
Is this always true when we multiply 2 fractions?
Use examples and diagrams to explain your answer.

3.4 Multiplying Mixed Numbers

Focus Apply knowledge of multiplying fractions to multiply mixed numbers.

Suppose  = 1.

How can you write the fraction representing

in 2 ways?

Investigate

Work with a partner.

During the salmon drift,
volunteers collect catch information from fisherpeople.
Akecheta volunteered for $3\frac{1}{2}$ h.

Onida volunteered for $\frac{2}{3}$ of the time
that Akecheta volunteered.
For how long did Onida volunteer?
How can you find out?
Show your work.
Use models or diagrams to justify your strategy.

Compare your strategy with that of another pair of classmates.
Do you think your strategy will work with all mixed numbers?
Test it with $\frac{3}{4} \times 2\frac{1}{3}$.

Connect

Here is an area model to show: $2\frac{1}{2} \times 1\frac{1}{3}$

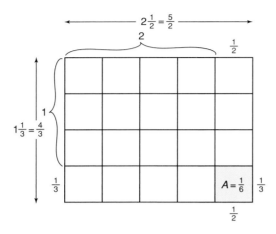

Write each mixed number as an improper fraction.
$2\frac{1}{2} = 2 + \frac{1}{2} = \frac{4}{2} + \frac{1}{2} = \frac{5}{2}$
$1\frac{1}{3} = 1 + \frac{1}{3} = \frac{3}{3} + \frac{1}{3} = \frac{4}{3}$

Each small rectangle has area: $\frac{1}{2} \times \frac{1}{3} = \frac{1}{6}$

There are 5×4, or 20 small rectangles.

So, the area is: $20 \times \frac{1}{6} = \frac{20}{6}$
$= \frac{10}{3}$
$= 3\frac{1}{3}$

After we write the mixed numbers as improper fractions, we can multiply the same way we multiplied proper fractions.

$2\frac{1}{2} \times 1\frac{1}{3} = \frac{5}{2} \times \frac{4}{3}$
$= \frac{5 \times 4}{2 \times 3}$
$= \frac{20}{6}$
$= \frac{20 \div 2}{6 \div 2}$
$= \frac{10}{3}$
$= 3\frac{1}{3}$

This is the same product as when we used the area model.
We can also use a rectangle model to multiply two mixed numbers.

Example 1

Multiply: $2\frac{1}{2} \times 1\frac{1}{3}$

▶ A Solution

Use a rectangle model.

$2\frac{1}{2} \times 1\frac{1}{3} = (2 \times 1) + (\frac{1}{2} \times 1) + (2 \times \frac{1}{3}) + (\frac{1}{2} \times \frac{1}{3})$

$\phantom{2\frac{1}{2} \times 1\frac{1}{3}} = 2 + \frac{1}{2} + \frac{2}{3} + \frac{1}{6}$ Add. Use common denominators.

$\phantom{2\frac{1}{2} \times 1\frac{1}{3}} = 2 + \frac{3}{6} + \frac{4}{6} + \frac{1}{6}$

$\phantom{2\frac{1}{2} \times 1\frac{1}{3}} = 2 + \frac{8}{6}$

$\phantom{2\frac{1}{2} \times 1\frac{1}{3}} = 2 + \frac{6}{6} + \frac{2}{6}$

$\phantom{2\frac{1}{2} \times 1\frac{1}{3}} = 2 + 1 + \frac{2}{6}$

$\phantom{2\frac{1}{2} \times 1\frac{1}{3}} = 3\frac{2}{6}$, or $3\frac{1}{3}$

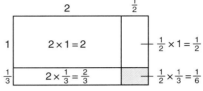

Remember to write the product in simplest form.

Example 2

Multiply. Estimate to check the product is reasonable.

$2\frac{1}{4} \times 3\frac{2}{5}$

▶ A Solution

$2\frac{1}{4} \times 3\frac{2}{5}$

$2\frac{1}{4} = 2 + \frac{1}{4}$ and $3\frac{2}{5} = 3 + \frac{2}{5}$ So, $2\frac{1}{4} \times 3\frac{2}{5} = \frac{9}{4} \times \frac{17}{5}$

$\phantom{2\frac{1}{4}} = \frac{8}{4} + \frac{1}{4}$ $\phantom{3\frac{2}{5}} = \frac{15}{5} + \frac{2}{5}$ $\phantom{So, 2\frac{1}{4} \times 3\frac{2}{5}} = \frac{153}{20}$

$\phantom{2\frac{1}{4}} = \frac{9}{4}$ $\phantom{3\frac{2}{5}} = \frac{17}{5}$ $\phantom{So, 2\frac{1}{4} \times 3\frac{2}{5}} = \frac{140}{20} + \frac{13}{20}$

$\phantom{So, 2\frac{1}{4} \times 3\frac{2}{5} aaaaa} = 7\frac{13}{20}$

Estimate to check.

$2\frac{1}{4}$ is between 2 and 3, but closer to 2.

$3\frac{2}{5}$ is between 3 and 4, but closer to 3.

So, the product is about $2 \times 3 = 6$.

Since $7\frac{13}{20}$ is close to 6, the product is reasonable.

Example 3

Multiply. Estimate to check the product is reasonable.
$3\frac{3}{8} \times 4\frac{2}{3}$

▶ **A Solution**

$3\frac{3}{8} \times 4\frac{2}{3}$

$3\frac{3}{8} = 3 + \frac{3}{8}$ and $4\frac{2}{3} = 4 + \frac{2}{3}$
$= \frac{24}{8} + \frac{3}{8}$ $\qquad\qquad = \frac{12}{3} + \frac{2}{3}$
$= \frac{27}{8}$ $\qquad\qquad\qquad = \frac{14}{3}$

So, $3\frac{3}{8} \times 4\frac{2}{3} = \frac{\overset{9}{\cancel{27}}}{\underset{4}{\cancel{8}}} \times \frac{\overset{7}{\cancel{14}}}{\underset{1}{\cancel{3}}}$ Divide by common factors.

$= \frac{9 \times 7}{4 \times 1}$ $\qquad\qquad\qquad$ $27 \div 3 = 9 \qquad 14 \div 2 = 7$
$\qquad\qquad\qquad\qquad\qquad\qquad\;\; 8 \div 2 = 4 \qquad 3 \div 3 = 1$
$= \frac{63}{4}$

$= \frac{60}{4} + \frac{3}{4}$

$= 15\frac{3}{4}$

Estimate to check.
$3\frac{3}{8}$ is between 3 and 4, but closer to 3.
$4\frac{2}{3}$ is between 4 and 5, but closer to 5.
So, the product is about $3 \times 5 = 15$.

Since $15\frac{3}{4}$ is close to 15, the product is reasonable.

Discuss the ideas

1. What is the difference between a proper fraction and an improper fraction?
2. How is multiplying two mixed numbers like multiplying two fractions?
3. How is the rectangle model useful when you multiply 2 mixed numbers?

Practice

Check

4. Write the mixed number and improper fraction represented by each picture.

a)

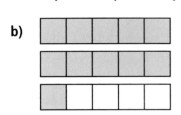

b)

c)

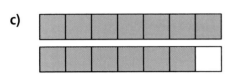

5. Write each mixed number as an improper fraction.
a) $2\frac{3}{10}$ b) $4\frac{1}{8}$ c) $3\frac{5}{6}$
d) $1\frac{2}{3}$ e) $3\frac{2}{5}$ f) $5\frac{1}{2}$
g) $2\frac{4}{7}$ h) $3\frac{5}{9}$ i) $6\frac{2}{3}$

6. Write each improper fraction as a mixed number.
a) $\frac{11}{3}$ b) $\frac{15}{4}$ c) $\frac{21}{5}$
d) $\frac{11}{8}$ e) $\frac{19}{6}$ f) $\frac{31}{7}$
g) $\frac{11}{2}$ h) $\frac{43}{10}$ i) $\frac{37}{8}$

7. Use estimation. Which number is each product closer to?
a) $2\frac{1}{8} \times 3\frac{3}{4}$ 6 or 8
b) $3\frac{5}{9} \times 1\frac{5}{6}$ 6 or 8
c) $7\frac{3}{8} \times 2\frac{4}{5}$ 21 or 24
d) $4\frac{7}{9} \times 3\frac{5}{12}$ 15 or 20

8. Multiply: $3\frac{3}{5} \times 2\frac{2}{9}$
a) Estimate the product.
b) Write each mixed number as an improper fraction.
c) Multiply the improper fractions. Simplify first.
d) Is the product reasonable? How do you know?

Apply

9. Multiply. Estimate to check the product is reasonable.
a) $3 \times 2\frac{1}{4}$ b) $4 \times 2\frac{1}{8}$
c) $1\frac{2}{3} \times 2$ d) $3\frac{1}{5} \times 3$

10. Use an area model to find each product.
a) $1\frac{1}{2} \times 1\frac{1}{3}$ b) $2\frac{3}{4} \times 2\frac{2}{3}$
c) $1\frac{1}{5} \times 3\frac{1}{3}$ d) $1\frac{1}{2} \times 2\frac{2}{5}$

11. Use improper fractions to find each product. Estimate to check the product is reasonable.
a) $1\frac{7}{8} \times 2\frac{2}{3}$ b) $4\frac{1}{6} \times 3\frac{2}{5}$
c) $2\frac{3}{7} \times 1\frac{5}{9}$ d) $3\frac{1}{2} \times 2\frac{2}{7}$
e) $2\frac{1}{4} \times 2\frac{2}{3}$ f) $1\frac{4}{5} \times 2\frac{1}{3}$

12. Multiply. Estimate to check the product is reasonable.
a) $1\frac{3}{4} \times 2\frac{1}{2}$ b) $3\frac{2}{3} \times 2\frac{1}{5}$
c) $4\frac{3}{8} \times 1\frac{1}{4}$ d) $3\frac{3}{4} \times 3\frac{3}{4}$
e) $4\frac{3}{10} \times \frac{4}{5}$ f) $\frac{7}{8} \times 2\frac{3}{5}$

13. A restaurant in Richmond, BC, lists the prices on its menu in fractions of a dollar. Three friends have lunch at the restaurant. Each of 3 friends orders a veggie mushroom cheddar burger for $11\frac{3}{4}$, with a glass of water to drink.
   a) What was the total bill before taxes, in fractions of a dollar?
   b) What was the total bill before taxes, in dollars and cents?

14. During the school year, the swim team practises $2\frac{3}{4}$ h per week.
During the summer, the weekly practice time is increased to $2\frac{1}{3}$ times the school-year practice time. How many hours per week does the team practise during the summer?

15. Write a story problem that can be represented by the expression $3\frac{1}{2} \times 2\frac{1}{8}$. Solve your problem. Trade problems with a classmate. Solve your classmate's problem. Check to see that your solutions are the same.

16. In a baseball game, the starting pitcher for the home team pitched $4\frac{2}{3}$ innings. The starting pitcher for the visiting team pitched $1\frac{1}{2}$ times as many innings. How many innings did the visiting team's pitcher pitch?

17. Assessment Focus Students baked cookies for a charity bake sale. Elsa baked $2\frac{1}{2}$ dozen cookies. Layton baked $2\frac{1}{6}$ times as many cookies as Elsa. Meghan and Josh together baked $5\frac{1}{3}$ times the number of cookies that Elsa baked.
   a) Estimate. About how many dozen cookies did Layton bake? About how many dozen cookies did Meghan and Josh bake altogether?
   b) Calculate how many dozen cookies Layton baked.
   c) Calculate how many dozen cookies Meghan and Josh baked.
   d) How many dozen cookies did these 4 students bake altogether?
   e) How many cookies did these 4 students bake altogether?
   Show your work.

18. Take It Further Use estimation. Which expression below has the greatest product? The least product? How do you know?
   a) $\frac{4}{3} \times \frac{8}{6}$ b) $2\frac{1}{8} \times 1\frac{1}{5}$
   c) $1\frac{3}{8} \times \frac{9}{4}$ d) $\frac{7}{2} \times 2\frac{3}{10}$

19. Take It Further Multiply. Estimate to check the product is reasonable.
   a) $2\frac{4}{9} \times 2\frac{2}{3} \times 2\frac{1}{2}$
   b) $3\frac{3}{5} \times 2\frac{3}{4} \times 1\frac{1}{4}$
   c) $4\frac{3}{8} \times 1\frac{1}{5} \times 2\frac{1}{4}$

Reflect

Describe 2 strategies you can use to multiply $3\frac{1}{2} \times 5\frac{1}{4}$.
Which strategy do you prefer? Why?

Spinning Fractions

HOW TO PLAY

Your teacher will give you a copy of the spinner.
Use an open paper clip as the pointer.
Use a sharp pencil to keep the pointer in place.
Record the scores in a chart.

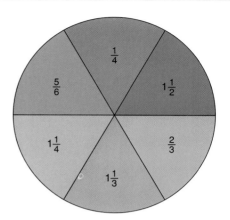

1. Player A spins the pointer twice.
   Player A adds the fractions.
   Player B multiplies the fractions.
   The player with the greater result gets one point.

2. Player B spins the pointer twice.
   Player A adds the fractions.
   Player B multiplies the fractions.
   The player with the greater result gets one point.

3. Players continue to take turns spinning the pointer.
   The first person to get 12 points wins.

REFLECT

- Do you think this game is fair?
  How many games do you need to play to find out?
  a) If this game is fair, explain how you know.
  b) If this game is not fair, how could you make the game a fair game?
- Without playing the game many times, how else could you find out if the game is fair?

YOU WILL NEED

A copy of the spinner; an open paper clip; a sharp pencil

NUMBER OF PLAYERS

2

GOAL OF THE GAME

To be the first to get 12 points

Mid-Unit Review

LESSON

3.1

1. Write each multiplication statement as repeated addition. Draw a picture to show each product.
   a) $4 \times \frac{1}{8}$
   b) $7 \times \frac{3}{5}$
   c) $\frac{5}{6} \times 3$
   d) $\frac{2}{9} \times 6$

2. Multiply. Draw a number line to show each product.
   a) $\frac{1}{4} \times 7$
   b) $8 \times \frac{3}{8}$
   c) $6 \times \frac{7}{10}$
   d) $\frac{5}{12} \times 3$

3. Sasha had 16 tomatoes in his garden. He gave: Samira $\frac{1}{8}$ of the tomatoes; Amandeep $\frac{1}{2}$ of the tomatoes; and Amina $\frac{1}{4}$ of the tomatoes.
   a) How many tomatoes did Sasha give away?
   b) How many tomatoes did Sasha have left?
   c) What fraction of the tomatoes did Sasha have left?

3.2

4. Draw a rectangle to find each product.
   a) $\frac{5}{8} \times \frac{1}{2}$
   b) $\frac{2}{3} \times \frac{3}{4}$
   c) $\frac{1}{2} \times \frac{4}{5}$
   d) $\frac{5}{6} \times \frac{3}{10}$

5. Use counters to find each product. Draw a diagram each time.
   a) $\frac{1}{2} \times \frac{4}{9}$
   b) $\frac{2}{3} \times \frac{6}{15}$
   c) $\frac{3}{4} \times \frac{8}{11}$
   d) $\frac{2}{5} \times \frac{10}{12}$

3.3

6. Multiply. Use benchmarks to estimate to check each product is reasonable.
   a) $\frac{1}{2} \times \frac{2}{3}$
   b) $\frac{4}{5} \times \frac{1}{4}$
   c) $\frac{3}{4} \times \frac{3}{8}$
   d) $\frac{4}{9} \times \frac{15}{18}$

7. Aiko says that $\frac{2}{3}$ of her stamp collection are Asian stamps. One-fifth of her Asian stamps are from India. What fraction of Aiko's stamp collection is from India? Estimate to check the solution is reasonable.

3.4

8. Use an area model to find each product.
   a) $2\frac{2}{3} \times 1\frac{7}{8}$
   b) $\frac{10}{3} \times \frac{5}{2}$
   c) $4\frac{3}{4} \times \frac{3}{8}$
   d) $1\frac{5}{6} \times 4\frac{1}{2}$

9. Multiply. Estimate to check each product is reasonable.
   a) $2\frac{1}{2} \times 3\frac{1}{4}$
   b) $4\frac{2}{5} \times \frac{1}{4}$
   c) $\frac{7}{3} \times \frac{6}{5}$
   d) $5\frac{1}{2} \times 2\frac{5}{8}$

10. Alek has a $1\frac{1}{4}$-h dance lesson every Saturday. The time he spends practising dance during the week is $3\frac{3}{5}$ times the length of his dance lesson. How long does Alek practise dance each week?

UNIT 3: Operations with Fractions

3.5 Dividing Whole Numbers and Fractions

Focus Use models to divide proper fractions and whole numbers.

When you first studied division, you learned two ways: sharing and grouping
For example, 20 ÷ 5 can be thought of as:

- Sharing 20 items equally among 5 sets
  There are 4 items in each set.

- Grouping 20 items into sets of 5
  There are 4 sets.

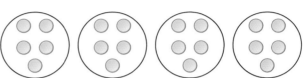

Recall that multiplication and division are inverse operations.
We know: 20 ÷ 5 = 4
What related multiplication facts do you know?

Investigate

Work with a partner.
You will need scissors.
Your teacher will give you 2 large copies of this diagram.

Square A models the whole number 1.
Figure B represents $\frac{3}{4}$ of Square A.
Which number does Rectangle C model?
Use paper cutting.

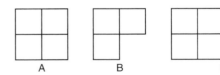

➤ Use one copy of the diagram.
  Find: $2 \div \frac{1}{4}$
➤ Use the other copy of the diagram.
  Find: $2 \div \frac{3}{4}$
  Illustrate your answers.

Compare your answers with those of another pair of classmates.
Did you deal with the leftover pieces the same way?
If not, explain your method to your classmates.

3.5 Dividing Whole Numbers and Fractions   **129**

Connect

➤ We can use a number line to divide a whole number by a fraction.
To find how many thirds are in 6, divide 6 into thirds.

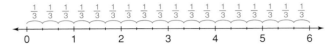

There are 18 thirds in 6.
Write this as a division equation.
$6 \div \frac{1}{3} = 18$

➤ Use the same number line to find how many two-thirds are in 6.

Arrange 18 thirds into groups of two-thirds.
There are 9 groups of two-thirds.
We write: $6 \div \frac{2}{3} = 9$

➤ Use the number line again to find how many four-thirds are in 6; that is, $6 \div \frac{4}{3}$.
Arrange 18 thirds into groups of four-thirds.

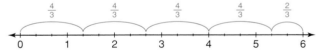

There are 4 groups of 4 thirds.
There are 2 thirds left over.
Think: What fraction of 4 thirds is 2 thirds?

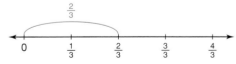

From the number line, $\frac{2}{3}$ is $\frac{1}{2}$ of $\frac{4}{3}$.
So, $6 \div \frac{4}{3} = 4\frac{1}{2}$

We can also use a number line to divide a fraction by a whole number.
This is illustrated in *Example 1*.

Example 1

Find each quotient.

a) Benny has one-half a litre of milk to pour equally among 3 glasses.
How much milk should he pour into each glass?

b) Chen and Luke equally shared $\frac{3}{4}$ of a pizza.
How much of the whole pizza was each person's share?

▶ **A Solution**

a) Find: $\frac{1}{2} \div 3$

Think: Share $\frac{1}{2}$ into 3 equal parts.

Use a number line. Mark $\frac{1}{2}$ on the line.

Divide the interval 0 to $\frac{1}{2}$ into 3 equal parts.

$\frac{1}{2} = \frac{3}{6}$

Each part is $\frac{1}{6}$.

So, $\frac{1}{2} \div 3 = \frac{1}{6}$

Benny should pour $\frac{1}{6}$ of a litre of milk into each glass.

b) Find: $\frac{3}{4} \div 2$

Think: Share $\frac{3}{4}$ into 2 equal parts.

Use a number line. Mark $\frac{3}{4}$ on the line.

Divide the interval 0 to $\frac{3}{4}$ into 2 equal parts.

To label this point, divide the fourths into eighths.

$\frac{3}{4} = \frac{6}{8}$

Each part is $\frac{3}{8}$.

So, $\frac{3}{4} \div 2 = \frac{3}{8}$

Each person's share was $\frac{3}{8}$ of the pizza.

Sometimes, when we divide fractions and whole numbers, there is a remainder.
This remainder is written as a fraction of the divisor.

Example 2

Use a model to divide: $5 \div \frac{3}{5}$

▶ **A Solution**

$5 \div \frac{3}{5}$

Think: How many $\frac{3}{5}$ are in 5 wholes?
Use fraction circles in fifths to model 5.

Count groups of three-fifths.
There are 8 groups of three-fifths.
There is 1 fifth left over.
The diagram shows that $\frac{1}{5}$ is $\frac{1}{3}$ of $\frac{3}{5}$.

So, $5 \div \frac{3}{5} = 8\frac{1}{3}$

Discuss the ideas

1. In *Example 1a*, the quotient is less than the dividend; that is, $\frac{1}{6} < \frac{1}{2}$.
   In *Example 2*, the quotient is greater than the dividend; that is, $8\frac{1}{3} > 5$.
   Why do you think this happens?
2. How can you use multiplication to check the quotient?

Practice

Check

3. Use each picture to find the quotient. Write the division equation each time.

   a) $4 \div \frac{1}{3}$

   b) $3 \div \frac{1}{6}$

   c) $4 \div \frac{2}{3}$

   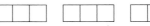 

   d) $3 \div \frac{3}{5}$

132 UNIT 3: Operations with Fractions

4. Use fraction circles to find: $4 \div \frac{5}{6}$
   a) Use fraction circles to model 4. How did you know which fraction circles to use?
   b) How many groups of five-sixths are in 4? What is the remainder?
   c) What fraction of $\frac{5}{6}$ does the remainder represent?
   d) Write the division equation.

5. Ioana wants to spend $\frac{4}{5}$ of an hour studying each subject. She has 4 h to study. How many subjects can she study?

6. Use fraction circles. Find each quotient.
   a) $2 \div \frac{1}{2}$ b) $3 \div \frac{1}{3}$ c) $4 \div \frac{1}{4}$
   d) $2 \div \frac{1}{6}$ e) $3 \div \frac{1}{2}$ f) $6 \div \frac{3}{4}$

Apply

7. Which division statement might each picture represent? How many different statements can you write each time? Use fraction circles if they help.
   a)
   b)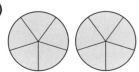
   c)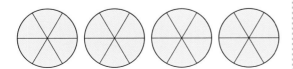

8. Use a number line to find each quotient.
   a) i) $2 \div \frac{1}{3}$ ii) $2 \div \frac{2}{3}$
   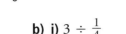
   b) i) $3 \div \frac{1}{4}$ ii) $3 \div \frac{2}{4}$ iii) $3 \div \frac{3}{4}$
   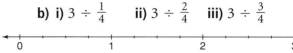
   c) i) $\frac{4}{8} \div 2$ ii) $\frac{4}{8} \div 4$ iii) $\frac{4}{8} \div 8$
   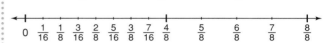

9. Use a model. Find each quotient.
   a) $5 \div \frac{2}{3}$
   b) $4 \div \frac{3}{4}$
   c) $\frac{1}{2} \div 5$
   d) $\frac{5}{8} \div 2$

10. Find each quotient. Use number lines to illustrate the answers.
   a) How many $\frac{1}{4}$-size sheets can be cut from 5 sheets of paper?
   b) How many $\frac{2}{3}$-cup servings are in 6 cups of fruit?
   c) Janelle feeds her cat $\frac{4}{5}$ of a tin of cat food each day. Janelle has 12 tins of cat food. How many days' supply of cat food does Janelle have?

11. Find each quotient. Use number lines to illustrate the answers.
   a) Three-quarters of a whole pizza is shared equally among 5 people. What fraction of the whole pizza does each person get?
   b) One-third of a carton of eggs is used to make a large omelette. How many large omelettes can be made from 4 cartons of eggs?
   c) Brandon planted trees for $\frac{11}{12}$ h. He planted 5 trees. Assume Brandon took the same amount of time to plant each tree. What fraction of an hour did it take to plant 1 tree?

12. Assessment Focus Copy these boxes.

   a) Write the digits 2, 4, and 6 in the boxes to find as many division expressions as possible.
   b) Which expression in part a has the greatest quotient? The least quotient? How do you know? Show your work.

13. Is $\frac{2}{3} \div 4$ the same as $4 \div \frac{2}{3}$? Use number lines in your explanation.

14. Take It Further
   a) Divide: $8 \div \frac{1}{3}$
   b) Divide: $\frac{1}{8} \div 3$
   c) Look at the quotients in parts a and b. What do you notice? How can you explain this?

15. Take It Further The numbers $\frac{9}{2}$ and 3 share this property: their difference is equal to their quotient. That is, $\frac{9}{2} - 3 = \frac{3}{2}$ and $\frac{9}{2} \div 3 = \frac{3}{2}$. Find other pairs of numbers with this property. Describe any patterns you see.

Math Link

Your World

Usually, a chef measures fractional amounts, such as $1\frac{1}{2}$ cups. Occasionally, a chef changes a recipe to serve more or fewer people. To do this, the amounts of ingredients are increased or decreased, often by multiplying or dividing fractions, mixed numbers, or whole numbers.

Reflect

When you divide a whole number by a proper fraction, is the quotient greater than or less than the whole number? Include an example in your explanation.

3.6 Dividing Fractions

Focus Develop algorithms to divide fractions.

You have used grouping to divide a whole number by a fraction: $4 \div \frac{2}{3} = 6$

You have used sharing to divide a fraction by a whole number: $\frac{2}{3} \div 4 = \frac{1}{6}$

You will now investigate dividing a fraction by a fraction.

Investigate

Work with a partner.
Use this number line. Find: $\frac{2}{3} \div \frac{1}{4}$

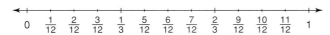

Use this number line. Find: $\frac{3}{5} \div \frac{1}{2}$

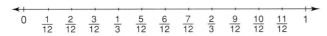

Use this number line. Find: $\frac{1}{3} \div \frac{1}{4}$

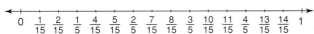

Use this number line. Find: $\frac{4}{5} \div \frac{2}{3}$

Look at the quotients.
How do the numbers in the numerators and denominators
relate to the quotients?
Try to find a strategy to calculate the quotient without using a number line.
Use a different division problem to check your strategy.

Compare your strategy with that of another pair of classmates.
Does your strategy work with their problem? Explain.
Does their strategy work with your problem? Explain.

Connect

Here are two ways to divide fractions.

➤ Use common denominators.
To divide: $\frac{4}{5} \div \frac{1}{10}$
Write each fraction with a common denominator.
Since 5 is a factor of 10, 10 is a common denominator.

So, $\frac{4}{5} \div \frac{1}{10} = \frac{8}{10} \div \frac{1}{10}$

When the denominators are the same, divide the numerators.

This means: How many 1 tenths are in 8 tenths?

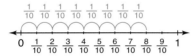

From the number line, this is the same as $8 \div 1 = 8$.
So, $\frac{4}{5} \div \frac{1}{10} = 8$

➤ Use multiplication.
Every whole number can be written as a fraction with denominator 1.
Here are some division equations from Lesson 3.5 and their related multiplication equations.

Each multiplication equation uses the reciprocal of the divisor.
This pattern is always true.

Division Equation		Related Multiplication Equation
$5 \div \frac{3}{5} = \frac{25}{3}$ or $\frac{5}{1} \div \frac{3}{5} = \frac{25}{3}$		$\frac{5}{1} \times \frac{5}{3} = \frac{25}{3}$
$4 \div \frac{3}{4} = \frac{16}{3}$ or $\frac{4}{1} \div \frac{3}{4} = \frac{16}{3}$		$\frac{4}{1} \times \frac{4}{3} = \frac{16}{3}$
$6 \div \frac{5}{3} = \frac{18}{5}$ or $\frac{6}{1} \div \frac{5}{3} = \frac{18}{5}$		$\frac{6}{1} \times \frac{3}{5} = \frac{18}{5}$
$\frac{1}{2} \div 3 = \frac{1}{6}$ or $\frac{1}{2} \div \frac{3}{1} = \frac{1}{6}$		$\frac{1}{2} \times \frac{1}{3} = \frac{1}{6}$
$\frac{3}{4} \div 2 = \frac{3}{8}$ or $\frac{3}{4} \div \frac{2}{1} = \frac{3}{8}$		$\frac{3}{4} \times \frac{1}{2} = \frac{3}{8}$
$\frac{4}{8} \div 4 = \frac{1}{8}$ or $\frac{4}{8} \div \frac{4}{1} = \frac{1}{8}$		$\frac{4}{8} \times \frac{1}{4} = \frac{1}{8}$

We can use the same patterns to divide two fractions.
For example, to find $\frac{3}{5} \div \frac{1}{4}$, write the reciprocal of the divisor, then multiply.
$\frac{3}{5} \div \frac{1}{4} = \frac{3}{5} \times \frac{4}{1}$
$= \frac{12}{5}$, or $2\frac{2}{5}$

Example 1

Divide. Estimate to check each quotient is reasonable.
$\frac{3}{4} \div \frac{5}{6}$

▶ **A Solution**

$\frac{3}{4} \div \frac{5}{6}$

Use multiplication.

$\frac{3}{4} \div \frac{5}{6}$ can be written as

Dividing by $\frac{5}{6}$ is the same as multiplying by $\frac{6}{5}$.

$\frac{3}{4} \times \frac{6}{5} = \frac{3 \times \cancel{6}^3}{\cancel{4}_2 \times 5}$ Simplify. Divide by common factor 2.

$= \frac{9}{10}$

Estimate to check.

$\frac{3}{4}$ is about 1. $\frac{5}{6}$ is about 1.

So, $\frac{3}{4} \div \frac{5}{6}$ is about $1 \div 1 = 1$.

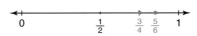

Since $\frac{5}{6}$ is greater than $\frac{3}{4}$,

$\frac{3}{4} \div \frac{5}{6}$ is less than 1.

Since $\frac{9}{10}$ is close to 1, and less than 1, the quotient is reasonable.

Example 2

Divide: $\frac{7}{8} \div \frac{1}{4}$

▶ **A Solution**

$\frac{7}{8} \div \frac{1}{4}$

Use common denominators.
Since 8 is a multiple of 4, 8 is a common denominator.
Multiply the numerator and denominator by 2: $\frac{1}{4} = \frac{2}{8}$

$\frac{7}{8} \div \frac{1}{4} = \frac{7}{8} \div \frac{2}{8}$

$= 7 \div 2$

$= \frac{7}{2}$, or $3\frac{1}{2}$

Since the denominators are the same, divide the numerators.

Estimate to check. $\frac{7}{8}$ is close to 1, but less than 1.

Since there are 4 quarters in one whole, $\frac{7}{8} \div \frac{1}{4}$ is close to 4, but less than 4.

Since $3\frac{1}{2}$ is close to 4, and less than 4, the quotient is reasonable.

Example 3

Use a number line to illustrate the quotient.

$\frac{3}{5} \div \frac{1}{4}$

▶ A Solution

$\frac{3}{5} \div \frac{1}{4}$

Use common denominators.
Write each fraction with a common denominator.
Since 5 and 4 have no common factors,
a common denominator is $5 \times 4 = 20$.

$\frac{3}{5} \div \frac{1}{4} = \frac{12}{20} \div \frac{5}{20}$

This means: How many 5 twentieths are in 12 twentieths?

From the number line, there are 2 groups of 5 twentieths with remainder 2 twentieths.
Write the remainder as a fraction of $\frac{5}{20}$.

2 twentieths is $\frac{2}{5}$ of 5 twentieths.

So, $\frac{3}{5} \div \frac{1}{4} = 2\frac{2}{5}$

Discuss the ideas

1. Without dividing, how do you know if the quotient of $\frac{5}{6} \div \frac{2}{3}$ is less than or greater than 1?
2. Why is it important to estimate to check the quotient?
3. What is a strategy for dividing two fractions?

Practice

Check

4. Write the reciprocal of each fraction.
  a) $\frac{5}{9}$ b) $\frac{3}{7}$ c) $\frac{7}{8}$ d) $\frac{14}{15}$

5. Use a copy of each number line to illustrate each quotient.
  a) $\frac{3}{4} \div \frac{3}{8}$

  b) $\frac{1}{2} \div \frac{1}{8}$

  c) $\frac{5}{6} \div \frac{5}{12}$

6. Divide: $\frac{3}{5} \div \frac{9}{10}$
  a) What is the reciprocal of $\frac{9}{10}$?
  b) Use multiplication. Simplify first.
  c) Estimate the quotient.
  d) Is the quotient reasonable? How do you know?

Apply

7. Use a copy of each number line to illustrate each quotient.
  a) $\frac{5}{6} \div \frac{1}{3}$

  b) $\frac{3}{4} \div \frac{1}{3}$

  c) $\frac{7}{9} \div \frac{1}{3}$

  d) $\frac{5}{8} \div \frac{1}{4}$
  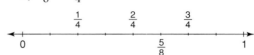

8. Find each quotient.
  a) $\frac{7}{10} \div \frac{3}{10}$ b) $\frac{5}{9} \div \frac{2}{9}$
  c) $\frac{3}{5} \div \frac{2}{5}$ d) $\frac{4}{5} \div \frac{2}{5}$

9. Use multiplication to find each quotient.
  a) $\frac{8}{5} \div \frac{3}{4}$ b) $\frac{9}{10} \div \frac{5}{3}$
  c) $\frac{7}{2} \div \frac{4}{3}$ d) $\frac{1}{2} \div \frac{7}{6}$

10. Use common denominators to find each quotient.
  a) $\frac{7}{12} \div \frac{1}{4}$ b) $\frac{3}{5} \div \frac{11}{10}$
  c) $\frac{5}{2} \div \frac{1}{3}$ d) $\frac{5}{6} \div \frac{9}{8}$

11. Divide. Estimate to check each quotient is reasonable.
  a) $\frac{5}{3} \div \frac{3}{5}$ b) $\frac{4}{9} \div \frac{4}{9}$ c) $\frac{1}{6} \div \frac{5}{2}$

12. Suppose you have $\frac{11}{12}$ of a cake. How many servings can you make of each size?
  a) $\frac{1}{4}$ of the cake   b) $\frac{1}{3}$ of the cake
  c) $\frac{1}{6}$ of the cake   d) $\frac{1}{2}$ of the cake

3.6 Dividing Fractions **139**

13. a) Find each quotient.
   i) $\frac{3}{4} \div \frac{5}{8}$ ii) $\frac{5}{8} \div \frac{3}{4}$
   iii) $\frac{7}{12} \div \frac{2}{5}$ iv) $\frac{2}{5} \div \frac{7}{12}$
   v) $\frac{5}{3} \div \frac{4}{5}$ vi) $\frac{4}{5} \div \frac{5}{3}$

   b) In part a, what patterns do you see in the division statements and their quotients? Write two more pairs of division statements that follow the same pattern.

14. As a busboy in a restaurant, Amiel takes $\frac{1}{12}$ h to clear and reset a table. How many tables can Amiel clear in $\frac{2}{3}$ h? Estimate to check the solution is reasonable.

15. Divide. Estimate to check each quotient is reasonable.
   a) $\frac{27}{28} \div \frac{9}{14}$ b) $\frac{15}{22} \div \frac{3}{11}$
   c) $\frac{32}{51} \div \frac{8}{17}$ d) $\frac{57}{69} \div \frac{19}{115}$

16. To conduct a science experiment, each pair of students requires $\frac{1}{16}$ cup of vinegar. The science teacher has $\frac{3}{4}$ cup of vinegar. How many pairs of students can conduct the experiment?

17. Assessment Focus
   a) Copy the boxes below. Write the digits 2, 3, 4, and 5 in the boxes to make as many different division statements as you can.

   $$\frac{\Box}{\Box} \div \frac{\Box}{\Box}$$

   b) Which division statement in part a has the greatest quotient? The least quotient? How do you know? Show your work.

18. Tahoe used the expression $\frac{7}{8} \div \frac{1}{4}$ to solve a word problem. What might the word problem be? Solve the problem. Estimate to check the solution is reasonable.

19. Take It Further Copy each division equation. Replace each $\Box$ with a fraction to make each equation true. Explain the strategy you used.
   a) $\frac{2}{3} \div \Box = \frac{8}{9}$ b) $\frac{3}{11} \div \Box = \frac{12}{55}$
   c) $\frac{1}{4} \div \Box = \frac{9}{20}$ d) $\frac{4}{5} \div \Box = \frac{28}{45}$

20. Take It Further Write as many division statements as you can that have a quotient between $\frac{1}{2}$ and 1. Explain the strategy you used.

Reflect

Explain how your knowledge of common denominators can help you divide two fractions. Include an example in your explanation.

3.7 Dividing Mixed Numbers

Focus Apply knowledge of dividing fractions to divide mixed numbers.

What strategies can you use to divide $\frac{5}{6} \div \frac{2}{3}$?

Investigate

Work with a partner.
Jeffrey wants to share some round Belgian waffles with his friends.
Suppose Jeffrey makes each portion $\frac{3}{4}$ of a waffle.

How many portions will he get from $4\frac{1}{2}$ waffles?
How can you find out?
Show your work.
Use models or diagrams to justify your strategy.

Compare your strategy with that of another pair of classmates.
Do you think your strategy will work with all mixed numbers?
Test it with $1\frac{1}{2} \div \frac{5}{8}$.

Connect

Here are three ways to divide mixed numbers.
In each method, the mixed numbers are first written as improper fractions.

➤ Use a number line.
To divide: $4\frac{2}{5} \div 1\frac{1}{2}$
$4\frac{2}{5} = \frac{22}{5}$ and $1\frac{1}{2} = \frac{3}{2}$
Write each fraction with a common denominator.
Since 2 and 5 have no common factors, a common denominator is $2 \times 5 = 10$.
$\frac{22}{5} = \frac{44}{10}$ and $\frac{3}{2} = \frac{15}{10}$
So, $4\frac{2}{5} \div 1\frac{1}{2} = \frac{44}{10} \div \frac{15}{10}$
This means: How many 15 tenths are in 44 tenths?
Use a number line divided in tenths.

From the number line, there are 2 groups of 15 tenths, with remainder 14 tenths.

Think: What fraction of 15 tenths is 14 tenths?

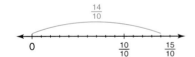

From the number line, 14 tenths is $\frac{14}{15}$ of 15 tenths.
So, $4\frac{2}{5} \div 1\frac{1}{2} = 2\frac{14}{15}$

➤ Use common denominators.
Divide: $4\frac{2}{5} \div 1\frac{1}{2}$
$4\frac{2}{5} = \frac{22}{5}$ and $1\frac{1}{2} = \frac{3}{2}$
Write each fraction with a common denominator.
$\frac{22}{5} = \frac{44}{10}$ and $\frac{3}{2} = \frac{15}{10}$
So, $4\frac{2}{5} \div 1\frac{1}{2} = \frac{44}{10} \div \frac{15}{10}$
$= 44 \div 15$ **Since the denominators are**
$= \frac{44}{15}$, or $2\frac{14}{15}$ **the same, divide the numerators.**

▶ Use multiplication.
$4\frac{2}{5} \div 1\frac{1}{2} = \frac{22}{5} \div \frac{3}{2}$

Recall that dividing by $\frac{3}{2}$ is the same as multiplying by $\frac{2}{3}$. $\frac{2}{3}$ **is the reciprocal of** $\frac{3}{2}$.

So, $\frac{22}{5} \div \frac{3}{2} = \frac{22}{5} \times \frac{2}{3}$

$= \frac{22 \times 2}{5 \times 3}$

$= \frac{44}{15}$

$= 2\frac{14}{15}$

Example 1

Divide. Estimate to check the quotient is reasonable.
$1\frac{7}{8} \div 1\frac{1}{4}$

▶ **A Solution**

$1\frac{7}{8} \div 1\frac{1}{4}$

Change the mixed numbers to improper fractions.

$1\frac{7}{8} = \frac{15}{8}$ and $1\frac{1}{4} = \frac{5}{4}$

So, $1\frac{7}{8} \div 1\frac{1}{4} = \frac{15}{8} \div \frac{5}{4}$

$= \frac{15}{8} \times \frac{4}{5}$

$= \frac{^3\cancel{15}}{_2\cancel{8}} \times \frac{\cancel{4}^1}{\cancel{5}_1}$

$= \frac{3 \times 1}{2 \times 1}$

$= \frac{3}{2}$, or $1\frac{1}{2}$

Estimate to check.
$1\frac{7}{8}$ is close to 2. $1\frac{1}{4}$ is close to 1.

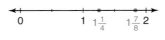

So, $1\frac{7}{8} \div 1\frac{1}{4}$ is about $2 \div 1 = 2$.

Since $1\frac{7}{8}$ is less than 2, and $1\frac{1}{4}$ is greater than 1, the quotient will be less than 2.

Since $1\frac{1}{2}$ is close to 2, the quotient is reasonable.

3.7 Dividing Mixed Numbers

Example 2

Brittany has a summer job in a bakery.
One day, she used $3\frac{3}{4}$ cups of chocolate chips to make chocolate-chip muffins.
A dozen muffins requires $\frac{3}{4}$ cup chocolate chips.
How many dozen chocolate-chip muffins did Brittany make that day?

▶ **A Solution**

$3\frac{3}{4} \div \frac{3}{4}$

Change the mixed number to an improper fraction.

$3\frac{3}{4} = \frac{15}{4}$

So, $3\frac{3}{4} \div \frac{3}{4} = \frac{15}{4} \div \frac{3}{4}$ Since the denominators are the same, divide the numerators.

$\phantom{So, 3\frac{3}{4} \div \frac{3}{4}} = 15 \div 3$

$\phantom{So, 3\frac{3}{4} \div \frac{3}{4}} = \frac{15}{3}$

$\phantom{So, 3\frac{3}{4} \div \frac{3}{4}} = 5$

Estimate to check.
$3\frac{3}{4}$ is close to 4. $\frac{3}{4}$ is close to 1.
So, $3\frac{3}{4} \div \frac{3}{4}$ is about $4 \div 1 = 4$.

Since 5 is close to 4, the solution is reasonable.
Brittany made 5 dozen chocolate-chip muffins.

Discuss the ideas

1. How is dividing mixed numbers similar to dividing fractions?
2. You have seen 3 methods for dividing mixed numbers. Which method are you likely to use most often? Justify your choice.
3. Why do we often write the quotient as a mixed number when we divide with mixed numbers?

Practice

Check

4. Write each mixed number as an improper fraction.
  a) $4\frac{3}{8}$ b) $3\frac{2}{7}$ c) $6\frac{1}{6}$ d) $2\frac{1}{4}$
  e) $1\frac{7}{10}$ f) $7\frac{2}{3}$ g) $2\frac{5}{9}$ h) $5\frac{2}{5}$

5. Write each improper fraction as a mixed number.
  a) $\frac{14}{9}$ b) $\frac{16}{7}$ c) $\frac{24}{5}$ d) $\frac{21}{10}$
  e) $\frac{15}{6}$ f) $\frac{23}{7}$ g) $\frac{17}{3}$ h) $\frac{25}{12}$

6. Use estimation. Which number is each quotient closer to?
  a) $6\frac{1}{8} \div 2\frac{3}{4}$ 2 or 3
  b) $7\frac{4}{5} \div 1\frac{3}{4}$ 3 or 4
  c) $3\frac{1}{8} \div 2\frac{3}{4}$ 1 or 2
  d) $9\frac{4}{7} \div 2\frac{1}{9}$ 4 or 5

7. Divide: $1\frac{4}{5} \div 2\frac{7}{10}$
  a) Estimate the quotient.
  b) Write each mixed number as an improper fraction.
  c) Divide the improper fractions. Simplify first.
  d) Is the quotient reasonable? How do you know?

Apply

8. Use common denominators to find each quotient. Estimate to check the quotient is reasonable.
  a) $3\frac{3}{4} \div 1\frac{1}{8}$ b) $1\frac{1}{6} \div 4\frac{1}{3}$
  c) $3\frac{1}{4} \div 3\frac{1}{4}$ d) $\frac{2}{3} \div 1\frac{1}{9}$

9. Use a copy of each number line to illustrate each quotient.
  a) $2\frac{1}{3} \div 1\frac{2}{3}$

  b) $1\frac{1}{8} \div \frac{3}{4}$

10. Use multiplication to find each quotient. Estimate to check the quotient is reasonable.
  a) $3\frac{2}{3} \div 5\frac{1}{4}$
  b) $4\frac{3}{8} \div 1\frac{5}{16}$
  c) $1\frac{3}{10} \div 3\frac{3}{5}$
  d) $3\frac{2}{3} \div 3\frac{2}{3}$

11. Divide. Estimate to check the quotient is reasonable.
  a) $1\frac{9}{10} \div 2\frac{2}{3}$ b) $2\frac{3}{4} \div 2\frac{1}{3}$
  c) $1\frac{4}{5} \div 3\frac{1}{2}$ d) $1\frac{3}{8} \div 1\frac{3}{8}$

12. Maxine took $12\frac{1}{2}$ h to build a model airplane. She worked for $1\frac{1}{4}$ h each evening. How many evenings did Maxine take to complete the model?

13. Glenn ran $3\frac{1}{3}$ laps in $11\frac{2}{3}$ min. Assume Glenn took the same amount of time to complete each lap. How long did Glenn take to run one lap?

3.7 Dividing Mixed Numbers

14. **Assessment Focus** Amelia has her own landscaping business. She ordered $10\frac{5}{8}$ loads of topsoil to fill large concrete planters. Each planter holds $1\frac{1}{2}$ loads of topsoil.
    a) Estimate the number of planters that Amelia can fill.
    b) Sketch a number line to illustrate the answer.
    c) Calculate the number of planters that Amelia can fill.
    d) What does the fraction part of the answer represent?

15. Write a story problem that could be solved using the expression $4\frac{2}{3} \div \frac{3}{5}$. Find the quotient to solve the problem. Estimate to check the solution is reasonable.

16. Use estimation. Which expression below has the greatest quotient? The least quotient? How do you know?
    a) $\frac{8}{5} \div \frac{4}{3}$
    b) $2\frac{3}{4} \div 1\frac{7}{8}$
    c) $4\frac{8}{9} \div 2\frac{1}{8}$
    d) $2\frac{1}{10} \div 1\frac{5}{6}$

17. **Take It Further**
    a) Which of these quotients is a mixed number? How can you tell without dividing?
       i) $4\frac{3}{8} \div 3\frac{2}{5}$
       ii) $3\frac{2}{5} \div 4\frac{3}{8}$
    b) Find the quotients in part a. What can you say about the order in which you divide mixed numbers?

18. **Take It Further** Which expression below has the greatest value? Give reasons for your answer. How could you find out without calculating each answer?
    a) $3\frac{1}{5} \times \frac{1}{2}$
    b) $3\frac{1}{5} \times \frac{2}{3}$
    c) $3\frac{1}{5} \div \frac{2}{3}$
    d) $3\frac{1}{5} \div \frac{2}{1}$
    e) $3\frac{1}{5} + \frac{2}{3}$
    f) $3\frac{1}{5} + \frac{3}{2}$

19. **Take It Further** One way to divide fractions is to use multiplication.
    a) How could you multiply fractions by using division? Explain.
    b) Do you think you would want to multiply fractions by using division? Why or why not?

Reflect

Suppose you divide one mixed number by another.
How can you tell, before you divide, if the quotient will be:
- greater than 1?
- less than 1?
- equal to 1?

Use examples in your explanation.

3.8 Solving Problems with Fractions

Focus Identify the operation required to solve a problem involving fractions.

A Grade 8 class is going to a canoe competition.
There are 24 students in the class.
Each canoe holds 4 people.
How many canoes does the class need?
How did you know which operation to use?

Investigate

Work with a partner.

Aidan wants to make this fruit punch recipe.
Answer each of these questions.
Show your work.
- How many cups of punch does the recipe make?
- Suppose Aidan makes the punch, then pours himself $\frac{3}{4}$ cup of punch.
  How much punch does he have left?
- Suppose Aidan only has $\frac{1}{3}$ cup of pineapple juice.
  How much of each of the other ingredients does he need to keep the flavour the same?
- Suppose Aidan decides to make one-third the recipe.
  How much soda will he need?

FRUIT PUNCH
$\frac{7}{8}$ cup orange juice
$\frac{2}{3}$ cup pineapple juice
$\frac{3}{4}$ cup lemon lime soda
$\frac{1}{4}$ cup cranberry juice
$\frac{1}{3}$ cup ice cubes

Compare your answers and strategies with those of another pair of classmates.
Did you solve the problems the same way?
How did you decide which operation to use each time?

Connect

When solving word problems, it is important to identify the operation or operations needed to solve the problem.
To identify the operation:
- Think about the situation.

- Make sense of the problem by explaining it in your own words, drawing a picture, or using a model.

- Think about what is happening in the problem. Sometimes, key words can help you identify the operation to use.

For example, "total" suggests adding,
"less than" suggests subtracting,
"times" suggests multiplying, and
"shared" suggests dividing.

Although key words may help identify an operation, the operation must make sense in the context of the problem.

Example 1

Kassie worked on her science project for $\frac{3}{4}$ h on Tuesday and $\frac{5}{6}$ h on Wednesday. She spent Thursday finishing her math homework.
a) How long did Kassie work on her science project altogether?
b) How much longer did Kassie work on the project on Wednesday than on Tuesday?
c) Altogether, Kassie spent 2 h on school work over the 3 days. How long did Kassie spend on her math homework?

▶ **A Solution**

a) The word "altogether" suggests addition.
Add: $\frac{3}{4} + \frac{5}{6}$

$$\frac{3}{4} + \frac{5}{6} = \frac{9}{12} + \frac{10}{12}$$
$$= \frac{19}{12}$$
$$= \frac{12}{12} + \frac{7}{12}$$
$$= 1\frac{7}{12}$$

Write an equivalent fraction with denominator 12 for each fraction.

$\frac{3}{4} = \frac{9}{12}$ $\frac{5}{6} = \frac{10}{12}$

Kassie worked on her science project for $1\frac{7}{12}$ h.

148   UNIT 3: Operations with Fractions

b) The words "longer … than" suggest subtraction.
To find how much longer Kassie worked on Wednesday than on Tuesday, subtract: $\frac{5}{6} - \frac{3}{4}$

$$\frac{5}{6} - \frac{3}{4} = \frac{10}{12} - \frac{9}{12}$$
$$= \frac{1}{12}$$

Kassie worked $\frac{1}{12}$ h longer on Wednesday than on Tuesday.

c) The word "altogether" suggests addition.
However, to find the time Kassie spent doing math homework, we subtract.
From part a, we know Kassie spent $1\frac{7}{12}$ h on her science project.
Altogether, Kassie spent 2 h on school work over the 3 days.
So, the time spent on her math homework is:

$$2 - 1\frac{7}{12} = 1\frac{12}{12} - 1\frac{7}{12}$$
$$= \frac{5}{12}$$

Kassie spent $\frac{5}{12}$ h on her math homework.

Example 2

Dakota volunteered at a gift-wrapping booth for a local charity.

He volunteered for $2\frac{3}{4}$ h and wrapped 11 gift boxes.

His friend Winona volunteered $1\frac{1}{3}$ times as long.

a) How long did Dakota spend wrapping each gift box?
   What assumptions do you make?
b) How many hours did Winona volunteer?

3.8 Solving Problems with Fractions

▶ **A Solution**

a) Think what the problem means:
We are given a time for many and we have to find a time for one.
This suggests division.
$2\frac{3}{4} \div 11$
Write the mixed number as an improper fraction.
$2\frac{3}{4} = \frac{11}{4}$
So, $2\frac{3}{4} \div 11 = \frac{11}{4} \div 11$ Write the reciprocal of the divisor, then multiply.
$\phantom{So, 2\frac{3}{4} \div 11} = \frac{11}{4} \times \frac{1}{11}$ Simplify. Divide by the common factor 11.
$\phantom{So, 2\frac{3}{4} \div 11} = \frac{{}^1\cancel{11} \times 1}{4 \times \cancel{11}_1}$
$\phantom{So, 2\frac{3}{4} \div 11} = \frac{1}{4}$

Dakota spent $\frac{1}{4}$ h wrapping each gift box.

It was assumed that Dakota wrapped for the entire time and that he spent the same amount of time wrapping each box.

b) The words "times as long" suggest multiplication.
So, Winona volunteered for $1\frac{1}{3}$ of the $2\frac{3}{4}$ h that Dakota volunteered.
$2\frac{3}{4} \times 1\frac{1}{3}$
Write the mixed numbers as improper fractions.
$2\frac{3}{4} = \frac{11}{4}$ and $1\frac{1}{3} = \frac{4}{3}$
So, $2\frac{3}{4} \times 1\frac{1}{3} = \frac{11}{4} \times \frac{4}{3}$
$\phantom{So, 2\frac{3}{4} \times 1\frac{1}{3}} = \frac{11 \times \cancel{4}^1}{{}_1\cancel{4} \times 3}$ Simplify. Divide by the common factor 4.
$\phantom{So, 2\frac{3}{4} \times 1\frac{1}{3}} = \frac{11}{3}$
$\phantom{So, 2\frac{3}{4} \times 1\frac{1}{3}} = 3\frac{2}{3}$

Winona volunteered for $3\frac{2}{3}$ h.

Discuss the ideas

1. What other words indicate addition, subtraction, multiplication, and division?
2. What other questions could you ask using the data in *Example 2*?
   What operation would you use to answer each question?

150   UNIT 3: Operations with Fractions

Practice

Check

3. Which operation would you use to solve each problem? How can you tell?
  a) Noel used $\frac{2}{3}$ cup of milk and $\frac{1}{4}$ cup of oil to make cookies. How much liquid did he use altogether?
  b) One-third of the cars in the parking lot are silver. There are 165 cars in the lot. How many cars are silver?
  c) Shania has $\frac{3}{8}$ cup of yogurt. She needs $\frac{3}{4}$ cup of yogurt to make a smoothie. How much more yogurt does she need?
  d) Part of a pizza was shared equally between two friends. Each friend got $\frac{5}{12}$ of the whole pizza. How much pizza was shared?

Solve each problem. For each problem, explain how you decided which operation or operations to use.

4. Chad mixed $\frac{2}{3}$ of one can of yellow paint and $\frac{1}{4}$ of one can of white paint to paint a wall in his bedroom. How much paint did he have altogether?

5. Vivi scored 5 goals in the Saskatoon Sticks lacrosse tournament. This was $\frac{1}{8}$ of her team's goals. How many goals did Vivi's team score altogether?

Apply

6. Parent-teacher interviews were held on Thursday. Of those parents who attended, $\frac{1}{6}$ attended in the morning, $\frac{1}{3}$ attended in the afternoon, and the rest attended in the evening.
  a) What fraction of the parents attended in the evening?
  b) Thirty parents attended the interviews. How many parents attended in the evening?

7. Patti works in a coffee shop. She usually takes $\frac{3}{4}$ h for lunch. One day the shop was very busy and Patti's manager asked her to shorten her lunch break by $\frac{1}{6}$ h. What fraction of an hour did Patti take for lunch that day?

8. Katrina's monthly salary is $2400. She uses $\frac{2}{5}$ of this money for rent. How much rent does Katrina pay?

9. A snail travelled 48 cm in $\frac{2}{3}$ h. Suppose the snail moved at a constant speed and made no stops. How far would the snail travel in 1 h?

3.8 Solving Problems with Fractions   151

10. Beven has a collection of 72 music CDs. One-sixth of the CDs are dance music, $\frac{1}{4}$ are hip hop, and $\frac{3}{8}$ are reggae. The rest of the CDs are rock music. What fraction of her CDs are rock music?

11. **Assessment Focus** A jug contains $2\frac{1}{2}$ cups of juice. Shavon pours $\frac{3}{8}$ cup of juice into each of three glasses, then $\frac{5}{6}$ cup of juice into a fourth glass.
    a) Estimate the fraction of the apple juice that remains in the jug.
    b) Calculate the total amount of juice in the 3 glasses that contain the same amount.
    c) Calculate the total amount of juice in the 4 glasses.
    d) How much juice remains in the jug after the 4 glasses have been poured?

12. The Hendersons went on a driving vacation. They decided to travel $\frac{1}{3}$ of the distance on the first day. Owing to bad weather, they had to stop for the night having gone only $\frac{1}{4}$ of the distance to the one-third point. What fraction of the total distance did they travel the first day?

13. Howie works at a petting zoo. He fed a piglet $\frac{1}{5}$ of a bottle of milk, then gave $\frac{3}{4}$ of what was left to a calf. How much of the bottle of milk did the calf drink?

14. Nathan used $2\frac{5}{6}$ loaves of bread to make sandwiches for a lunch. He made equal numbers of 4 different types of sandwiches. What fraction of a loaf of bread did Nathan use for each type of sandwich?

15. **Take It Further** A steward reported to an airport official before takeoff that three-fifths of the passengers were women, three-eighths were men, and one-twentieth were children. After thinking for a moment, the official seemed puzzled and asked the steward to repeat the fractions. Why do you think the official was puzzled? Explain.

Reflect

Use the fractions $\frac{1}{2}$ and $\frac{7}{8}$. Write 4 problems. Each problem should require a different operation (addition, subtraction, multiplication and division). Solve each problem. How did you decide which operation to use?

3.9 Order of Operations with Fractions

Focus Use the order of operations to evaluate expressions.

Suppose you had to answer this skill-testing question to win a contest. What answer would you give?
$5 + 20 \times 2 - 36 \div 9$
Explain the strategy you used.

Every fraction can be written as a decimal.
So, we use the same order of operations for fractions as for whole numbers and decimals.

Investigate

Work with a partner.
Use these fractions: $\frac{9}{4}, \frac{3}{8}, \frac{15}{16}$
Use any operations or brackets.
Write an expression that has value 4.
Show your work.

Compare your expression with that of another pair of classmates.
If the expressions are different, check that both expressions have value 4.
What strategies did you use to arrive at your expression?

Connect

In *Investigate*, you used the order of operations to write an expression with a specific value. To make sure everyone gets the same value for a given expression, we add, subtract, multiply, and divide in this order:
- Do the operations in brackets first.
- Then divide and multiply, in order, from left to right.
- Then add and subtract, in order, from left to right.

Example 1

Evaluate: $\frac{5}{16} - \frac{3}{8} \times \frac{2}{3}$

▶ **A Solution**

$\frac{5}{16} - \frac{3}{8} \times \frac{2}{3}$ Multiply. Simplify first.

$= \frac{5}{16} - \frac{\cancel{3}^1}{\cancel{8}_4} \times \frac{\cancel{2}^1}{\cancel{3}_1}$

$= \frac{5}{16} - \frac{1}{4}$ Use common denominators to subtract.

$= \frac{5}{16} - \frac{4}{16}$

$= \frac{1}{16}$

Example 2

Evaluate: $\frac{3}{4} - \frac{2}{3} \div \frac{4}{5} \times (\frac{1}{8} + \frac{1}{4})$

▶ **A Solution**

$\frac{3}{4} - \frac{2}{3} \div \frac{4}{5} \times (\frac{1}{8} + \frac{1}{4})$ First do the operation in brackets.
Use common denominators to add.

$= \frac{3}{4} - \frac{2}{3} \div \frac{4}{5} \times (\frac{1}{8} + \frac{2}{8})$

$= \frac{3}{4} - \frac{2}{3} \div \frac{4}{5} \times (\frac{3}{8})$ Divide and multiply from left to right.

$= \frac{3}{4} - \frac{\cancel{2}^1}{3} \times \frac{5}{\cancel{4}_2} \times (\frac{3}{8})$ To divide by $\frac{4}{5}$, multiply by $\frac{5}{4}$. Simplify first.

$= \frac{3}{4} - \frac{5}{\cancel{6}_2} \times \frac{\cancel{3}^1}{8}$ Multiply. Simplify first.

$= \frac{3}{4} - \frac{5}{16}$ Use common denominators to subtract.

$= \frac{12}{16} - \frac{5}{16}$

$= \frac{7}{16}$

Discuss the ideas

1. A student suggested that brackets should be put around $\frac{3}{8} \times \frac{2}{3}$ in *Example 1*. What is your response to this suggestion?
2. Why are the brackets necessary in *Example 2*?
3. Do you think most people would get the skill-testing question in the introduction correct? If not, what answer do you think they would give?

Practice

Check

4. Which operation do you do first?
   a) $\frac{1}{3} \times (\frac{7}{8} - \frac{3}{4})$
   b) $\frac{7}{8} \div (\frac{1}{3} \times \frac{1}{8})$
   c) $\frac{9}{5} \times (\frac{3}{5} \div \frac{1}{10})$
   d) $(\frac{5}{3} + \frac{7}{12}) \times \frac{4}{9}$

5. Raj and Rena evaluated this expression:
   $\frac{5}{9} + \frac{2}{3} \times \frac{1}{2}$
   Raj got $\frac{8}{9}$. Rena got $\frac{11}{18}$. Who is correct? What mistake did the other person make?

6. Evaluate. Which operation is done first?
   a) $\frac{1}{2} \times \frac{3}{5} + \frac{1}{4}$
   b) $\frac{2}{3} + \frac{5}{6} \div \frac{1}{2}$
   c) $\frac{4}{5} \div \frac{7}{10} + \frac{1}{3}$
   d) $\frac{1}{4} \times (\frac{11}{12} - \frac{5}{6})$
   e) $\frac{1}{2} \times (\frac{4}{5} \div \frac{3}{10})$
   f) $(\frac{3}{5} + \frac{7}{15}) \times \frac{5}{6}$

Apply

7. Evaluate. Show all steps.
   a) $\frac{1}{8} \times \frac{3}{4} \times \frac{7}{5} \div \frac{7}{10}$
   b) $\frac{14}{15} \div \frac{2}{3} \times \frac{5}{8} + \frac{3}{4}$
   c) $\frac{2}{3} - \frac{1}{4} + \frac{1}{2} \div \frac{2}{5}$
   d) $\frac{5}{6} - \frac{1}{5} \times \frac{5}{8} + \frac{2}{3}$

8. Emma thinks that the expressions $1\frac{1}{2} \div \frac{1}{4} \times \frac{2}{3}$ and $1\frac{1}{2} \div (\frac{1}{4} \times \frac{2}{3})$ have the same value. Is Emma correct? Explain.

9. Evaluate.
   a) $\frac{7}{10} - (\frac{1}{5} + \frac{1}{4}) \times \frac{2}{3}$
   b) $(\frac{1}{4} + \frac{5}{6} - \frac{1}{3}) \times \frac{8}{5}$
   c) $(\frac{6}{5} + \frac{4}{10}) \times (\frac{3}{8} - \frac{1}{16})$

10. Evaluate.
    a) $\frac{5}{2} + \frac{1}{4} \times \frac{4}{5} \div \frac{1}{10} - \frac{1}{2}$
    b) $\frac{4}{9} \times (\frac{2}{3} - \frac{1}{6}) - \frac{1}{8} \times \frac{4}{3}$

11. **Assessment Focus** Robert, Myra, and Joe evaluated this expression:
    $4 \times (\frac{3}{4} - \frac{1}{2}) + \frac{13}{6} \times \frac{1}{2}$
    Robert's answer was $5\frac{1}{3}$, Myra's answer was $2\frac{1}{12}$, and Joe's answer was $4\frac{5}{6}$.
    a) Who had the correct answer? How do you know?
    b) Show and explain how the other two students got their answers. Where did they go wrong?

12. **Take It Further** Evaluate.
    a) $\frac{14}{15} \div 4\frac{2}{3} \times \frac{5}{8} + 2\frac{3}{4}$
    b) $3\frac{2}{3} - 2\frac{1}{4} + \frac{1}{2} \div 2\frac{2}{5}$
    c) $8\frac{7}{10} - (2\frac{1}{5} + 2\frac{1}{4}) \times \frac{2}{3}$

Reflect

Write an expression that contains fractions and three operations.
Talk to a partner. Discuss the steps you would follow to evaluate the expression.

Checking and Reflecting

Have you ever finished your math homework and thought "What was that all about?"

One purpose of homework is to increase your understanding of the math you are learning.
When you have finished your homework, how do you know you have understood everything you are supposed to understand?

It is important to reflect on what you have done and to check your understanding.

Solve this problem:

> Chris used $\frac{2}{3}$ of a roll of stamps.
>
> His sister then used $\frac{1}{4}$ of the stamps left on the roll. What fraction of the roll of stamps did Chris' sister use? What fraction of the roll of stamps is left?

When you have solved the problem, ask yourself these questions.
- What math idea did this problem involve?
- How difficult was this problem for me?
- Did I need help with this problem?
- What could I say to help someone else understand the problem?
- What would make this problem easier to solve?
- What would make this problem more difficult to solve?
- What other math ideas does this problem remind me of?

Strategies for Success

Using a Traffic Light Strategy

You can use "traffic light" signals to help you reflect on your understanding. It is a good way to signal what you *know* you can do, what you *think* you can do, and what you need help with.

When you have finished your homework, reflect on all the questions. Label each question with a green dot, a yellow dot, or a red dot.

A green dot means: I understand this. I can answer this question and am confident about my answer. If someone else is struggling with this question, I can help.

A yellow dot means: I am not sure I understand this. I might be able to answer this question with a little help but I am not confident about my answer. I might need to ask a question or two to get started.

A red dot means: I do not understand this. I cannot answer the question. I need some help.

Answer these questions.
When you have finished, use the "Traffic Light" strategy to label each question in your notebook with a green, yellow, or red dot.

1. Evaluate.
   a) $\frac{3}{5} \times 5$
   b) $\frac{3}{7} \times \frac{14}{15}$
   c) $\frac{5}{2} \div \frac{6}{7}$
   d) $\frac{5}{6} - \frac{1}{9} \times \frac{3}{4}$

2. A glass holds $\frac{2}{3}$ cup of juice. A jug contains 8 cups of juice. How many glasses can be filled from the juice in the jug?

3. A soccer practice lasted $1\frac{1}{4}$ h. Drills took up $\frac{2}{3}$ of the time. How much time was spent on drills?

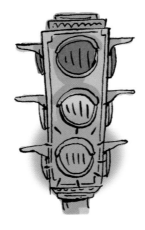

When you have completed the *Unit Review* and *Practice Test*, use the "Traffic Light" strategy to check your understanding. Get help with questions labelled with yellow and red dots.

Strategies for Success: Checking and Reflecting

Unit Review

What Do I Need to Know?

☑ **To multiply two fractions:**
Multiply the numerators and multiply the denominators.
$$\frac{2}{3} \times \frac{1}{5} = \frac{2 \times 1}{3 \times 5} = \frac{2}{15}$$

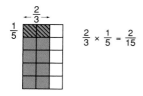

To multiply two mixed numbers:
Write each number as an improper fraction, then multiply.
$$1\frac{1}{2} \times 2\frac{5}{6} = \frac{\cancel{3}^1}{2} \times \frac{17}{\cancel{6}_2}$$
$$= \frac{17}{4}, \text{ or } 4\frac{1}{4}$$

Simplify first, when you can.

☑ **To divide two fractions:**

Method 1:
Use common denominators.
$$\frac{4}{5} \div \frac{3}{2} = \frac{8}{10} \div \frac{15}{10} = \frac{8}{15}$$

Method 2:
Use multiplication.
$\frac{4}{5} \div \frac{3}{2}$ is the same as $\frac{4}{5} \times \frac{2}{3} = \frac{8}{15}$

To divide two mixed numbers:
Write each number as an improper fraction, then divide.
$$3\frac{1}{2} \div 1\frac{2}{3} = \frac{7}{2} \div \frac{5}{3}$$
$$= \frac{7}{2} \times \frac{3}{5}$$
$$= \frac{21}{10}, \text{ or } 2\frac{1}{10}$$

☑ **To identify the operation:**
- Think about the situation.
- Make sense of the problem.
- Think about what is happening in the problem. Use key words to help.

☑ **The order of operations with whole numbers and decimals applies to fractions.**
- Do the operations in brackets first.
- Then divide and multiply, in order, from left to right.
- Then add and subtract, in order, from left to right.

What Should I Be Able to Do?

LESSON

3.1

1. Write the multiplication equation each number line represents.

   a)
   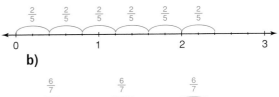

   b)
   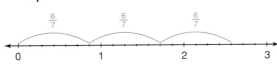

2. Multiply. Draw a picture or a number line to show each product.
   a) $\frac{1}{3} \times 3$ b) $7 \times \frac{1}{2}$ c) $8 \times \frac{2}{5}$

3. Solve each problem.
   a) There are 30 students in a class. Three-fifths of the students are girls. How many girls are in the class?
   b) Six glasses are $\frac{2}{3}$ full. How many full glasses could be made?
   c) There are 75 cars in the parking lot of a car dealership. Two-thirds of the cars are new. How many of the cars are new?
   d) One serving is $\frac{1}{12}$ of a cake. How many cakes are needed for 18 servings?

3.2

4. Draw an area model to find each product.
   a) $\frac{2}{3} \times \frac{3}{8}$ b) $\frac{4}{5} \times \frac{3}{10}$
   c) $\frac{7}{10} \times \frac{3}{4}$ d) $\frac{3}{7} \times \frac{1}{3}$

5. Fasil donated $\frac{3}{5}$ of $\frac{1}{4}$ of his allowance to a charity. What fraction of his allowance did Fasil donate?

3.3

6. Multiply. Use benchmarks to estimate to check each product is reasonable.
   a) $\frac{1}{2} \times \frac{3}{10}$ b) $\frac{3}{5} \times \frac{1}{8}$
   c) $\frac{7}{8} \times \frac{2}{5}$ d) $\frac{3}{11} \times \frac{44}{63}$

7. Twenty Grade 8 students are going on a school trip. They pre-order sandwiches. Three-quarters of the students order a turkey sandwich, while $\frac{1}{4}$ of the students order a roasted vegetable sandwich. Of the $\frac{3}{4}$ who want turkey, $\frac{2}{5}$ do not want mayonnaise. What fraction of the students do not want mayonnaise?

8. Write a story problem that could be solved using the expression $\frac{5}{7} \times \frac{3}{8}$. Find the product to solve the problem. Estimate to check the solution is reasonable.

3.4

9. Write each mixed number as an improper fraction.
   a) $7\frac{1}{2}$ b) $2\frac{7}{8}$ c) $10\frac{7}{10}$

10. Use an area model to find each product.
    a) $1\frac{1}{2} \times 2\frac{1}{3}$ b) $\frac{19}{3} \times \frac{6}{5}$
    c) $3\frac{1}{5} \times \frac{1}{4}$ d) $2\frac{1}{4} \times 3\frac{1}{3}$

Unit Review 159

LESSON

11. Multiply. Estimate to check the product is reasonable.
   a) $1\frac{2}{3} \times 1\frac{9}{10}$
   b) $4\frac{1}{2} \times \frac{5}{8}$
   c) $\frac{9}{5} \times \frac{14}{8}$
   d) $1\frac{3}{10} \times 6\frac{2}{3}$

12. Jonathan works for a landscape maintenance company. It took Jonathan $1\frac{3}{4}$ h to mow Mr. Persaud's lawn. The lawn he will mow next is $2\frac{1}{3}$ times as large as Mr. Persaud's lawn. How long will it take Jonathan to mow the next lawn? What assumptions do you make?

3.5

13. Find each quotient. Use number lines to illustrate the answers.
   a) One-half of a cake is shared equally among 5 people. What fraction of the whole cake does each person get?
   b) Nakkita's dog eats $\frac{3}{4}$ of a can of dog food each day. Nakkita has 9 cans of dog food. How many days' supply of dog food does Nakkita have?

14. Find each quotient.
   a) $3 \div \frac{4}{5}$
   b) $4 \div \frac{5}{6}$
   c) $\frac{3}{10} \div 2$
   d) $2\frac{5}{8} \div 3$

15. A glass holds $\frac{3}{4}$ cup of milk. A jug contains 12 cups of milk. How many glasses can be filled from the milk in the jug?

16. Kayla uses $\frac{2}{3}$ of a scoop of detergent to do one load of laundry. Kayla has 9 scoops of detergent. How many loads of laundry can Kayla do?

17. When you divide a fraction by a whole number, is the quotient greater than or less than 1? Include examples in your explanation.

3.6

18. Use a copy of each number line to illustrate each quotient.
   a) $\frac{9}{10} \div \frac{3}{5}$

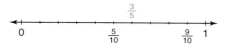

   b) $\frac{3}{4} \div \frac{1}{2}$

19. Divide. Estimate to check each quotient is reasonable.
   a) $\frac{3}{4} \div \frac{3}{8}$
   b) $\frac{1}{4} \div \frac{7}{8}$
   c) $\frac{5}{12} \div \frac{1}{3}$
   d) $\frac{1}{2} \div \frac{3}{5}$

20. Midori lives on a farm. Midori has $\frac{7}{8}$ of a tank of gas. Each trip to town and back uses $\frac{1}{6}$ of a tank of gas. How many trips to town and back can Midori make until she needs more gas? Estimate to check the solution is reasonable.

21. When you divide a proper fraction by its reciprocal, is the quotient less than 1, greater than 1, or equal to 1? Use examples in your explanation.

LESSON 3.7

22. Write each mixed number as an improper fraction.
    a) $3\frac{7}{11}$ b) $5\frac{1}{6}$
    c) $4\frac{8}{9}$ d) $2\frac{5}{12}$

23. Divide. Estimate to check the quotient is reasonable.
    a) $1\frac{3}{4} \div 2\frac{1}{8}$
    b) $3\frac{5}{6} \div 2\frac{1}{5}$
    c) $3\frac{1}{2} \div 1\frac{3}{8}$
    d) $2\frac{1}{5} \div 4\frac{2}{5}$

24. A recipe for cereal squares calls for $1\frac{1}{4}$ bags of regular marshmallows. The recipe makes a cookie sheet of squares. Marcus has $\frac{3}{4}$ of a bag of marshmallows. He buys 5 more bags. How many sheets of cereal squares can Marcus make?

3.8

25. A cookie recipe calls for $\frac{3}{4}$ cup of rolled oats. Norma has $\frac{5}{8}$ cup of rolled oats. How much more rolled oats does she need to make the cookies? How did you decide which operation to use?

26. In a lottery for a local charity, 1000 tickets are sold. Of these tickets, $\frac{1}{1000}$ will win $1000, $\frac{1}{500}$ will win $50, $\frac{1}{200}$ will win $25, $\frac{1}{100}$ will win $10, and $\frac{1}{10}$ will win $5. How many tickets will not win a prize? How did you decide which operations to use?

27. There are 30 students in a Grade 8 class. One-third of the students take a school bus, $\frac{1}{5}$ take public transportation, $\frac{1}{6}$ are driven by family, and the rest walk to school.
    a) What fraction of the students in the class walk to school?
    b) How many of the students in the class walk to school? How did you decide which operations to use?

3.9

28. Evaluate. State which operation you do first.
    a) $\frac{1}{5} + \frac{2}{3} \times \frac{3}{5}$ b) $\frac{4}{5} \div (\frac{2}{3} - \frac{3}{10})$
    c) $\frac{7}{3} + \frac{1}{6} \times \frac{2}{5}$ d) $\frac{7}{8} \div \frac{5}{6} \times \frac{4}{7}$

29. Evaluate.
    a) $\frac{2}{3} + \frac{1}{4} - \frac{1}{6}$ b) $\frac{3}{2} \times (\frac{4}{3} - \frac{1}{6})$
    c) $\frac{9}{8} \div (\frac{3}{4} + \frac{3}{2})$ d) $\frac{2}{3} \times (\frac{1}{8} + \frac{5}{6} - \frac{3}{4})$

30. Carlton evaluated this expression:
    $2\frac{4}{5} \div (\frac{2}{3} + \frac{1}{12})$
    His work is shown below.
    Where did Carlton go wrong?
    What is the correct answer?

    $2\frac{4}{5} \div (\frac{2}{3} + \frac{1}{12}) = 2\frac{4}{5} \div (\frac{8}{12} + \frac{1}{12})$
    $= 2\frac{4}{5} \div (\frac{9}{12})$
    $= \frac{14}{5} \div \frac{9}{12}$
    $= \frac{14}{5} \times \frac{9}{12}$
    $= \frac{\overset{7}{14}}{5} \times \frac{\overset{3}{9}}{\underset{2}{12}}$
    $= \frac{21}{10}$
    $= 2\frac{1}{10}$

Practice Test

1. Use a number line to divide: $5 \div \frac{5}{6}$

2. Copy this rectangle.
   Shade the rectangle to find the product:
   $\frac{5}{8} \times \frac{2}{3}$

   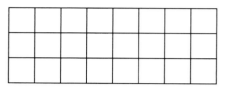

3. Find each product and quotient.
   a) $\frac{1}{4} \times 28$ b) $\frac{3}{8} \times \frac{1}{2}$ c) $\frac{5}{6} \div 2$ d) $\frac{1}{5} \div \frac{2}{3}$

4. Find each product and quotient.
   Estimate to check each solution is reasonable.
   a) $\frac{5}{8} \times 3\frac{1}{4}$ b) $3\frac{1}{3} \times 2\frac{1}{10}$ c) $2\frac{1}{4} \div \frac{7}{8}$ d) $1\frac{2}{5} \div 1\frac{1}{2}$

5. Three-fifths of the Grade 8 class are in the band.
   a) On Tuesday, only $\frac{1}{3}$ of these students went to band practice.
      What fraction of the class went to band practice on Tuesday?
   b) How many students might be in the class? How do you know?

6. Predict without calculating. Which statement below has the greatest value? How do you know?
   a) $\frac{7}{3} \times \frac{3}{4}$ b) $\frac{7}{3} - \frac{3}{4}$ c) $\frac{7}{3} \div \frac{3}{4}$ d) $\frac{7}{3} + \frac{3}{4}$

7. Multiply a fraction by its reciprocal. What is the product?
   Use an example and a model to explain.

8. Evaluate.
   a) $\frac{1}{2} - \frac{3}{5} \times \frac{1}{6}$ b) $\frac{1}{4} \div \frac{1}{8} + (\frac{1}{2} - \frac{3}{8})$

9. A dog groomer takes $\frac{5}{6}$ h to groom a poodle.
   a) Estimate the number of poodles the groomer can groom in $4\frac{1}{2}$ h.
   b) Sketch a number line to illustrate the answer.
   c) Calculate the number of poodles the groomer can groom in $4\frac{1}{2}$ h.
   d) What assumptions do you make?

10. Solve each problem.
Explain how you decided which operation or operations to use.
   a) Haden is making a milkshake.
   He has $\frac{1}{3}$ cup of milk in a glass.
   How much milk must Haden add so there will be $\frac{4}{5}$ cups of milk in the glass?
   b) Lacy, Lamar, and Patti own a dog-walking business.
   Lacy owns $\frac{5}{12}$ of the business and Lamar owns $\frac{1}{3}$ of the business.
   How much of the business does Patti own?
   c) The science teacher has $2\frac{1}{4}$ cups of baking soda and $1\frac{1}{3}$ cups of salt.
   To conduct an experiment, each student needs $\frac{1}{8}$ cup of baking soda
   and $\frac{1}{12}$ cup of salt. There are 18 students in the class.
   i) Does the teacher have enough baking soda for each student?
   If not, how much more baking soda does the teacher need?
   ii) Does the teacher have enough salt for each student?
   If not, how much more salt does the teacher need?

11. Which of these statements is always true?
Use examples to support your answer.
   a) Division always results in a quotient that is less than the dividend.
   b) Division always results in a quotient that is greater than the dividend.
   c) Division sometimes results in a quotient that is less than the dividend and sometimes in a quotient that is greater than the dividend.

Unit Problem Sierpinski Triangle

A *fractal* is a shape that can be subdivided into parts, each of which is a reduced-size copy of the original. Many natural objects – blood vessels, lungs, coastlines, galaxy clusters – can be modelled by fractals.

The Sierpinski Triangle is a famous fractal.

Use triangular dot paper.
Draw an equilateral triangle
with side length 16 units.

Step 1

Find the midpoint of each side.
Connect each midpoint to form a triangle
inside the original triangle.
Shade the new triangle.
Label this diagram Figure 1.

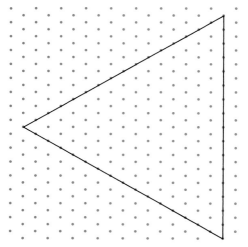

Suppose the area of the original
triangle is 1 square unit.
Find the total area *not* shaded in Figure 1.
Record your results in a copy of
the table below.

Area of Sierpinski Triangle						
Figure	Original	1	2	3	4	5
Area Not Shaded (square units)	1					

Step 2

Use Figure 1. Repeat *Step 1* with each triangle that is not shaded.
Label this diagram Figure 2.
Repeat this process 3 more times to get Figures 3, 4, and 5.

Step 3

Describe any patterns you see in the table you completed in *Step 1*.
Write a pattern rule for the areas not shaded.
Suppose the pattern continues.
Use your rule to find the areas not shaded in Figure 6 and Figure 7.
How did you use multiplication of fractions to help?

Step 4

Suppose the side length of the original triangle is 1 unit.
Record the perimeter of this triangle in a copy of the table below.
Find the total perimeter of the triangles not shaded in Figure 1.
Record your results in the table.
Explain how you found the perimeter.

Use the figures drawn in *Step 2* to complete the table.

Perimeter of Sierpinski Triangle						
Figure	Original	1	2	3	4	5
Perimeter of Triangles Not Shaded (units)						

Describe any patterns you see in the table.
Write a pattern rule for the perimeters of the triangles not shaded.
Use your rule to find the perimeters of the triangles not shaded in Figure 6 and Figure 7.
How did you use multiplication of fractions to help?

Check List

Your work should show:
✓ all diagrams, tables, and calculations in detail
✓ clear explanations of the patterns you found
✓ how you used multiplication of fractions to find area and perimeter

Reflect on Your Learning

What do you now know about fractions that you did not know before this unit?
Use examples in your explanation.

Units 1–3 Cumulative Review

UNIT 1

1. a) List the factors of each number in ascending order.
   i) 84 ii) 441
   iii) 236 iv) 900
   b) Which numbers in part a are square numbers? How can you tell?

2. Estimate each square root to 1 decimal place. Show your work.
   a) $\sqrt{52}$ b) $\sqrt{63}$
   c) $\sqrt{90}$ d) $\sqrt{76}$

3. Find each length indicated. Sketch and label the triangle first. Give your answers to one decimal place where needed.
   a)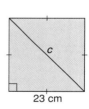
   b)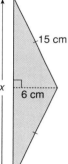

4. The area of the square on each side of a triangle is given. Is the triangle a right triangle? How do you know?
   a) 16 cm², 8 cm², 30 cm²
   b) 16 cm², 8 cm², 24 cm²

5. Identify the sets of numbers that are Pythagorean triples. How did you decide?
   a) 2, 5, 6 b) 6, 10, 8
   c) 12, 9, 7 d) 18, 30, 24

6. The dimensions of a rectangle are 3 cm by 4 cm. What is the length of a diagonal? Explain your reasoning.

7. Jacobi wants to install an underground sprinkler system in her backyard. The backyard is rectangular with side lengths 17 m and 26 m. The water pipe will run diagonally across the yard. About how many metres of water pipe does Jacobi need?

UNIT 2

8. Evaluate.
   a) $(-9) \times (+8)$ b) $(+14) \times (+8)$
   c) $(-18) \times (-1)$ d) $(+21) \times (-6)$

9. The Brandon Birdies junior golf team has 4 golfers. Each golfer is –5 on her round for the day. What is the team score for the day?
   a) Write this problem as a multiplication expression using integers.
   b) Solve the problem.

10. Write a word problem that could be solved using the expression $(+8) \times (-5)$. Solve the problem.

11. Divide.
    a) $(-77) \div (+7)$ b) $(+63) \div (+3)$
    c) $(-30) \div (-1)$ d) $\dfrac{-24}{+6}$
    e) $\dfrac{+51}{-3}$ f) $\dfrac{0}{-8}$

UNIT

12. Divide.
   a) $(+84) \div (-12)$ b) $(-75) \div (-15)$
   c) $(-78) \div (+13)$ d) $(+98) \div (+14)$

13. Four friends had lunch at a restaurant. They share the bill equally. The bill is $52. How much does each friend owe?
   a) Write this problem as a division expression using integers.
   b) Solve the problem.

14. For each number below, find two integers for which that number is:
   i) the sum
   ii) the difference
   iii) the product
   iv) the quotient
   a) -8 b) -2 c) -12 d) -3

15. Evaluate.
   a) $(-3)[5 - (-3)]$
   b) $[8 \div (-4)] - 10(3)$
   c) $\dfrac{(-6) + 12 \div (-4) \times (-5)}{(-18) \div 6}$

16. Write each improper fraction as a mixed number.
   a) $\dfrac{7}{3}$ b) $\dfrac{15}{2}$ c) $\dfrac{21}{8}$ d) $\dfrac{19}{5}$

17. Multiply. Estimate to check each product is reasonable.
   a) $\dfrac{7}{9} \times 36$ b) $\dfrac{5}{6} \times 21$
   c) $\dfrac{5}{6} \times \dfrac{5}{6}$ d) $\dfrac{9}{4} \times \dfrac{8}{3}$
   e) $1\dfrac{1}{2} \times 2\dfrac{3}{4}$ f) $3\dfrac{1}{5} \times 4\dfrac{1}{8}$

18. Divide. Estimate to check each quotient is reasonable.
   a) $9 \div \dfrac{3}{5}$ b) $\dfrac{5}{6} \div 3$
   c) $\dfrac{5}{3} \div \dfrac{3}{4}$ d) $\dfrac{5}{3} \div \dfrac{4}{3}$
   e) $1\dfrac{2}{3} \div 1\dfrac{1}{3}$ f) $2\dfrac{1}{4} \div 3\dfrac{1}{2}$

19. A recipe for one bowl of punch calls for $1\dfrac{1}{8}$ bottles of sparkling water. For a party, Aaron wants to make $5\dfrac{1}{2}$ bowls of punch. How many bottles of sparkling water does Aaron need to make the punch?

20. Evaluate.
   a) $\dfrac{3}{20} \div \dfrac{2}{5} \div \dfrac{1}{2} \times \dfrac{3}{4}$
   b) $\dfrac{7}{8} - \dfrac{7}{12} + 2$
   c) $\dfrac{5}{9} \div (\dfrac{1}{2} \times \dfrac{7}{9})$
   d) $4 \div \dfrac{2}{3} - 3\dfrac{1}{4} + \dfrac{7}{12}$

21. Solve each problem. Explain how you decided which operation to use.
   a) Justine usually runs $7\dfrac{1}{2}$ laps of the track before breakfast. This morning, she had to be at school early so she only ran $3\dfrac{1}{4}$ laps. By how many laps did Justine cut her run short?
   b) Lyle slept for $6\dfrac{1}{2}$ h on Monday night. On Tuesday night, Lyle slept $1\dfrac{1}{4}$ times as long as he did on Monday night. How many hours of sleep did Lyle get on Tuesday night?

Cumulative Review

UNIT 4

Measuring Prisms and Cylinders

Most products are packaged in boxes or cans. Think about how a box or can is made. How do you think the manufacturer chooses the shape and style of package? Why do you think tennis balls are sold in cylinders but golf balls are sold in rectangular prisms?

Look at the packages on these pages. Choose one package. Why do you think the manufacturer chose that style of packaging?

What You'll Learn

- Draw and construct nets of 3-D objects.
- Determine the surface areas of prisms and cylinders.
- Develop formulas for the volumes of prisms and cylinders.
- Solve problems involving prisms and cylinders.

Why It's Important

- We need measurement and calculation skills to design and build objects to meet our needs.
- Calculating the surface area and volume of prisms and cylinders is an extension of the measuring you did in earlier grades.

Key Words

- net
- polyhedron
- regular prism
- regular pyramid
- regular dodecagon
- surface area
- volume
- capacity

4.1 Exploring Nets

Focus Identify, draw, and construct nets of objects.

Suppose this diagram was cut out and folded along the dotted lines to make an object.

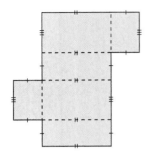

Which object do you think would be made?
Why do you think so?

Investigate

Work in groups of 3.
You will need 3 different objects, one for each group member, scissors, tape, and markers.

Name each face of your object.
Label each face with a letter.
➤ Trace each face of the object.
  Label each tracing with its letter.
➤ Cut out your tracings.
  Tape them together at the edges
  to make a diagram.
  Arrange the faces so the diagram can be
  folded to build a model of your object.
➤ Fold the diagram to build the model.
➤ How many different ways can you make
  a diagram to form the model?
➤ Trade diagrams with another group member.
  Check that each diagram folds to build the model.

Share your objects and diagrams with another group of classmates.
How did you decide how to arrange the faces?
How are the diagrams for the prisms different?
How are they alike?

170 UNIT 4: Measuring Prisms and Cylinders

Connect

A **net** is a diagram that can be folded to make an object.
A net shows all the faces of an object.

A net can be used to make an object called a **polyhedron**.
A polyhedron has faces that are polygons.
Two faces meet at an *edge*.
Three or more edges meet at a *vertex*.

➤ A *prism* has 2 congruent bases and is named
   for its bases.
   When all its faces, other than the bases,
   are rectangles and they are perpendicular
   to the bases, the prism is called a **right prism**.
   Here is a right pentagonal prism and its net.

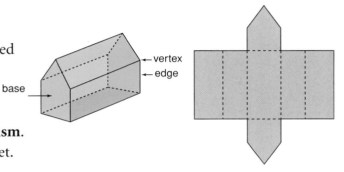

A **regular prism** has
regular polygons as bases.

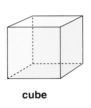 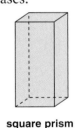

cube   square prism

A **regular pyramid** has
a regular polygon as its base.
Its other faces are triangles.

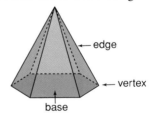

regular hexagonal pyramid

A regular polygon has all sides equal and all angles equal.

➤ Here is a right cylinder and its net.
   The line joining the centres of the circular bases is perpendicular to the bases.

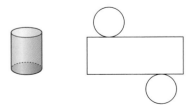

The two congruent circles are on opposite sides of the rectangle.

4.1 Exploring Nets

Example 1

Which diagram is the net for this right square prism?

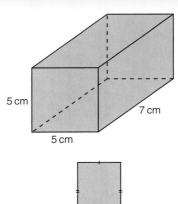

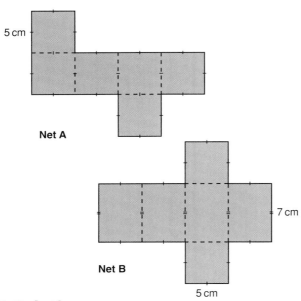

▶ A Solution

This square prism has 2 square bases and 4 rectangular faces that are not squares.
Net A has 6 square faces, so it is not the correct net.
Net B has 2 square faces and 4 rectangular faces, so it is the correct net.
Net C has all rectangular faces, so it is not the correct net.

Example 2

Use a ruler and compass.
Construct a net of this right triangular prism.

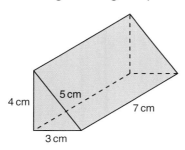

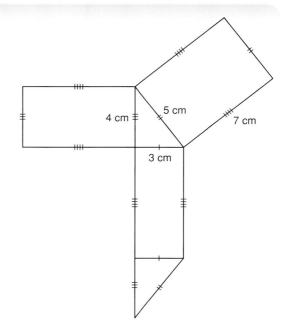

▶ A Solution

The prism has 2 congruent triangular bases and
3 rectangular faces.
Sketch the net of the triangular prism.

172   UNIT 4: Measuring Prisms and Cylinders

Construct the net of the prism.

➤ Start by constructing a base, △ABC.
Use a ruler to draw BC = 3 cm.
With the compass point and pencil 5 cm apart,
put the compass point on C and draw an arc.
With the compass point and pencil 4 cm apart,
put the compass point on B and draw an arc.
Mark point A where the arcs intersect.
Join AB and AC.
Label each side with its length.

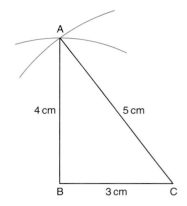

➤ Use what you know about parallel and
perpendicular lines to:
- Construct a rectangle on AB with length 7 cm.
- Construct a rectangle on AC with length 7 cm.
- Construct a rectangle on BC with length 7 cm.
- Construct a triangle congruent to △ABC
  on the rectangle on BC.

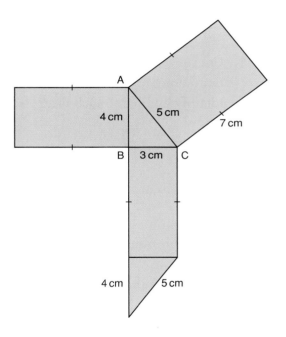

1. Look at the net of a cylinder in *Connect*. Does it matter where the circles are placed along the sides of the rectangle? Justify your answer.
2. In *Example 2*, the base of the right prism is a right triangle. What other types of triangles could be the base of a right prism? How would you name each prism?
3. Why is it useful to be able to draw nets?

4.1 Exploring Nets 173

Practice

Check

4. Use square dot paper. Draw a net of this right rectangular prism. Identify and name each face.

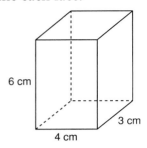

5. Use square dot paper. Draw a net of this cube. Identify and name each face.

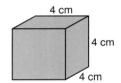

6. One diagram below is the net of this object. Choose the correct net. Justify your choice.

   a)

   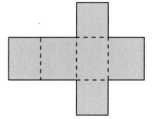

   b)

   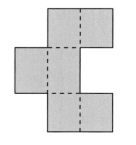

7. Use square dot paper. Draw a net of this square prism. Identify and name each face.

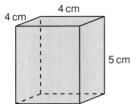

Apply

8. Construct a net of this right isosceles triangular prism. Identify and name each face.

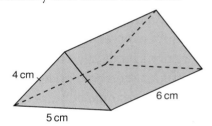

9. a) Match each object to its net.
   b) Identify and name each face of each object.

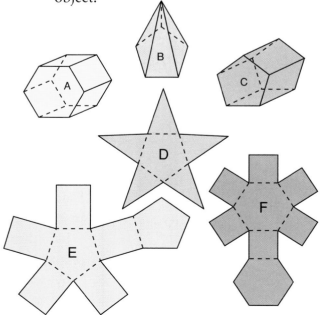

174 UNIT 4: Measuring Prisms and Cylinders

10. Which of the diagrams below are nets of a square pyramid? Justify your choices.

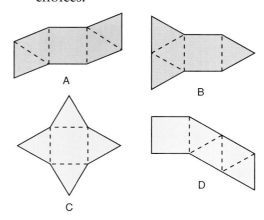

11. The Saamis Teepee in Medicine Hat, Alberta is shaped like a dodecagonal pyramid. Which diagram below is a net for a regular dodecagonal pyramid? How do you know?

A **regular dodecagon** is a polygon with 12 equal sides and 12 equal angles.

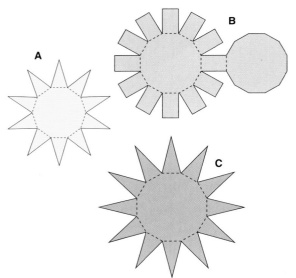

12. **Assessment Focus**
    Name each object and identify its faces. Draw or construct as many different nets as you can for each object. Use square dot paper or triangular dot paper if it helps. Use hatch marks to show which sides have the same length.
    a)   b)

    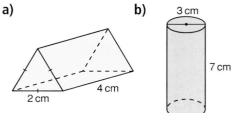

13. The sums of the numbers on opposite faces of a die are equal.
    a) Copy each net onto grid paper. Label each face with a number from 1 to 6, so the net can fold to make a die.

    i)

    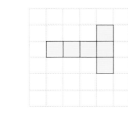

    ii)

    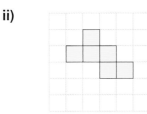

    iii)

    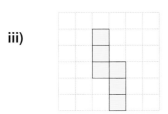

    b) Draw a different net of a die. Label its faces.

4.1 Exploring Nets 175

14. Use the descriptions below. Identify an object that has each set of faces.
   a) four equilateral triangles and one square
   b) two congruent squares and four congruent rectangles
   c) one rectangle and two pairs of congruent triangles
   d) five congruent triangles and one regular pentagon
   e) four congruent equilateral triangles

15. Take It Further Decorative wrapping paper is often used to wrap a gift.
How is wrapping paper similar to a net? How is it different?

16. Take It Further You will need triangular dot paper.
   a) A triangular pyramid with all faces congruent is a regular tetrahedron. Draw all possible nets for a regular tetrahedron.
   b) Draw all possible nets for a pyramid with an equilateral triangular base and three isosceles triangular faces.
   c) Compare your nets in parts a and b. Explain any differences.

17. Take It Further
Rubik's cube, a popular and challenging puzzle, is made of 27 small cubes.

   a) Draw a net of one of the small cubes. Indicate which cube you are drawing and label each face you see.
   b) Draw a net of the Rubik's cube. Label each face you see.
   c) How are the nets in parts a and b alike? How are they different?
   d) The Rubik's cube was turned and twisted. How would you change the net you drew in part b to reflect the changes shown?

Reflect

Choose a package in the shape of a polyhedron.
Sketch the package. Include appropriate dimensions.
Cut along the edges to make the net.
Why do you think the manufacturer used this shape for the package?
Can you find a better shape? Explain.

4.2 Creating Objects from Nets

Focus Build objects from nets.

In Vincent Massey Junior High School in Calgary, Grade 8 students taking Fashion Studies work with a pattern to make a pair of pyjama pants. How is a pattern similar to a net? How is it different?

Investigate

Work with a partner.
You will need scissors and tape.
Your teacher will give you large copies of the nets below.

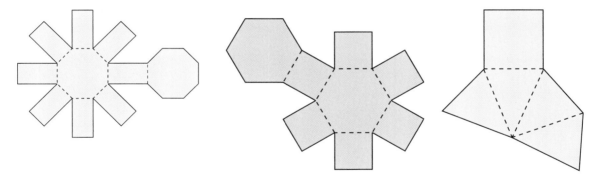

➤ Predict the object that can be formed from each net above.
➤ Use a copy of each net.
  Fold, then tape the net to verify your prediction.

Compare the three objects you made with those of another pair of classmates. What do you notice? What does this tell you about nets of different objects?

Connect

To determine if a diagram is a net for an object,
look at each shape and at how the shapes are arranged.

➤ This is *not* a net for a rectangular prism.
If this diagram was folded, it would form a box that is
open at one end.
At the opposite end, two rectangles would overlap.

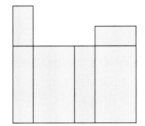

Example 1

Look at the diagrams below.
Is each diagram the net of an object?
If your answer is yes, name and describe the object.
If your answer is no, what changes could you make so it could be a net?

a)

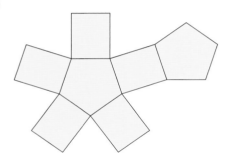

b)

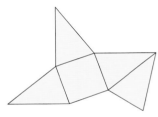

c)

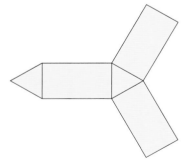

d)

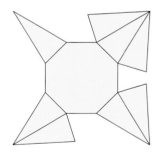

▶ **A Solution**

a) This diagram has 2 congruent regular pentagons
and 5 congruent rectangles.
When it is folded, congruent sides join to form edges.
The diagram is a net of a right pentagonal prism.

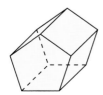

178 UNIT 4: Measuring Prisms and Cylinders

b) This diagram has 1 square and
4 congruent isosceles triangles.
The diagram is a net of a
square pyramid.

c) This diagram has 2 congruent equilateral triangles
and 3 congruent rectangles.
The diagram is a net of a right triangular prism.
It has equilateral triangular bases.

d) This diagram is not a net. When it is folded,
2 triangular faces overlap, and the
opposite face is missing.
To make a net, move one triangular face
from the top right to the top left.
The diagram is now a net of an octagonal pyramid.

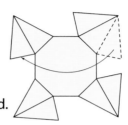

Example 2

a) Predict the object this net will form.
b) Fold a copy of the net to verify your prediction.
c) Describe the object.

▶ **A Solution**

a) The net has one regular hexagon and
6 congruent triangles.
So, the net probably makes a hexagonal pyramid.

b) When the net is folded, it forms
a hexagonal pyramid.

c) The object is a polyhedron because its faces are polygons.
The object is a regular hexagonal pyramid.
This means that the base of the pyramid is a regular hexagon.
The pyramid has 6 congruent triangular faces.

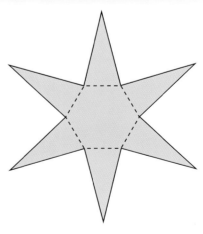

Discuss the ideas

1. How can you tell whether a given net is for a prism or a pyramid?
2. How can you change the net in *Connect* to create a net of a rectangular prism?
   How many different ways can you do this?

Practice

Check

Your teacher will give you a large copy of each net.

3. One diagram below is the net of a right cylinder. Predict which diagram is a net. Cut out the diagrams. Fold them to confirm your prediction.

   a)

   b)

   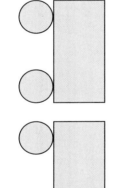

4. a) Predict the object this net will form.
   b) Fold the net to verify your prediction.
   c) Describe the object.

   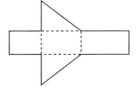

5. a) Predict the object in Set B that each net in Set A will form.
   b) Fold each net to verify your prediction.

   Set A

   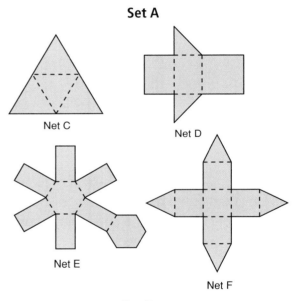

   Net C

   Net D

   Net E

   Net F

   Set B

   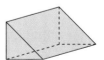

   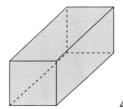

180   UNIT 4: Measuring Prisms and Cylinders

Apply

6. a) Describe the object this net will form.
   b) Fold the net to verify your prediction.
   c) Is the object a polyhedron? If it is, describe the polyhedron.
   d) Identify parallel faces and perpendicular faces.

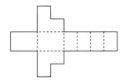

7. Assessment Focus
   a) Is this diagram a net of a right octagonal prism? If your answer is yes, tell how you know.
   b) If your answer is no, describe and sketch any changes you would make to correct the diagram. How many different ways could you correct the diagram? Explain.

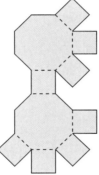

8. A soccer ball is not a sphere. It is a polyhedron. Explain which polygons are joined to make the ball. How are the polygons joined?

9. Which diagrams are nets? How do you know? For each net, identify and describe the object it will form. Verify your answer. Explain how you did this.
   a)   **b)**
   c)   **d)**

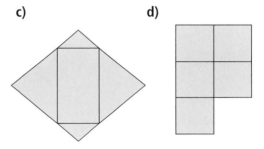

10. One diagram below is the net of this decagonal pyramid. Predict which diagram is a net. Cut out the diagrams. Fold them to confirm your prediction.

   a)   **b)**

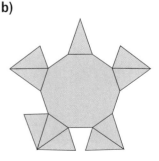

4.2 Creating Objects from Nets

11. Which diagrams below are nets of a cube? Explain how you know.

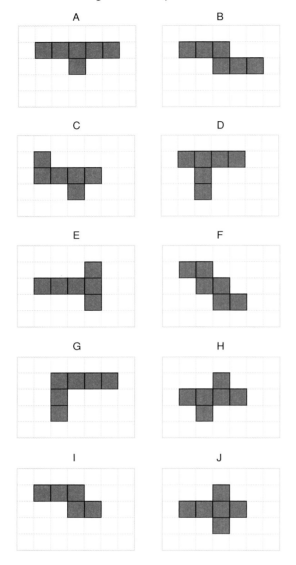

12. Take It Further Fold 2 of these nets to make 2 square pyramids with no base.

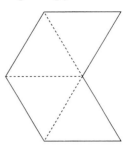

Tape the 2 pyramids together at their missing bases. You have made a regular octahedron.
  a) Why does it have this name?
  b) Describe the octahedron. How do you know it is regular?

13. Take It Further
  a) Fold this net to make an object. Describe the object.

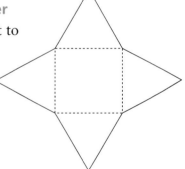

  b) Suppose the object in part a is cut and part of it is removed to form the object below.
  What changes would you make to the net so that it folds to make the new object?
  Show your work.

Reflect

When you see a diagram that may be a net of an object, how do you find out if it is a net?
Include an example in your explanation.

4.3 Surface Area of a Right Rectangular Prism

Focus Find the surface area of a right rectangular prism.

This rectangular prism is made from 1-cm cubes.
What is the surface area of the prism?

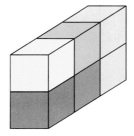

Investigate

Work with a partner.
You will need an empty cereal box, scissors, and a ruler.

➤ Open the bottom of the box without tearing the edges.
  Then cut along one edge to make a net.
➤ Find the area of the net.
  What measurements did you make?
➤ Did you find any shortcuts? Explain.
  How does the area of the net relate to the area of the
  surface of the cereal box?
➤ Describe a method to find the area of the surface
  of any right rectangular prism.

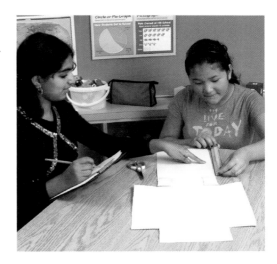

Compare your method with that of another pair of classmates.
How were your methods similar?
How were they different?
Do both methods work? How could you check?

4.3 Surface Area of a Right Rectangular Prism   **183**

Connect

Here is a right rectangular prism and its net.
The net has 6 rectangles,
labelled A to F.
The area of the net is the sum of
the areas of the 6 rectangles.

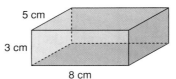

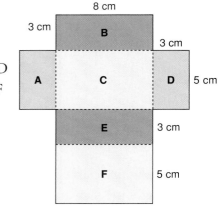

The area of the net = area of Rectangle A + area of Rectangle B
  + area of Rectangle C + area of Rectangle D
  + area of Rectangle E + area of Rectangle F
  = (3 × 5) + (8 × 3) + (8 × 5) + (3 × 5)
    + (8 × 3) + (8 × 5)
  = 15 + 24 + 40 + 15 + 24 + 40
  = 158

The area of the net is 158 cm².

We say that the **surface area** of the rectangular prism is 158 cm².

The surface area of an object is the sum of the areas of its faces.
We can use the net of a rectangular prism to find its surface area.

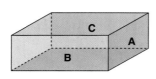

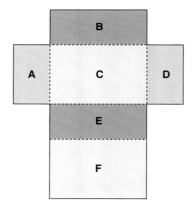

There are 3 pairs of congruent rectangles in the net
of a rectangular prism.
So, we can write the surface area a shorter way.
Surface area = 2 × area of Rectangle A
  + 2 × area of Rectangle B
  + 2 × area of Rectangle C

Example 1

Find the surface area of this right rectangular prism.

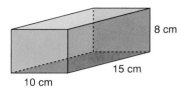

▶ **A Solution**

Identify each rectangle with a letter.
Rectangle A has area: 15 × 8 = 120
Rectangle B has area: 15 × 10 = 150
Rectangle C has area: 8 × 10 = 80

So, surface area = 2(120) + 2(150) + 2(80)
 = 240 + 300 + 160
 = 700

The surface area of the rectangular prism is 700 cm².

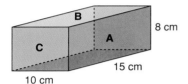

Example 2

The school is holding elections for student council.
The ballot box is to be painted.
The ballot box is a cube with edge length 30 cm.
There is a slot on the top.
The slot has length 16 cm and width 1 cm.
What is the total surface area to be painted?
Assume the base is to be painted.

▶ **A Solution**

Draw a labelled picture.
The cube has 6 congruent square faces.
Each face has area: 30 cm × 30 cm = 900 cm²
So, the surface area of the cube is:
6 × 900 cm² = 5400 cm²

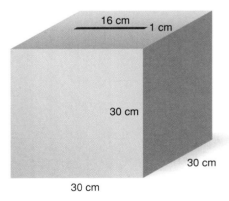

The area of the slot is: 16 cm × 1 cm = 16 cm²

The surface area of the ballot box is:
Surface area of cube − area of slot = 5400 cm² − 16 cm²
 = 5384 cm²

The surface area to be painted is 5384 cm².

Discuss the ideas

1. Explain how the net of a rectangular prism can help you find the surface area of the prism.
2. Suppose the rectangular prism in *Example 1* was open at the top. How would this affect its surface area?
3. In *Example 2*, suppose it was decided not to paint the base. What is the total surface area to be painted?

Practice

Check

4. Here is the net of a right rectangular prism. The area of each face is given. What is the surface area of the prism? How did you find out?

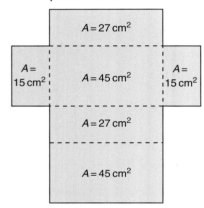

5. Sketch a net of this right rectangular prism. What is its surface area?

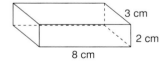

6. Find the surface area of each right rectangular prism.

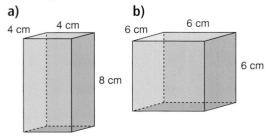

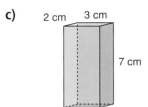

c)

7. Find the surface area of a right rectangular prism with these dimensions.
   a) 4 m by 3 m by 10 m
   b) 3 cm by 5 cm by 8 cm

Apply

8. Find a right rectangular prism in the classroom. Measure its faces. Find its surface area.

9. Tanya paints the walls of her family room. The room measures 7 m by 4 m by 3 m. The walls need 2 coats of paint. A 4-L can of paint covers 40 m².
   a) How much paint should Tanya buy?
   b) What assumptions do you make? Explain.

10. The surface area of a cube is 54 cm².
    a) What is the area of one face of the cube?
    b) What is the length of one edge of the cube?

11. A window washing company is hired to wash the windows in a condominium. The building is 50 m by 30 m by 300 m. Windows cover about one-quarter of the building. What is the total surface area of the windows to be washed? What assumptions do you make?

12. The Sandberg Institute building in Amsterdam generates revenue by selling advertising space on the exterior of the building. The building is a rectangular prism with dimensions 50 m by 40 m by 75 m. Suppose it costs 1 Euro per month to rent an advertising space of 50 cm². Each of the 4 walls of the building is covered with advertisements. How much money will the institute earn in one month?

13. Which prism has the greatest surface area? The least surface area?

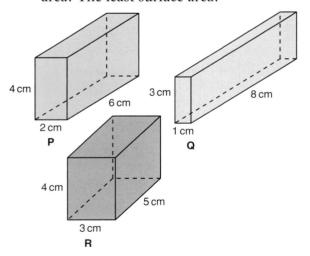

14. Assessment Focus Sketch a right rectangular prism. Label its dimensions. Answer the question below. Justify your answer. What do you think happens to the surface area of a prism in each case?
  i) Its length is doubled.
  ii) Its length is halved.

15. Each object has the shape of a rectangular prism, but one face or parts of faces are missing. Find each surface area.

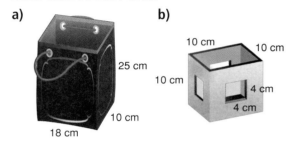

16. Take It Further A right rectangular prism has a square base with area 4 m². The surface area of the prism is 48 m². What are the dimensions of the prism?

17. Take It Further A right rectangular prism has faces with these areas: 12 cm², 24 cm², and 18 cm²
What are the dimensions of the prism? How did you find out?

Reflect

Explain how you would find the surface area of a rectangular prism. Include a diagram in your explanation.
How does a net help you find the surface area?

4.4 Surface Area of a Right Triangular Prism

Focus Find the surface area of a right triangular prism.

What is the area of this rectangle?

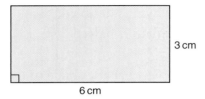

What is the area of this triangle?

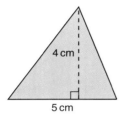

You will use these measures to find the surface area of a right triangular prism.

Investigate

Work with a partner.
Your teacher will give you a right triangular prism.

➤ Find the surface area of the triangular prism.

➤ Describe your strategy for finding the surface area of the prism.

➤ Do you think your strategy would work for any prism?
Trade prisms with a pair of classmates who used a different prism. Use your strategy to find its surface area.

Compare your strategy and results with the same pair of classmates.
Did you use the same strategy?
If not, do both strategies work? Explain.

Connect

A right triangular prism has 5 faces: 3 rectangular faces and 2 triangular bases.
The two triangular bases of the prism are congruent.
Here is a triangular prism and its net.
The measurements are given to the nearest centimetre.

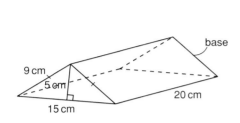
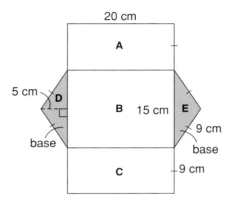

The surface area of a triangular prism is the sum of the areas of its faces.
Rectangle A has area: $20 \times 9 = 180$
Rectangle B has area: $20 \times 15 = 300$
Rectangle C has area: $20 \times 9 = 180$
Triangle D has area: $\frac{1}{2} \times 15 \times 5 = \frac{75}{2} = 37.5$
Triangle E has area: $\frac{1}{2} \times 15 \times 5 = \frac{75}{2} = 37.5$

The area measurements are approximate because the measurements are given to the nearest centimetre.

Surface area = area of Rectangle A + area of Rectangle B + area of Rectangle C
             + area of Triangle D + area of Triangle E
     = 180 + 300 + 180 + 37.5 + 37.5
     = 735

The surface area of the prism is 735 cm².

Since the bases of a triangular prism are congruent,
we can say:
Surface area = sum of the areas of the 3 rectangular faces
             + 2 × area of one triangular base

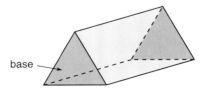

4.4 Surface Area of a Right Triangular Prism

Example 1

Find the surface area of this prism.

Each dimension is given to the nearest centimetre.

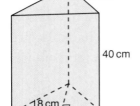

▶ A Solution

Draw a net. Label its dimensions.
The prism is 40 cm tall, so each rectangle has height 40 cm.
The width of each rectangle is a side length of the triangular base.
One rectangle has area: $40 \times 20 = 800$
Another rectangle has area: $40 \times 31 = 1240$
The third rectangle has area: $40 \times 29 = 1160$
The triangular base has area: $\frac{1}{2} \times 31 \times 18 = 279$

The surface area $= 800 + 1240 + 1160 + 2 \times 279$
$= 3758$
The surface area of the prism is 3758 cm².

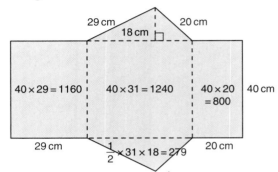

Instead of drawing a net, we can visualize each face as we calculate its area.

Example 2

A wooden doorstop is a triangular prism. It is to be painted. The bottom rectangular face is covered with rubber and will not be painted.
Find the total surface area to be painted.

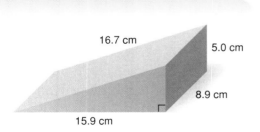

▶ A Solution

Each dimension is given to one decimal place.

There are 4 faces to be painted: 2 triangular bases, the slanted rectangular face, and the vertical rectangular face
The area of each triangular base is: $\frac{1}{2} \times 15.9 \times 5.0 = 39.75$
The slanted rectangular face has area: $8.9 \times 16.7 = 148.63$
The vertical rectangular face has area: $5.0 \times 8.9 = 44.5$
Total area to be painted is: $(2 \times 39.75) + 148.63 + 44.5 = 272.63$
The surface area to be painted is 272.6 cm² to one decimal place.

1. Explain how the net of a triangular prism can help you find the surface area of the prism.
2. What do you know about the rectangular faces in an equilateral triangular prism?
   How would you find the surface area of the prism?
3. What do you know about the rectangular faces in an isosceles triangular prism?
   How would you find the surface area of the prism?

Practice

Check

Use a calculator when you need to.

4. Here is the net of a right triangular prism. The area of each face is given. What is the surface area of the prism? How did you find out?

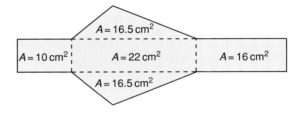

5. Here is a right isosceles triangular prism. Which faces are congruent and share the same area? How do you know?

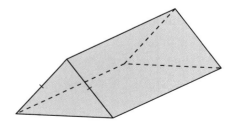

6. Sketch a net of this triangular prism. What is its surface area?

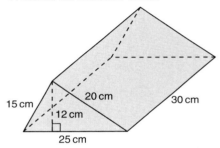

7. a) Calculate the area of each net.
   i)
   ii)

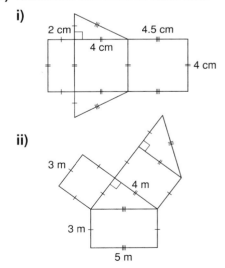

b) How does the area of each net compare to the surface area of the prism formed by the net?

4.4 Surface Area of a Right Triangular Prism   191

Apply

8. Calculate the surface area of each prism. Order the prisms from greatest to least surface area. Show your work.

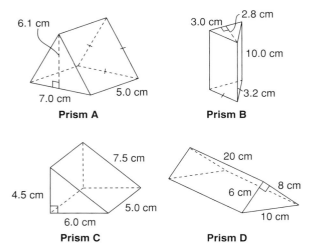

Prism A / **Prism B**

9. Find the surface area of each triangular prism.

a)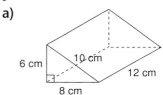

b)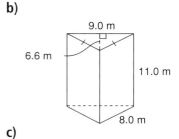

c)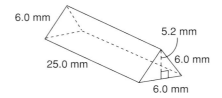

10. The 3 rectangular faces of a triangular prism have areas 30 cm^2, 40 cm^2, and 50 cm^2. The 2 triangular bases have a combined area of 12 cm^2. What are the dimensions of the triangular prism? Explain your thinking using diagrams, numbers, and words.

11. Suppose you want to construct a right triangular prism 15 cm long with the greatest surface area. Which of these triangles should you choose for its base? Explain your choice.

a)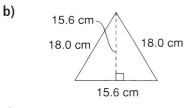

b)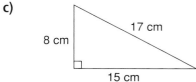

c)
8 cm, 17 cm, 15 cm triangle

12. Assessment Focus A student said, "If you double all the dimensions of a triangular prism, you will double its surface area." Is the student correct? Use words, numbers, and diagrams to explain your answer.

13. How much metal is needed to build this water trough?

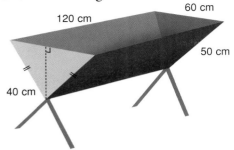

14. Daniel wants to cover the outside of an empty 3-ring binder with plastic. Each dimension of the binder has been written to one decimal place.
How much plastic is needed to cover the outside of the binder?
What assumptions do you make?

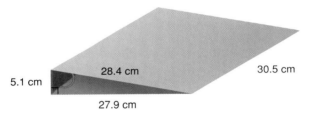

15. A rectangular prism is cut as shown to form two congruent triangular prisms. Is the surface area of one triangular prism one-half the surface area of the rectangular prism? Justify your answer.

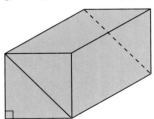

16. Take It Further This triangle is one base of a right triangular prism. What should the length of the prism be so its surface area is between 100 cm² and 150 cm²? Show your work.

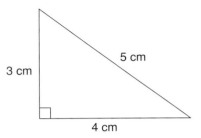

17. Take It Further
a) Use the Pythagorean Theorem. Find the height of a triangular base of this prism.
b) What is the surface area of the prism? Give your answer to the nearest square centimetre.

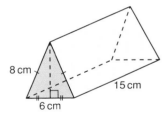

Reflect

How is the strategy for finding the surface area of a triangular prism similar to the strategy for finding the surface area of a rectangular prism? How is it different?

Mid-Unit Review

LESSON

4.1

1. Sketch two different nets for each object.
   a)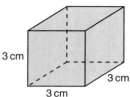
   b)

4.2

2. a) Which diagrams are nets? How do you know?
   b) For each net, identify the object it forms.

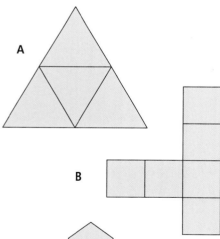

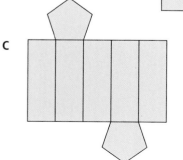

4.3
4.4

3. Find the surface area of each prism.
   a)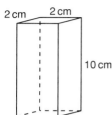
   b)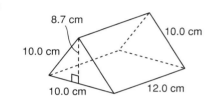

4. A skyscraper is shaped like a rectangular prism. The outside of the building is almost entirely glass. The base of the building is 58 m by 75 m. The height of the skyscraper is 223 m. What is the area of glass needed to cover the building?

5. A clerk gift wrapped some golf clubs in a box shaped like a right triangular prism. The base of the prism is an equilateral triangle with side length 28 cm and height about 24 cm. The prism is 120 cm long.
   a) What is the surface area of the box?
   b) How much wrapping paper does the clerk need? What assumptions do you make?

4.5 Volume of a Right Rectangular Prism

Focus Develop and use a formula to calculate the volume of a right rectangular prism.

This rectangular prism is made from 1-cm cubes. What is the volume of the prism?

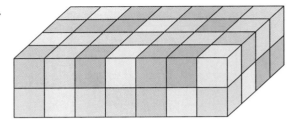

Investigate

Work with a partner.
You will need 2 empty cereal boxes, a ruler, and a calculator.

➤ Compare the two boxes.
   Which box do you think holds more cereal?
   Why do you think so?
➤ Find the volume of each box.
   Which box has the greater volume?
   How does this compare with your prediction?
➤ Work together to write a formula you can use to find the volume of any rectangular prism.
➤ Suppose you know the area of one face of a cereal box. What else do you need to know to find the volume of the box?
➤ Work together to write a formula for the volume of a right rectangular prism in terms of the area of its base.

Compare your formula with that of another pair of classmates.
Did you write the same formula?
If not, do both formulas work? Explain.
Compare the formulas for the volume of a right rectangular prism and the area of a rectangle. What do you notice?
How can you explain this?

Connect

This box is a right rectangular prism.
The volume of the box is the number of centimetre cubes the box holds.

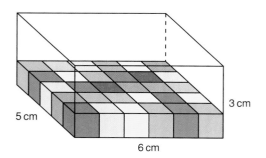

Recall that, when the dimensions are measured in centimetres (cm), the volume is measured in cubic centimetres (cm³).

One layer of cubes will be 5 cm wide and 6 cm long.
So, 5 × 6, or 30 cubes fit in one layer.
The box is 3 cm high, so 3 layers can fit.
The total number of cubes is 30 × 3 = 90.
So, the volume of the box is 90 cm³.

One way to write the volume of the box:
Volume = the number of cubes in one layer × the number of layers
This is the same as:
Volume = the area of the base of the box × its height
This is true for all rectangular prisms.

We can use variables to write a formula for
the volume of a rectangular prism.
Let A represent the base area and h represent the height.
Then, the volume of a rectangular prism is:
$V = A \times h$ or $V = Ah$

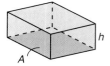

Example 1

The area of the base of a fish tank is 2013 cm².
The height of the tank is 30 cm.
Find the volume of the fish tank.

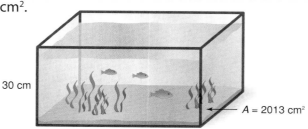

▶ **A Solution**

The fish tank is a right rectangular prism with base area 2013 cm² and height 30 cm.
Volume = base area × height
   = 2013 × 30
   = 60 390

The volume of the fish tank is 60 390 cm³.

Example 2

A deck of 54 cards fits in a box shaped like a right rectangular prism.
The box has dimensions 6.5 cm by 9.0 cm by 1.6 cm.
What is the volume of the box?
Give the answer to the nearest cubic centimetre.

▶ **A Solution**

Draw a diagram.
Label each dimension.
Let the base be one rectangle with length 9.0 cm
and width 6.5 cm.
$A = 9.0 \times 6.5$
  $= 58.5$
The area of the base is 58.5 cm².
The height of the box is 1.6 cm.
Use the formula: $V = Ah$
$V = 58.5 \times 1.6$ Use a calculator.
  $= 93.6$
The volume is 94 cm³, to the nearest cubic centimetre.

Discuss the ideas

1. Suppose the rectangular prism in *Connect* holds 210 centimetre cubes. How high is the box? Assume the area of the base is unchanged.
2. When you find the volume of a right rectangular prism, does it matter which face you use as the base?
3. For *Example 2*, suggest a different way to find the volume of the box.

Practice

Use a calculator when it helps.

Check

4. The base area and height of each prism are given. Find the volume of each prism.

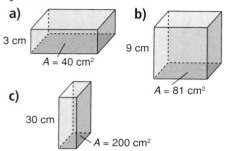

5. A box of laundry detergent has dimensions 28 cm by 16 cm by 25 cm.
   a) Sketch the box. Label each dimension.
   b) What volume of detergent will fill the box?

6. a) Find the volume of each prism.

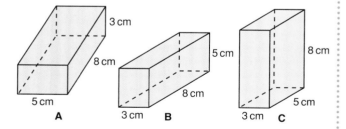

   b) What do you notice about the volumes in part a?
   c) Does the volume of a rectangular prism change when you place the prism on a different base? Justify your answer.

Apply

7. Find the volume of each rectangular prism.

a)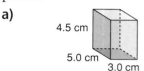

b)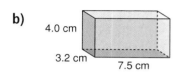

c)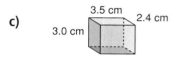

8. Find a right rectangular prism in the classroom. Measure its dimensions. Find its volume.

9. Each dogsled team that enters the Iditarod has a portable doghouse for each sled dog. Two mushers are comparing the sizes of their doghouses. Each of Rick's doghouses is 94 cm by 63 cm by 71 cm. Each of Susan's doghouses is 109 cm by 71 cm by 81 cm.
   a) What is the volume of each doghouse?
   b) About now many times as great as the volume of Rick's doghouse is the volume of Susan's doghouse?

10. Suppose a milk carton is 10 cm wide and 10 cm long. How tall must the carton be to hold 1 L of milk?

    Recall 1 cm³ = 1 mL.

11. Large trucks often tow trailers that are shaped like right rectangular prisms. A standard trailer is 2.74 m by 2.43 m by 6.1 m.
    a) What is the greatest volume of cargo a standard trailer can hold?
    b) How many trailers would it take to transport 100 m³ of goods? What assumptions do you make?

12. A rectangular swimming pool is to be filled with water. The pool has a uniform depth of 2 m and is surrounded by a wooden deck. The pool is 20 m wide and 50 m long. How much water is needed in each case?
    a) The pool is filled to the level of the deck.
    b) The pool is filled to within 20 cm of the level of the deck.
    c) The pool is half filled.

13. **Assessment Focus**
    a) Sketch all possible right rectangular prisms with volume 36 cm³. Label each prism with its dimensions in centimetres. How do you know you have found all possible prisms?
    b) Use the prisms you sketched.
       i) Which prism has the greatest surface area?
       ii) Which prism has the least surface area?
       How did you find out?

14. Philip made fudge that filled a 20-cm by 21-cm by 3-cm pan.
    a) What is the volume of the fudge?
    b) Philip shares the fudge with his classmates.
       There are 30 people in the class. How much fudge will each person get?
    c) How could Philip cut the fudge so each person gets an equal sized piece?
       Sketch the cuts Philip could make.
    d) What are the dimensions of each piece of fudge in part c?

15. Sketch a right rectangular prism. Label its dimensions.
What do you think happens to the volume of the prism when:
a) its length is doubled?
b) its length and width are doubled?
c) its length, width, and height are doubled?
Investigate to find out.
Show your work.
Will the results be true for all rectangular prisms?
Why do you think so?

16. Take It Further
How can you double the volume of a right rectangular prism?
Does its surface area double, too? Explain.

17. Take It Further
Students in a Grade 8 class are filling shoeboxes with toys for children in other countries. A shoebox measures 30 cm by 18 cm by 16 cm.
a) Find the volume of a shoebox.
b) The students fill 24 shoeboxes. Eight shoeboxes are packed into a larger box. What could the dimensions of this larger box be?
c) What are the most likely dimensions of the larger box? Justify your choice.

18. Take It Further
a) Sketch 3 different right rectangular prisms with volume 24 cm³.
b) Which prism has the greatest surface area?
Which prism has the least surface area?
c) Sketch a prism with a greater surface area but the same volume. Describe the shape of this prism.
d) Sketch a prism with a lesser surface area but the same volume. Describe the shape of this prism.

Reflect

Suppose you know the area of one face of a rectangular prism.
What else do you need to know to find the volume of the prism? Explain.
Suppose you know the volume of a rectangular prism.
Can you find its dimensions? Use words and diagrams to explain.

Largest Box Problem

HOW TO PLAY

1. Cut congruent squares from the four corners of the grid. Think about what size the squares should be to make a box with the greatest volume.

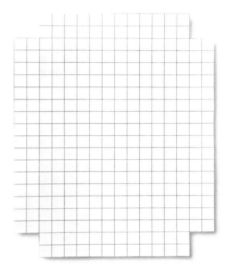

YOU WILL NEED
20-cm by 16-cm grids; scissors; tape; rulers

NUMBER OF PLAYERS
4

GOAL OF THE GAME
To make a box with the greatest volume

2. Fold, then tape the sides to form an open box.

3. Measure the length, width, and height of your box. Find its volume.

4. Compare the volume of your box to the volumes of the boxes the other players in your group made. The player whose box has the greatest volume wins.

Reflect

➤ What strategies did you use to make the box with the greatest volume?
➤ Compare results with another group of students. How do you know you cannot make a box with a greater volume?

4.6 Volume of a Right Triangular Prism

Focus Develop and use a formula to find the volume of a right triangular prism.

Here is another way to visualize a triangular prism. A triangle is translated in the air so that each side of the triangle is always parallel to its original position.

How could you use this model to find the volume of the prism?

Investigate

Work with a partner.
You will need scissors and tape.
Your teacher will give you a large copy of these nets.

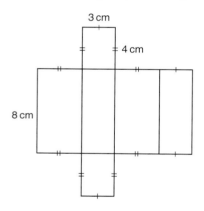
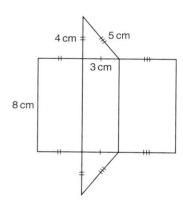

➤ Identify the prism each net will form.
➤ Cut out the nets and construct the right prisms.
➤ Visually compare the volumes of the two prisms.
  How are they related?
➤ What is the volume of the rectangular prism?
  How can you use this volume to find the volume of the triangular prism?
➤ What is a formula for the volume of a rectangular prism?
  How can you use this formula to write a formula for the volume of a triangular prism?

Reflect & Share

Combine your prisms with those of another pair of classmates. How can you arrange the prisms to verify the relationship you found in *Investigate*?

Connect

The volume of a right rectangular prism can be written as:
$V = $ base area $\times$ length

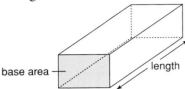

To avoid confusion between the height of a triangle and the height of the prism, use *length* to describe the height of the prism.

Suppose we draw a triangle on the base of the prism so that the base of the triangle is one edge, and the third vertex of the triangle is on the opposite edge.

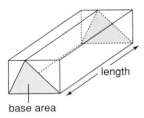

The volume of a triangular prism with this base, and with length equal to the length of the rectangular prism, is one-half the volume of the rectangular prism.

The volume of a right triangular prism
is also: $V = $ base area $\times$ length
The base is a triangle, so the base area is the area of the triangle.

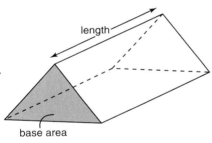

We can use variables to write a formula for the volume of a triangular prism.

For the triangular prism below:

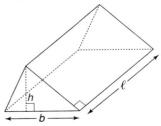

The length of the prism is ℓ.
Each triangular face has base b and height h.
The volume of the prism is:
$V = $ base area $\times$ length, or $A \times \ell$, or $A\ell$, where $A = \frac{1}{2}bh$

Example 1

Find the volume of the prism.

▶ **A Solution**

The area of the base of a triangular prism is 12 cm².
The length of the prism is 9 cm.
Volume of triangular prism = base area × length
$$V = 12 \times 9$$
$$= 108$$

The volume of the triangular prism is 108 cm³.

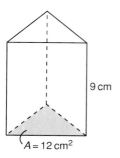

Example 2

Here is a diagram of Renee's new house.
What is the volume of the attic?

▶ **A Solution**

The attic is a triangular prism.
Sketch the prism.
Use a variable to represent each dimension.
The base of the triangle is: $b = 8$
The height of the triangle is: $h = 3$
The length of the prism is: $\ell = 10$

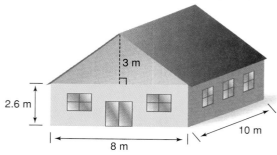

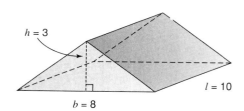

Use: $V = A\ell$
First find A.
$A = \frac{1}{2}bh$
Substitute: $b = 8$ and $h = 3$
$A = \frac{1}{2} \times 8 \times 3$
$= 12$

Now find V.
Substitute: $A = 12$ and $\ell = 10$ into $V = A\ell$
$V = 12 \times 10$
$= 120$

The volume of the attic is 120 m³.

1. A rectangular prism is cut in half to make 2 congruent triangular prisms.
What do you know about the volume of each triangular prism?

2. Any face can be used as the base of a rectangular prism.
Can any face be used as the base of a triangular prism? Explain.

Practice

Check

3. Each rectangular prism is divided into 2 congruent triangular prisms along the diagonal shown. The volume of each rectangular prism is given. Find the volume of each triangular prism.

a)
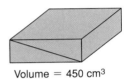
Volume = 450 cm³

b)
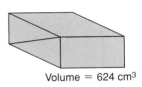
Volume = 624 cm³

4. The base area and length of each triangular prism are given.
Find the volume of each prism.

a)
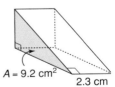
$A = 9.2$ cm² 2.3 cm

b)

5 cm
$A = 43.5$ cm²

c)
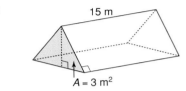
15 m
$A = 3$ m²

5. Find the volume of each triangular prism.

a)
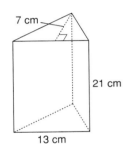
7 cm
21 cm
13 cm

b)
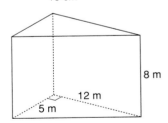
8 m
12 m
5 m

c)
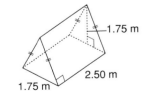
1.75 m
2.50 m
1.75 m

4.6 Volume of a Right Triangular Prism   **205**

Apply

6. Find the volume of each prism.

a)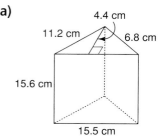

b)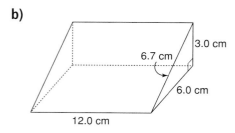

7. What is the volume of glass in this glass prism?

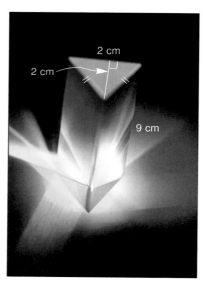

8. The volume of a right triangular prism is 30 cm³. Each triangular face has area 4 cm².
How long is the prism?

9. Assessment Focus
a) Find possible values for A and ℓ for each volume of a right triangular prism. Sketch one possible triangular prism for each volume.
  i) 5 cm³
  ii) 9 m³
  iii) 8 m³
  iv) 18 cm³
b) How many different prisms can you find in each case?

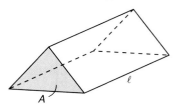

10. Chico has a wedge of cheddar cheese. He plans to serve the cheese as an appetizer before dinner.

a) What volume of cheese does Chico have?
b) Suppose each person eats 20 cm³ of cheese.
How many people will the cheese serve?

11. The volume of a triangular prism is 50 m³. The length of the prism is 5 m. What is the area of each triangular face?

12. Jackie uses this form to build a concrete pad.

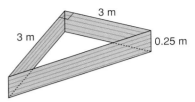

a) How much concrete will Jackie need to mix to fill the form?
b) Suppose Jackie increases the lengths of the equal sides of the form from 3 m to 6 m.
How much more concrete will Jackie need to mix?
Include a diagram.

13. a) Predict which triangular prism has the greater volume.

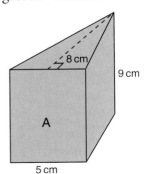

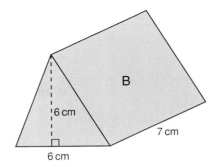

b) Find the volume of each prism. Was your prediction correct?
c) How could you change one dimension of Prism B so the two prisms have the same volume?

14. a) Find the volume of this prism.

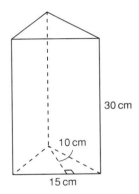

b) Suppose the prism contains 1350 mL of water.
What is the depth of the water?
c) What percent of the volume of the prism is water?

15. The volume of a right triangular prism is 198 cm³. Each triangular face is a right triangle with area 18 cm². Find as many dimensions of the prism as you can.

4.6 Volume of a Right Triangular Prism   **207**

16. Take It Further

A chocolate company produces different sizes of chocolate bars that are packaged in equilateral triangular prisms.
Here is the 100-g chocolate bar.

a) Calculate the surface area and volume of the box.
b) The company produces a 400-g chocolate bar. It has the same shape as the 100-g bar.
   i) What are the possible dimensions for the 400-g box? How many different sets of dimensions can you find?
   ii) How are the dimensions of the two boxes related, in each case?

17. Take It Further

a) Find the surface area and volume of this triangular prism.

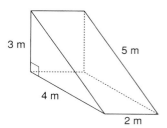

b) What do you think happens to the surface area and volume when the length of the prism is doubled? Justify your prediction.
c) What do you think happens to the surface area and volume when the base and height of the triangular face are doubled? Justify your prediction.
d) What do you think happens to the surface area and volume when all the dimensions are doubled? Justify your prediction.
e) For parts b to d, find the surface area and volume to verify your predictions.

Reflect

How did you use what you know about the volume of a right rectangular prism in this lesson?

4.7 Surface Area of a Right Cylinder

Focus Find the surface area of a right cylinder.

What is the area of this circle?

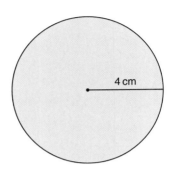

What is the circumference of this circle?

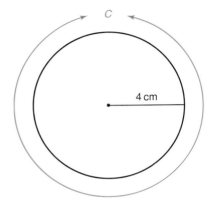

Investigate

Work with a partner.

You will need a cardboard tube, scissors, and tape.
Cut out two circles to fit the ends of the tube.
Hold a circle at each end of the tube.
You now have a right cylinder.

Find a way to calculate the surface area of the cylinder.

Share your strategy for finding the surface area with another pair of classmates.
Did you use the same strategy?
If not, do both strategies work?
How could you check?

Connect

The bases of a right cylinder are 2 congruent circles.
The curved surface of a cylinder is a rectangle when laid flat.
These 3 shapes make the net of a cylinder.

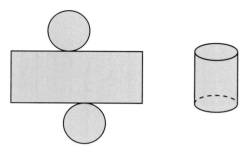

The surface area of a cylinder = 2 × area of one circular base + area of a rectangle
Label the cylinder and its net.

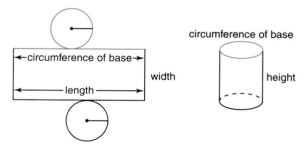

The width of the rectangle is equal to the height of the cylinder.
The length of the rectangle is equal to the circumference
of the base of the cylinder.

So, the area of the rectangle = circumference of base × height of cylinder

When a cylinder is like a cardboard tube and has no circular bases,
its surface area is the curved surface only:
Curved surface area = circumference of base × height of cylinder

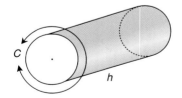

Example 1

Find the surface area of this cylinder.

A Solution

Sketch the net.

Surface area = 2 × area of one circle + area of the rectangle

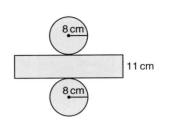

- The area of the circle is: $A = \pi r^2$

  Substitute: $r = 8$

  So, area of circle is: $A = \pi \times 8^2$
  $\doteq 201.06$

- The area of the rectangle = circumference × height
  $= 2\pi r \times h$

  Substitute: $r = 8$ and $h = 11$
  Use a calculator. For π, press the π key.
  The area of the rectangle = $2\pi \times 8 \times 11$
  $\doteq 552.92$

Surface area $\doteq 2 \times 201.06 + 552.92$
$= 955.04$

The surface area of the cylinder is about 955 cm².

Example 2

A manufacturer produces a can with height 7 cm and diameter 5 cm. What is the surface area of the label, to one decimal place?

A Solution

Sketch the can.

The label does not cover the circular bases. So, the surface area of the label is equal to the curved surface area of the can.

Curved surface area = circumference of base × height of cylinder
$= \pi d \times h$

Substitute: $d = 5$ and $h = 7$

Use a calculator. For π, press the π key.
Curved surface area = $\pi \times 5 \times 7$
$\doteq 109.956$

The surface area of the label is 110.0 cm², to one decimal place.

1. In *Example 2*, what is the surface area of the can? The can is opened and one end removed. What is the surface area of the open can?
2. What is an algebraic formula for the surface area of a right cylinder with height *h* and radius *r*?
3. Why is the surface area of a cylinder always approximate?

Practice

Check

Give each area to the nearest square unit.

4. Find the area of each net.
   a)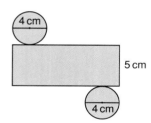
   b)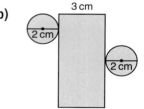
   c)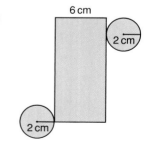

5. Describe the cylinder that each net in question 4 forms.

6. Calculate the curved surface area of each tube.
   a)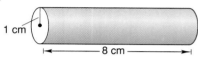
   b)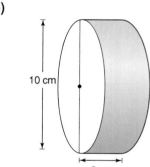
   c)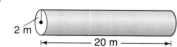

7. Find a right cylinder in the classroom. Use thin string to find its circumference. Use a ruler to measure its radius and height. Calculate the surface area of the cylinder.

Apply

8. Calculate the surface area of each cylinder.

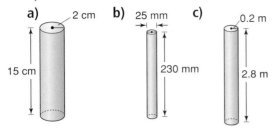

9. A cylindrical tank has diameter 3.8 m and length 12.7 m. What is the surface area of the tank?

10. Cylindrical paper dryers are used in pulp and paper mills. One dryer has diameter 1.5 m and length 2.5 m. What is the area of the curved surface of this dryer?

11. A wooden toy kit has different painted solids. One solid is a cylinder with diameter 2 cm and height 14 cm.

   a) What is the surface area of the cylinder?
   b) One can of paint covers 40 m². Each cylinder is painted with one coat of paint.
   How many cylinders can be painted with one can of paint?

12. **Assessment Focus**
   A soup can has diameter 6.6 cm. The label on the can is 8.8 cm high. There is a 1-cm overlap on the label. What is the area of the label?

13. A hot water tank is cylindrical. Its interior is insulated to reduce heat loss. The interior has height 1.5 m and diameter 65 cm. What is the surface area of the interior of the tank? Give the answer in two different square units.

14. A tom-tom hoop drum is made of stretched membranes, called heads, which are held tightly across a tubular shell. The drum has diameter 30 cm and height 30 cm. The shell of the drum is made of 5 layers of birch sheathing.

   a) How much sheathing is needed to make the shell?
   b) Suppose the drum has two heads. How much membrane would you need to make the heads?
   What assumptions do you make?

4.7 Surface Area of a Right Cylinder 213

15. A candy company can sell fruit gums in rectangular boxes or in cylindrical tubes. Each box is 8 cm by 3 cm by 7 cm. Each cylinder has radius 3 cm and height 6 cm. The company wants the packaging that uses less material. Which packaging should the company choose? Justify your choice.

16. Take It Further
The curved surface area of a solid cylinder is 660 cm².
The cylinder has height 10 cm.
 a) What is the circumference of the cylinder?
 b) What is the radius of the cylinder?
 c) What is the area of one circular base?
 d) What is the surface area of the cylinder?

17. Take It Further Benny places a glass cylinder, open at one end, over a rose cutting in his garden. The cylinder has diameter 9 cm and height 20 cm.
To make sure animals cannot knock the cylinder over, Benny covers the bottom 5 cm of the cylinder with soil. What is the surface area of the cylinder exposed to the sun?

Math Link

History

In the late 1800s, Thomas Edison developed the earliest method for recording sound, the phonograph cylinder. The open cylinder was made of wax. Audio recordings were etched on its outside surface. The sounds were reproduced when the cylinder was played on a mechanical phonograph. There are about 3312 original wax cylinders recorded by First Nations and French Canadian people on display at the Canadian Museum of Civilization in Gatineau, Quebec.

The standard wax cylinder had diameter about 5.5 cm and height 10.5 cm. What surface area was available to be etched?

Reflect

How is the formula for the surface area of a cylinder related to the net of the cylinder? Include a diagram in your explanation.

4.8 Volume of a Right Cylinder

Focus Develop and use a formula to find the volume of a right cylinder.

Here is a way to visualize a right cylinder.
A circle is translated through the air so that
the circle is always parallel to its
original position.

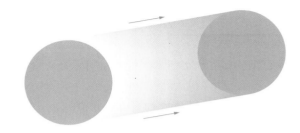

How does this relate to the triangular
prism in Lesson 4.6, page 202?

Investigate

Work with a partner.
You will need 2 identical rectangular sheets of construction paper, rice, and tape.

➤ Roll one sheet of paper lengthwise to create a tube.
  Tape the edges together.
  Repeat with the second sheet of paper.
  This time roll the paper widthwise.
➤ Predict which tube has the greater volume.
  Use rice to check your prediction.
  Do the results match your prediction? Explain.
➤ Calculate the volume of the taller tube.
  How did you use the diameter and radius in your
  calculations?
  How did you use π?

Share your strategy for calculating the volume with another
pair of classmates.
Work together to write a formula for
the volume of a right cylinder.
Use any of diameter, radius, height, and π in your formula.
Use your formula to find the volume of the shorter tube.

Connect

The volume of a right prism is: base area × height
We can use this formula to find the volume of a right cylinder.

Example 1

The area of the base of a cylinder is about 154 cm².
The height of the cylinder is 24 cm.
Find the volume of the cylinder.

▶ **A Solution**

Volume of a cylinder = base area × height
≐ 154 × 24
= 3696

The volume of the cylinder is about 3696 cm³.

We can write an algebraic formula for the volume.
The base of a cylinder is a circle with radius r.
The area of a circle is: $A = \pi r^2$
Let the height of the cylinder be h.

So, the volume of a cylinder is: V = base area × height
= area of circle × height
= $\pi r^2 \times h$
= $\pi r^2 h$

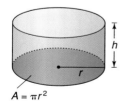

So, a formula for the volume of a cylinder is $V = \pi r^2 h$, where r is the radius of its base, and h its height.

Example 2

In 2002, nine Pennsylvania miners were trapped in a flooded coal mine. Rescue workers drilled a hole about 90 cm wide and 73 m deep into the ground to make an escape shaft. The soil from the hole was removed and piled on the ground.
What volume of soil did the rescue workers remove?
Give your answer to the nearest cubic metre.

▶ **A Solution**

The hole is shaped like a cylinder.
Draw a picture. Label the cylinder.

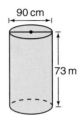

The radius of the base is: $\frac{90 \text{ cm}}{2} = 45 \text{ cm} = 0.45 \text{ m}$
The height of the cylinder is 73 m.

Use the formula for the volume of a cylinder:
$V = \pi r^2 h$
Substitute: $r = 0.45$ and $h = 73$
$V = \pi (0.45)^2 \times 73$ Use a calculator.
$\doteq 46.44$

Both dimensions must have the same units.

The rescue workers removed 46 m³ of soil, to the nearest cubic metre.

1. A student measured a can of beans. The height was 10.5 cm. The diameter was 7.4 cm.
   The student calculated the volume to be about 452 cm³.
   The label on the can shows the capacity as 398 mL.
   How is this possible?
2. Why was the base radius in *Example 2* converted from centimetres to metres?
   What would the volume be if the height was converted to centimetres?
3. In *Example 2*, why do you think the volume was asked for in cubic metres?

Practice

Check

Give each volume to the nearest cubic unit.

4. The base area and height of each cylinder are given to one decimal place. Calculate the volume of each cylinder.

   a)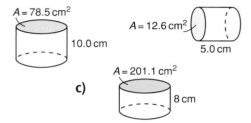
   b)
   c)

5. Calculate the volume of each cylinder.

   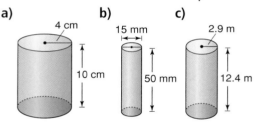
   a)   b)   c)

6. A candle mould is cylindrical. Its radius is 5 cm and its height is 20 cm. What volume of wax will fit in the mould?

Apply

7. Find a right cylinder in the classroom.
   a) Measure its height and diameter.
   b) Calculate its base area.
   c) Calculate its volume.

8. A hockey puck is a solid piece of rubber with the dimensions shown. How much rubber is used to make a hockey puck?

9. How do the volumes of these cylinders compare? How can you tell without calculating each volume?

   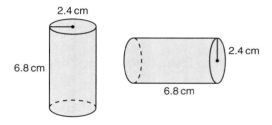

10. Kari has 125 mL of water. She wants to pour it into one of these cylindrical bottles. Which bottle will hold all the water? How do you know?
    Bottle A: $d = 7$ cm, $h = 3$ cm
    Bottle B: $r = 2$ cm, $h = 6$ cm
    Bottle C: $r = 3.5$ cm, $h = 7$ cm
    Bottle D: $d = 3$ cm, $h = 4$ cm

11. **Assessment Focus** Frozen apple juice comes in cylindrical cans. A can is 12 cm high with radius 3.5 cm.
    a) What is the capacity of the can?
    b) What happens to the capacity of the can if the dimensions of the radius and height are switched? Why does this happen?

12. A core sample of soil is cylindrical. The length of the core is 300 mm. Its diameter is 15 cm. Calculate the volume of soil.

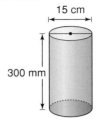

13. Carol and Tom are drilling a well for water at their cottage in Lac La Hache, B.C. The drill is about 15 cm wide. Carol and Tom found water at a depth of 25 m. About how much soil did they remove before they found water?

14. A farmer has 3 cylindrical containers to hold feed. Each container has radius 91 cm and height 122 cm. What is the total volume of the three containers? How did you find out?

15. Orange juice concentrate is poured into cylindrical cans with diameter 7 cm and height 12 cm. A space of 1.5 cm is left at the top of the can to allow for expansion when the concentrate freezes. What volume of concentrate is poured into each can?

16. **Take It Further** Which right cylinder do you think has the greater volume?
    - a cylinder with radius 1 m and height 2 m, or
    - a cylinder with radius 2 m and height 1 m

    How can you find out without using a calculator? Explain.

17. **Take It Further**
    A concrete column in a parkade is cylindrical. The column is 10 m high with diameter 3.5 m.
    a) What is the volume of concrete in one column?
    b) There are 127 columns in the parkade. What is the total volume of concrete?
    c) Suppose the concrete in part a is made into a cube. What would the dimensions of the cube be?

18. **Take It Further** A study shows that consumers think the diameter of a large can of coffee is too wide. The study suggests that a narrower can would increase sales. The original can has diameter 20 cm and height 18 cm. Suppose the diameter of the can is decreased by 20% without changing the volume. What is the height of the new can?

Reflect

How did your knowledge of circles help you in this lesson?

Choosing the Correct Answer

Have you ever written a multiple-choice test?
Many students like multiple-choice tests because they know the correct answer is one of the choices.

The other choices – called *distractors* – are created by making common mistakes.

Answer these multiple-choice questions.
Try to answer each question before you look at the choices.
What mistakes lead to the other choices?

1. A rectangular prism has dimensions 4 m by 6 m by 3 m. What is the surface area of the prism?

    a) 54 m²   **b)** 72 m²   **c)** 108 m²   **d)** 144 m²

2. What is the volume of this triangular prism?

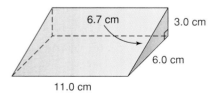

    a) 110.6 cm³
    b) 99 cm³
    c) 190.7 cm³
    d) 198 cm³

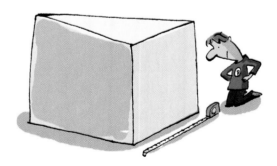

Strategies for Success

Here are some strategies you can use to help choose the correct answer for a multiple-choice question.

Before you start

➤ Make sure you understand what you are supposed to do.
- Is it okay to guess?
- Is there only one correct answer?
- Where should you record your answer?
- Are you supposed to show your work?

For each question

➤ Read the question carefully. Underline the key words.
➤ Draw a sketch or make a calculation if it helps.
➤ Try to answer the question before you look at the choices.
➤ Read all the choices.
➤ If your answer doesn't appear as a choice, read the question again. Look for any mistakes you might have made.
➤ If you still have trouble deciding, read each choice again.
  – Cross out any choices you know are not correct.
  – If two choices appear to be similar, identify any differences.

Organizing your time and checking

➤ Leave questions that you are unsure of until the end.
➤ If it is okay to guess, make your best guess. Do so after you have eliminated the choices you know are not correct.
➤ Read all the questions and your choices. Check that you have not missed any questions.

Unit Review

What Do I Need to Know?

✓ Right Rectangular Prism

Surface area = 2 × area of Rectangle A
             + 2 × area of Rectangle B
             + 2 × area of Rectangle C

Volume = base area × height
$V = Ah$,
where A represents the area of the base

✓ Right Triangular Prism

The length of the prism is ℓ. Each triangular base has height h and base b.
Surface area = sum of the areas of 3 rectangular faces + 2 × area of one triangular base
Volume = area of triangular base × length of prism
$V = A\ell$ where $A = \frac{1}{2}bh$

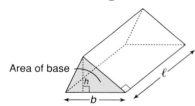

✓ Right Cylinder

The height of a cylinder is h and its radius r.
Surface area = 2 × area of one circular base
             + area of a rectangle
Curved surface area = circumference of base
             × height of cylinder

Volume = base area × height
$V = \pi r^2 h$

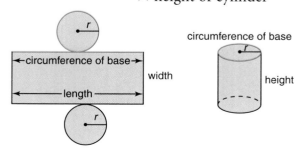

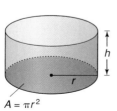

What Should I Be Able to Do?

LESSON

4.1

1. Draw three different nets for the same right rectangular prism.
   What must be true for a net to be that of a rectangular prism?

2. For each net, identify the object it folds to form.
   a)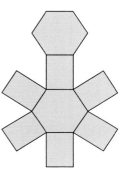

   b)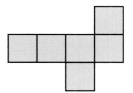

   c)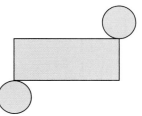

   d)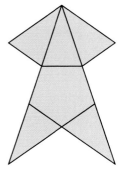

4.2

3. One diagram is a net of a triangular prism. Predict which net is correct.
   Your teacher will give you a large copy of each net. Cut out the nets.
   Fold them to confirm your prediction.
   How might the incorrect net be fixed?
   a)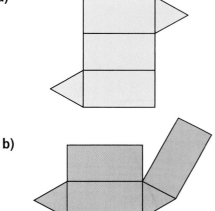

   b)

4. Which diagrams are nets?
   For each net, identify the object.
   For each diagram that is not a net, explain how to change it so it is a net.
   a)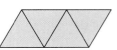

   b)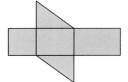

   c)

Unit Review 223

LESSON

4.3

5. A cube has edge length 4 cm.
   a) What is its surface area?
   b) What shortcut could you use to find the surface area?

4.3
4.5

6. Find the surface area and volume of each rectangular prism.

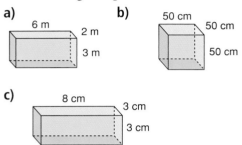

7. Elizabeth wallpapers 3 walls of her bedroom. She paints the 4th wall. This is one of the smaller walls. The room has length 4 m, width 6 m, and height 3 m. A roll of wallpaper covers about 5 m². A 4-L can of paint covers about 40 m².
   a) How much wallpaper and paint should Elizabeth buy?
   b) What assumptions do you make?

8. a) Sketch all possible right rectangular prisms with volume 28 m³. Each edge length is a whole number of metres. Label each prism with its dimensions.
   b) Calculate the surface area of each prism.

9. The base area, A, and height, h, of a right rectangular prism are given. Find the volume of each prism.
   a) $A = 6$ m², $h = 4$ m
   b) $A = 15$ cm², $h = 3$ cm

4.6

10. Here is a net of a triangular prism.

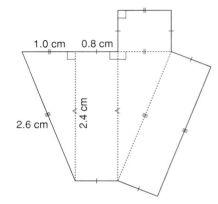

   a) Calculate the surface area of the prism in square centimetres.
   b) Calculate the volume of the prism in cubic centimetres.

11. a) Calculate the surface area of this prism. Sketch a net first, if it helps.

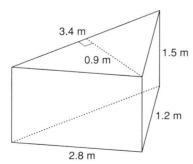

   b) Calculate the volume of the prism.
   c) Suppose you sit the prism on one of its rectangular faces. How does this affect the volume?

224 UNIT 4: Measuring Prisms and Cylinders

LESSON

12. The horticultural society is building a triangular flower bed at the intersection of two streets.
The edges of the bed are raised 0.25 m. How much soil is needed to fill this flower bed?
Justify your answer.

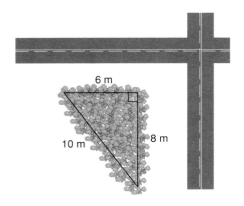

13. Alijah volunteers with the horticultural society. He wants to increase the size but not the depth of the flower bed in question 12.
  a) How can Alijah change the dimensions so that:
    • the flower bed remains triangular, and
    • the area of the ground covered by the bed doubles?
  b) Sketch the new flower bed. Label its dimensions.
  c) How does the change in size affect the volume of soil needed? Explain.

4.7
4.8

14. The label on a can of soup indicates a capacity of 398 mL. The height of the can is 10.5 cm. The diameter of the can is 7.2 cm.
  a) Find the actual capacity of the can in millilitres.
  b) Give a reason why the answer in part a is different from the capacity on the label.

15. A sculpture comprises 3 cylindrical columns. Each column has diameter 1.2 m. The heights of the columns are 3 m, 4 m, and 5 m. The surfaces of the cylinders are to be painted. Calculate the area to be painted. (The base each column sits on will *not* be painted.)

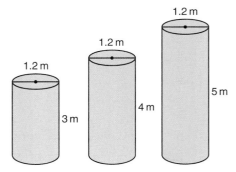

16. A building is to be built in the shape of a cylinder. It will have height 155 m and diameter 25 m. The outside of the building will be made of zinc panels. What area of zinc panels is needed to cover the vertical surface of the building?

Practice Test

1. a) Describe the object this net folds to form.
   b) Fold a copy of the net to check your prediction.
   c) Sketch the object.

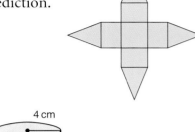

2. Draw a net for each object. Identify and name each face.
   a)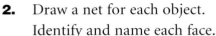
   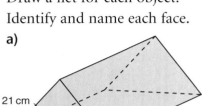
   b)

3. Which diagrams are nets of a cylinder? How do you know?
   a) 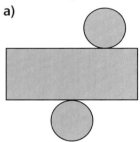 b) 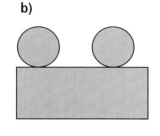 c)

4. Here are the nets of a rectangular prism, a triangular prism, and a cylinder. What is the surface area of each object?
   a) 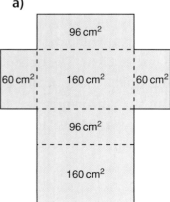 b) 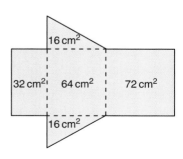 c)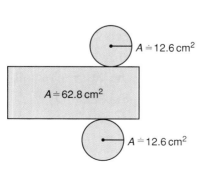

226 UNIT 4: Measuring Prisms and Cylinders

5. The base area and height of each prism are given.
   Calculate the volume of each prism.
   a)
   b)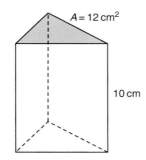

6. Find the surface area and volume of each prism.
   a)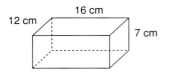
   b)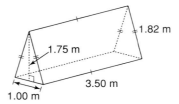

7. Find the volume of each cylinder.
   a)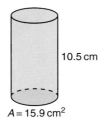
   b)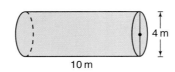

8. Look at the triangular prism in question 6.
   Suppose the base and height of the triangular faces are tripled.
   a) How does this affect the volume of the prism? Explain.
   b) Sketch the larger prism.
   c) Calculate the volume of the larger prism.

9. The dimensions of a wooden sandbox for a local playground are 2 m by 3 m by 25 cm. The sandbox is a rectangular prism.
   a) Calculate the area of wood needed to build the sandbox.
   b) Calculate the volume of sand it will hold.

10. Which has the greater volume?
    • a piece of paper rolled into a cylinder lengthwise, or
    • the same piece of paper rolled into a cylinder widthwise
    Justify your answer. Include diagrams in your answer.

Unit Problem Prism Diorama

Here are some structures that represent the prisms studied in this unit.

Chan Centre, Vancouver

Winnipeg Art Gallery

Husky Oil Building, Calgary

A *diorama* uses objects to create a landscape or a cityscape against a realistic background.
Your task is to create your own cityscape to resemble part of the downtown core of a major city.

Part A

Plan your diorama on grid paper.
Your design must include at least:
- 1 rectangular prism
- 1 triangular prism
- 1 cylinder
- 1 different prism (hexagonal, pentagonal, and so on)
- 1 water storage facility, such as a reservoir, water tank, swimming pool, and so on.
  Record the dimensions of each prism you use.

Part B

Build a model of your diorama.
Use any materials available: cardboard tubes, boxes, plastic objects, coloured pencils, coloured paper, paint, and so on.
Make your model appealing.

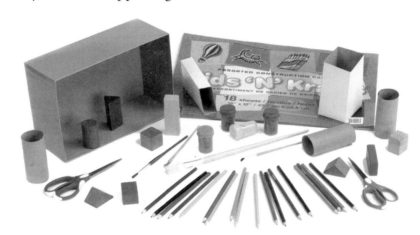

Check List

Your work should show:

✓ a plan of your diorama on grid paper

✓ a model of your plan

✓ each painted surface area and the volume of water

✓ detailed and accurate calculations

Part C

Each object in your diorama will be painted.
You must order the correct amount of each colour of paint.
To do this, you need to know the area of each surface that is to be painted.
Calculate each surface area.

Assume each water storage facility is full.
Calculate the volume of water in your diorama.

Reflect on Your Learning

How are the volumes of prisms and cylinders related?
How can you use nets to calculate surface areas?

Investigation   Pack It Up!

Work with a partner.

Imagine that you work for a packaging company.
You have a sheet of thin card that measures 28 cm by 43 cm.
You use the card to make a triangular prism with the
greatest volume. The triangular faces of the prism
are right isosceles triangles.
As you complete this *Investigation*, include all your work in a
report that you will hand in.

Materials:
- sheets of paper measuring 28 cm by 43 cm
- piece of thin card measuring 28 cm by 43 cm
- centimetre ruler
- 0.5-cm grid paper
- scissors
- tape

Part 1

Here is one prism and its net.

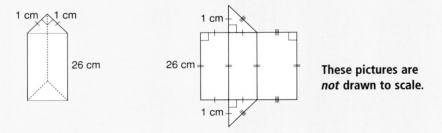

These pictures are *not* drawn to scale.

Each triangular face of this prism has base and height 1 cm.
Increase the base and height of the triangular faces by 1 cm
each time.
Calculate the length of the prism and its volume.
Copy and complete this table.

Triangular Face			Length of Prism (cm)	Volume of Prism (cm³)
Base (cm)	Height (cm)	Area (cm²)		
1	1			
2	2			

Continue to increase the base and height of the triangular face.
Use a sheet of paper to draw each net if you need to.
The length of the prism decreases each time, so the net always fits on the piece of paper.

➤ When do you know that the table is complete?
➤ What patterns do you see in the table? Explain.
➤ From the table, what is the greatest volume of the prism? What are its dimensions?

Part 2

Use 0.5-cm grid paper.
Draw a graph of *Volume* against *Base of triangular face*.

➤ What inferences can you make from the graph?
➤ How can you find the greatest volume from the graph?

Part 3

Use a piece of thin card that measures 28 cm by 43 cm.
Construct the net for your prism.
Cut out, then fold and tape the net to make the prism.

Take It Further

Does the prism with the greatest volume also have the greatest surface area? Write about what you find out.

UNIT 5
Percent, Ratio, and Rate

The animal kingdom provides much interesting information. We use information to make comparisons. What comparisons can you make from these facts?

The amount of fruits and vegetables a grizzly bear eats a day represents about 25% of its body mass.

A Great Dane can eat up to 4 kg of food a day.

A cheetah can reach a top speed of 110 km/h.

A human can run at 18 km/h.

The heart of a blue whale is the size of a small car.

One in 5000 North Atlantic lobsters is born bright blue.

What You'll Learn

- Understand percents greater than or equal to 0%.
- Solve problems that involve percents.
- Understand ratios.
- Understand rates.
- Solve problems that involve ratios and rates.

Why It's Important

You use ratios and rates to compare numbers and quantities; and to compare prices when you shop. You use percents to calculate sales tax, price increases, and discounts.

Key Words

- percent increase
- percent decrease
- discount
- two-term ratio
- three-term ratio
- part-to-whole ratio
- part-to-part ratio
- equivalent ratios
- proportion
- rate
- unit rate

5.1 Relating Fractions, Decimals, and Percents

Focus Given a fraction, decimal, or percent, calculate the other two forms.

The students in a First Nations school were asked which of 5 events at the Northern Manitoba Trappers' Festival they would most like to attend. The circle graph shows the results.

Which event was the favourite? How do you know? How else can you write that percent?

Northern Manitoba Trappers' Festival Event Preference
- Pole Climbing, 25%
- Goose Calling, 20%
- Bannock Baking, 7%
- Fish Filleting, 13%
- Log Balance, 35%

Investigate

Work with a partner.
The Grades 7 and 8 students in 2 schools in Winnipeg were asked which of these cultural or historical attractions they would most like to visit.

Attraction	Number of Students	
	School A	School B
Winnipeg Aboriginal Centre	48	18
Chinese Cultural Centre	30	24
Upper Fort Gary Gate	39	3
Gabrielle Roy House	33	15

➤ How many students were surveyed in each school?
➤ Which school had the greater percent of students choosing the Winnipeg Aboriginal Centre? The Chinese Cultural Centre? Upper Fort Gary Gate? Gabrielle Roy House?

Reflect & Share

What strategies did you use to find the percents?
Compare your answers with those of another pair of classmates.
If the answers are different, how do you know which answers are correct?
At each school, what percent of students did not choose Upper Fort Gary Gate? How did you find out?

234   UNIT 5: Percent, Ratio, and Rate

Connect

To write a fraction as a percent, we first write the fraction with a denominator that is a power of 10; such as 10, 100, 1000, or 10 000. Some fractions cannot be written this way. Then, we can use a calculator to divide.

A *power* of 10 is a product of any number of 10s.

➤ We can use a *hundred chart* to represent one whole, or 100%.
Each small square represents 1%.

We can describe the shaded part of the hundred chart in 3 ways: as a percent, a decimal, and as a fraction.

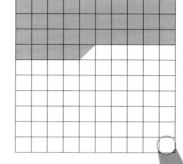

There are $34\frac{1}{2}$ blue squares in 100 squares.
So, 34.5% of the squares are blue.
As a decimal: $\frac{34.5}{100} = 0.345$

$\frac{1}{2} = 0.5$

As a fraction: Since the decimal has 3 digits after the decimal point, write a fraction with denominator 1000.

$0.345 = \frac{345}{1000}$

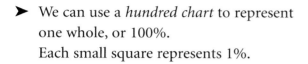

$= \frac{345 \div 5}{1000 \div 5}$

Reduce to simplest form. 5 is a factor of both the numerator and the denominator.

$= \frac{69}{200}$

➤ We can use a *hundredths chart* to represent 1%.
Each small square represents $\frac{1}{100}$ of 1%, which we write as $\frac{1}{100}$%, or 0.01%.
To represent $\frac{1}{5}$ of 1%, or $\frac{1}{5}$% on the hundredths chart, shade $\frac{1}{5}$ of the chart, which is 20 squares.
Since 1 small square is 0.01%, then 20 small squares are 0.20%, or 0.2%.

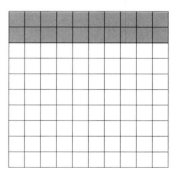

We can write this percent as a decimal.
$0.2\% = \frac{0.2}{100} = \frac{2}{1000} = 0.002$

Be careful not to confuse a decimal percent, such as 0.2%, with the decimal 0.2, which is 20%.

Example 1

Write each percent as a fraction and as a decimal.
a) 7% b) 7.75% c) $7\frac{1}{4}$%

▶ **A Solution**

a) $7\% = \frac{7}{100}$
$= 0.07$

b) $7.75\% = \frac{7.75}{100}$ Multiply the numerator and the denominator by 100.
$= \frac{775}{10\,000}$
$= 0.0775$

Write the fraction in simplest form.
$\frac{775}{10\,000} = \frac{775 \div 25}{10\,000 \div 25}$ 25 is a factor of both the numerator and the denominator. So, divide by 25.
$= \frac{31}{400}$

c) $7\frac{1}{4}\% = \frac{7.25}{100}$ $\frac{1}{4} = 0.25$
$= \frac{725}{10\,000}$
$= 0.0725$

Write the fraction in simplest form.
$\frac{725}{10\,000} = \frac{725 \div 25}{10\,000 \div 25}$
$= \frac{29}{400}$

We can show each decimal in *Example 1* in a place-value chart.

Ones	Tenths	Hundredths	Thousandths	Ten Thousandths
0	0	7	0	0
0	0	7	7	5
0	0	7	2	5

UNIT 5: Percent, Ratio, and Rate

Example 2

Write each fraction as a decimal and as a percent.
a) $\frac{5}{8}$ b) $\frac{5}{6}$ c) $\frac{5}{1000}$

▶ **A Solution**

a) $\frac{5}{8}$ means $5 \div 8$. Use a calculator.

$\frac{5}{8} = 0.625$

$0.625 = \frac{625}{1000}$ Divide the numerator and the denominator by 10 to get an equivalent fraction with denominator 100.

$= \frac{625 \div 10}{1000 \div 10}$

$= \frac{62.5}{100}$

$= 62.5\%$

b) $\frac{5}{6} = 5 \div 6$ Use a calculator.

$= 0.8\overline{3}$

This is a repeating decimal.

Recall that the bar over the 3 indicates that the digit repeats.

To write an equivalent fraction with denominator 100, first write $0.8\overline{3}$ as $0.83\overline{3}$.

$0.83\overline{3} = \frac{83.\overline{3}}{100}$

$= 83.\overline{3}\%$

c) $\frac{5}{1000} = 5 \div 1000$

$= 0.005$

Divide the numerator and the denominator by 10 to get an equivalent fraction with denominator 100.

$\frac{5}{1000} = \frac{5 \div 10}{1000 \div 10}$

$= \frac{0.5}{100}$

$= 0.5\%$

We can use decimals or percents to compare two test marks when the total marks are different.

Example 3

Buffy had $23\frac{1}{2}$ out of 30 on her first math test.

She had $31\frac{1}{2}$ out of 40 on her second math test.

On which test did Buffy do better?

▶ **A Solution**

Write each test mark as a percent.

First test:
$23\frac{1}{2}$ out of $30 = \frac{23.5}{30}$ Divide. Use a calculator.
$= 0.78\overline{3}$

Since the decimal has 3 digits after the decimal point, write a fraction with denominator 1000.

$0.78\overline{3} = \frac{783.\overline{3}}{1000}$

$= \frac{783.\overline{3} \div 10}{1000 \div 10}$ Divide the numerator and the denominator by 10 to get an equivalent fraction with denominator 100.

$= \frac{78.\overline{3}}{100}$

$= 78.\overline{3}\ \%$

Second test:
$31\frac{1}{2}$ out of $40 = \frac{31.5}{40}$ Use a calculator.
$= 0.7875$

Since the decimal has 4 digits after the decimal point, write a fraction with denominator 10 000.

$0.7875 = \frac{7875}{10\ 000}$ Divide the numerator and the denominator by 100.

$= \frac{7875 \div 100}{10\ 000 \div 100}$

$= \frac{78.75}{100}$

$= 78.75\%$

Since $78.75\% > 78.\overline{3}\ \%$, Buffy did better on the second test.

Discuss the ideas

1. How can we use a grid with 100 squares to show 100% and also to show 1%?
2. How can we use a grid with 100 squares to show 0%? Justify your answer.
3. Explain why $\frac{1}{5}$ and $\frac{1}{5}$% represent different numbers.
4. In *Example 1*, we simplified $\frac{775}{10\,000}$ by dividing the numerator and denominator by 25.
   a) Why did we choose 25?
   b) Could we have simplified the fraction a different way? Explain.
5. In *Example 3*, how could we solve the problem without finding percents?

Practice

Check

6. Each hundred chart represents 100%. What fraction of each hundred chart is shaded? Write each fraction as a decimal and as a percent.

   a)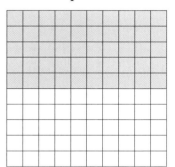

   b)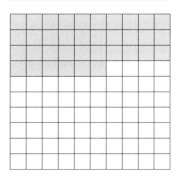

   c)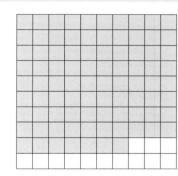

   d)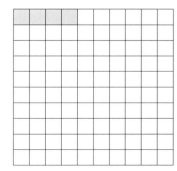

7. Write each percent as a fraction and as a decimal.
   a) 3%   b) 51%   c) 98%   d) 29%

5.1 Relating Fractions, Decimals, and Percents   239

8. Each hundred chart represents 100%. What fraction of each hundred chart is shaded? Write each fraction as a decimal and as a percent.

a)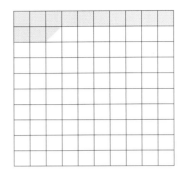

b)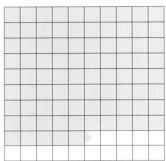

c)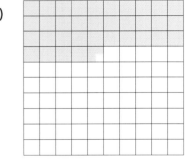

Apply

9. Write each percent as a fraction and as a decimal.
   a) 73.5%
   b) 21.25%
   c) $8\frac{3}{4}$%
   d) $1\frac{1}{5}$%

10. Use a hundredths chart to represent 1%. Shade the chart to represent each percent.
    a) 0.75%
    b) 0.4%
    c) 0.07%
    d) 0.95%

11. Use a hundredths chart to represent 1%. Shade the chart to represent each percent.
    a) 0.655%
    b) 0.0225%
    c) $\frac{2}{3}$%
    d) $\frac{2}{5}$%

12. Write each percent as a fraction and as a decimal.
    a) 0.25%
    b) 0.6%
    c) 0.5%
    d) 0.38%

13. Write each fraction as a decimal and as a percent.
    a) $\frac{2}{300}$
    b) $\frac{18}{400}$
    c) $\frac{7}{500}$
    d) $\frac{8}{250}$

14. Write each decimal as a fraction and as a percent.
    a) 0.345
    b) 0.0023
    c) 0.1825
    d) 0.007

15. A hundredths chart represents 1%. Forty-five of its squares are shaded. Arjang says the shaded squares represent $\frac{45}{100}$. Fiona says the shaded squares represent 0.0045.
    Who is correct? Write to explain where the other student went wrong.

16. Vince scored 82.5% on a math test. Junita had 15 out of 18 on the same test. Who did better? How do you know?

17. Suppose you were asked to tutor another student.
   a) i) How would you explain $\frac{5}{8}$ as a fraction?
   ii) What real-life example could you use to help?
   b) i) How would you explain $\frac{5}{8}$ as a quotient?
   ii) What real-life example could you use to help?

18. Assessment Focus You will need 1-cm grid paper and coloured pencils.
   a) Draw a 6-cm by 8-cm rectangle. Shade:
   - $33.\overline{3}$ % of the grid squares in the rectangle red
   - 0.25 of the grid squares green
   - $\frac{3}{8}$ of the grid squares blue

   Explain how you did this.
   b) What fraction of the rectangle is not shaded? Write this fraction as a decimal and as a percent.
   c) Do you think you could have completed part a with a 6-cm by 9-cm rectangle? With a square of side length 7 cm? Explain.

19. A student council representative is elected from each homeroom class in the school. Joanna received 23 of 30 votes in her class. Kyle received 22 of 28 votes in his class. Who received the greater percent of votes, Joanna or Kyle? How did you find out?

20. A student used this strategy to write $6\frac{1}{4}$% as a fraction.

$$6\frac{1}{4}\% = 6.25\%$$
$$= \frac{625}{100}$$
$$= \frac{625 \div 25}{100 \div 25}$$
$$= \frac{25}{4}$$

   a) Check the student's work. Is the strategy correct?
   b) If your answer is yes, write the fraction as a decimal. If your answer is no, describe the error then correct it.

21. Take It Further Replace each ☐ with <, >, or = to make each statement true.
   a) 3.21 ☐ 321%
   b) $1\frac{5}{8}$ ☐ 158%
   c) 0.76 ☐ 7.6%
   d) 0.9% ☐ 0.9
   e) $0.\overline{3}$% ☐ $\frac{1}{3}$%
   f) 125% ☐ $1\frac{1}{4}$

Reflect

What did you know about fractions, decimals, and percents before you began this lesson? What do you know about fractions, decimals, and percents now?

5.2 Calculating Percents

Focus Calculate percents from 0% to greater than 100%.

Have you ever used a photocopier to reduce or enlarge a picture?
To choose the size of the image picture, you select a percent.
Which percents might you choose if you want to reduce the picture?
Which percents might you choose if you want to enlarge the picture?

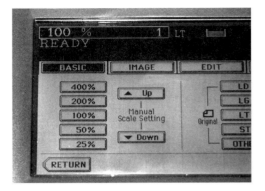

Investigate

Work with a partner.
Copy this shape.

➤ Redraw the shape so that each line segment is 150% of the length shown.

➤ Draw your own shape.
Choose a different percent between 100% and 200%.
Repeat the activity above.

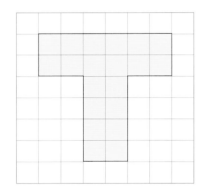

Compare your drawings with those of another pair of classmates.
What strategies did you use to create your enlargements?
What do you notice about the lengths of corresponding line segments on the original shape and the enlarged shape?

Connect

Recall that when the whole is 1.0, you know that:
- 100% = 1.0
- 10% = 0.10
- 1% = 0.01

We can extend the pattern to write percents less than 1% as decimals:
- 0.1% = 0.001
- 0.5% = 0.005

We can use number lines to show percents between 0% and 1%.
For example, this number line shows 0.2%.

We can also extend the pattern to write percents greater than 100% as decimals.
- 101% = 1.01
- 110% = 1.10, or 1.1
- 150% = 1.50, or 1.5
- 200% = 2.00, or 2.0

We can use a number line to show percents greater than 100%.

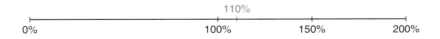

Percents greater than 100% are used by store owners to calculate the prices of items they sell. A store has to make a profit; that is, to sell goods for more than the goods cost to buy.

A store manager buys merchandise from a supplier.
The price the manager pays is called the *cost* price.
The manager *marks up* the cost price to arrive at the *selling price* for the customer.
The markup is the *profit*.
Cost price + Profit = Selling price

Example 1

a) Write 210% as a decimal.
b) Shade hundred charts to show 210%.

▶ **A Solution**

a) $210\% = \frac{210}{100}$
   $= 2.10$, or 2.1

b) $210\% = 100\% + 100\% + 10\%$
   Use a hundred chart to represent 100%.
   To show 200%, shade all the squares in 2 hundred charts.

   Each small square represents 1%.
   So, to show 10%, shade 10 squares of a third hundred chart.

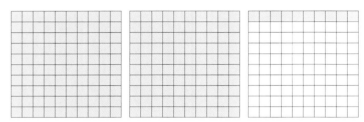

Example 2

The cost price of a winter coat is $80.
The selling price of the coat is 230% of the cost price.
What is the selling price of the coat?
Illustrate the answer with a number line.

▶ **A Solution**

To find the selling price of the coat, find 230% of $80.
First, write 230% as a decimal.
$230\% = \frac{230}{100}$
$= 2.30$, or 2.3
Then, 230% of $80 = 2.3 \times \$80$
$= \$184$
The selling price of the coat is $184.
We can show this answer on a number line.

Example 3

In 2004, the population of First Nations people living on reserves in Alberta was 58 782.
About 0.28% of these people belonged to the Mikisew Cree band.
a) About how many people belonged to the Mikisew Cree band?
b) Estimate to check the answer is reasonable.
c) Illustrate the answer with a diagram.

▶ **A Solution**

a) Find 0.28% of 58 782.
First write 0.28% as a decimal.
$0.28\% = \frac{0.28}{100}$ Multiply the numerator and the denominator by 100.
$= \frac{28}{10\,000}$
$= 0.0028$
Then, 0.28% of 58 782 $= 0.0028 \times 58\,782$ Use a calculator.
$= 164.5896$
About 165 people belonged to the Mikisew Cree band.

b) 0.28% is approximately 0.25%.
0.25% is $\frac{1}{4}$%.
1% of 58 782 is: $0.01 \times 58\,782 = 587.82$
587.82 is about 600.
$600 \div 4 = 150$
This estimate is close to the calculated answer, 165.

c) To illustrate 0.28%, first show 1% on a number line.
Then, 0.28% is about $\frac{1}{4}$ of 1%.

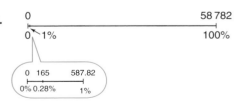

Discuss the ideas

1. As a decimal, 100% = 1.
   What decimals correspond to percents greater than 100%?
   What decimals correspond to percents less than 1%?
2. In *Example 2*, how could you use the number line to find the profit?
3. In *Example 2*, how could you estimate to check the answer?

Practice

Check

4. A hundred chart represents 100%. Shade hundred charts to show each percent.
   a) 150%   **b)** 212%   **c)** 300%   **d)** 198%

5. Write each percent as a decimal. Draw a diagram or number line to illustrate each percent.
   a) 120%   **b)** 250%   **c)** 475%
   d) 0.3%   **e)** 0.53%   **f)** 0.75%

6. Write each decimal as a fraction and as a percent.
   a) 1.7   **b)** 3.3   **c)** 0.003   **d)** 0.0056

7. The cost price of a baseball cap is $9. The selling price of the cap is 280% of the cost price. What is the selling price of the baseball cap? Illustrate the answer with a number line.

Apply

8. What does it mean when someone states, "She gave it 110%"? How can this comment be explained using math? Is it possible to give 110%? Explain.

9. a) Describe two situations when a percent may be greater than 100%.
   b) Describe two situations when a percent may be between 0% and 1%.

10. a) Write each fraction as a percent.
   i) $\frac{1}{3}$ ii) $\frac{2}{3}$ iii) $\frac{3}{3}$
   iv) $\frac{4}{3}$ v) $\frac{5}{3}$ vi) $\frac{6}{3}$
   b) What patterns do you see in your answers in part a?
   c) Use these patterns to write each fraction as a percent.
   i) $\frac{7}{3}$ ii) $\frac{8}{3}$ iii) $\frac{9}{3}$
   iv) $\frac{10}{3}$ v) $\frac{11}{3}$ vi) $\frac{12}{3}$

11. a) Find each percent of the number. Draw a diagram to illustrate each answer.
   i) 200% of 360   ii) 20% of 360
   iii) 2% of 360   iv) 0.2% of 360
   b) What patterns do you see in your answers in part a?
   c) Use the patterns in part a to find each percent. Explain your work.
   i) 2000% of 360   ii) 0.02% of 360

12. A marathon had 618 runners registered. Of these runners, about 0.8% completed the race in under 2 h 15 min.
   a) How many runners completed the race in this time?
   b) Estimate to check your answer.

13. a) This shape represents 100%. Draw a shape that represents 375%.
   b) Repeat part a using a shape of your own choice.

14. The population of a small town in Alberta was 2600. The population increased by 5% one year and by 15% the next year. What was the town's population after the 2 years?
   a) To solve this problem, Juan calculated the population after a 5% increase. He then used his number to find the population after a 15% increase. What was Juan's answer?
   b) To solve this problem, Jeremy calculated the population after a 20% increase. What was Jeremy's answer?
   c) Compare your answers to parts a and b. Are Juan and Jeremy's answers the same? If your answer is yes, explain why both strategies work. If your answer is no, who is correct? Justify your choice.

15. At the local theatre, 120 people attended the production of *Romeo and Juliet* on Friday. The attendance on Saturday was 140% of the attendance on Friday.
   a) How many people went to the theatre on Saturday?
   b) Estimate to check your answer is reasonable.

16. Assessment Focus During the 1888 Gold Rush, a British Columbia town had a population of about 2000. By 1910, the town had become a ghost town. The population was 0.75% of its population in 1888.
   a) Estimate the population in 1910. Justify your estimate.
   b) Calculate the population in 1910.
   c) Find the decrease in population from 1888 to 1910. Show your work.

17. Take It Further Twenty boys signed up for the school play. The number of girls who signed up was 195% of the number of boys. At the auditions, only 26 girls attended. What percent of the girls who signed up for the play attended the auditions?

18. Take It Further At an auction, a painting sold for $148 500. This was 135% of what it sold for 3 years ago. What was the selling price of the painting 3 years ago? Justify your answer.

19. Take It Further The perimeter of a rectangular window is 280% of its length. The length of the window is 145 cm. What is the width of the window? Show your work.

Reflect

How do you find a percent of a number in each case?
• The percent is less than 1%.   • The percent is greater than 100%.
Use an example to explain each case. Include diagrams.

5.3 Solving Percent Problems

Focus Find the whole, when given a percent, and find the percent increase and decrease.

Investigate

Work with a partner.

Tasha conducted a survey of the students in her school.

➤ From the results, Tasha calculated that 60% of the students go to school by bus.

➤ Liam knows that 450 students go to school by bus. How can Liam use these data to find the number of students in the school?

➤ Tasha also found that 50% more students go by bus than walk or drive.
About how many students walk or drive to school?

Sketch number lines to illustrate your work.

Compare your results with those of another pair of classmates. Discuss the strategies you used to solve the problems.

Connect

Grady is 13 years old and 155 cm tall.
His height at this age is about
90% of his final height.

To estimate Grady's final height:
90% of Grady's height is 155 cm.
So, 1% of his height is: $\frac{155 \text{ cm}}{90}$

And, 100% of his height is: $\frac{155 \text{ cm}}{90} \times 100 \doteq 172.2$ cm

So, Grady's final height will be about 172 cm.

When we know a percent of the whole, we divide to find 1%, then multiply by 100 to find 100%, which is the whole.

Example 1

Find the number in each case.
a) 70% of a number is 63.
b) 175% of a number is 105.

▶ **A Solution**

a) 70% of a number is 63.
   So, 1% of the number is:
   $\frac{63}{70} = 0.9$
   And, 100% of the number is:
   $0.9 \times 100 = 90$
   The number is 90.

b) 175% of a number is 105.
   So, 1% of the number is:
   $\frac{105}{175} = 0.6$
   And, 100% of the number is:
   $0.6 \times 100 = 60$
   The number is 60.

5.3 Solving Percent Problems

Example 2

a) A length of 30 cm increased by 40%. What is the new length?
b) A mass of 50 g decreased by 17%. What is the new mass?

▶ **A Solution**

a) The length increased by 40%.
So, the increase in length is 40% of 30.
First, write 40% as a decimal.
$40\% = \frac{40}{100}$
$= 0.4$
Then, 40% of $30 = 0.4 \times 30$
$= 12$
The length increased by 12 cm.
So, the new length is: $30 \text{ cm} + 12 \text{ cm} = 42 \text{ cm}$

b) The mass decreased by 17%.
So, the decrease in mass is 17% of 50.
First, write 17% as a decimal.
$17\% = \frac{17}{100}$, or 0.17
Then, 17% of $50 = 0.17 \times 50$
$= 8.5$
The mass decreased by 8.5 g.
So, the new mass is: $50 \text{ g} - 8.5 \text{ g} = 41.5 \text{ g}$

▶ **Example 2**
Another Solution

a) The length increased by 40%.
So, the new length is $100\% + 40\% = 140\%$ of the original length.
Find 140% of 30.
First, write 140% as a decimal.
$140\% = \frac{140}{100}$, or 1.4
Then, 140% of $30 = 1.4 \times 30$
$= 42$
The new length is 42 cm.

We can illustrate this answer on a number line.

```
0 cm              30 cm    42 cm
|------------------|--------|----------|
0%               100%     140%       200%
```

b) The mass decreased by 17%.
So, the new mass is 100% − 17% = 83% of the original mass.
Find 83% of 50.
First, write 83% as a decimal.
$83\% = \frac{83}{100}$
$= 0.83$
Then, 83% of 50 = 0.83 × 50
$= 41.5$
The new mass is 41.5 g.

We can illustrate this answer on a number line.

```
0 g                        41.5 g  50 g
|──────────────────────────┼──────|
0%                          83%   100%
```

Another type of problem involving percents is to find the **percent increase** or **percent decrease**.

Example 3

a) The price of a carton of milk in the school cafeteria increased from 95¢ to $1.25.
What was the percent increase in price?
b) The price of a green salad decreased from $2.50 to $1.95.
What was the percent decrease in price?

▶ A Solution

Write $1.25 in cents: $1.25 = 125¢
a) The increase in price was: 125¢ − 95¢ = 30¢
To find the percent increase, write the increase as a fraction of the original price: $\frac{30¢}{95¢}$

To write this fraction as a percent: $\frac{30}{95} \doteq 0.32$ Use a calculator.
$= \frac{32}{100}$
$= 32\%$

The price of a carton of milk increased by about 32%.

```
0¢                        95¢   $1.25
|──────────────────────────┼──────|
0%                         100%  132%
```

5.3 Solving Percent Problems   251

b) The decrease in price is: $2.50 - $1.95 = 0.55, or 55¢
To find the percent decrease, write the decrease as a fraction of the original price: $\frac{55¢}{\$2.50} = \frac{55¢}{250¢}$

$2.50 = 250¢

To write this fraction as a percent: $\frac{55}{250} = 0.22$ Use a calculator.
$$= \frac{22}{100}$$
$$= 22\%$$
The price of a green salad decreased by 22%.

Discuss the ideas

1. In *Example 2*, you learned two methods to find the new length and mass. Which method do you think you would prefer to use in the *Practice* questions that follow? Explain.
2. In *Example 3*, how could you check that the percent increase and percent decrease are correct?

Practice

Check

3. Use the number line to find each number.
   a) 50% of a number is 15.

   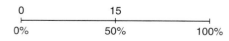

   b) 75% of a number is 12.

   c) 30% of a number is 60.

   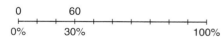

   d) 80% of a number is 120.

4. Find the number in each case. Illustrate each answer with a number line.
   a) 25% of a number is 5.
   b) 75% of a number is 18.
   c) 4% of a number is 32.
   d) 120% of a number is 48.

5. Write each increase as a percent. Illustrate each answer with a number line.
   a) The elastic band stretched from 5 cm to 10 cm.
   b) The price of a haircut increased from $8.00 to $12.00.

6. Write each decrease as a percent. Illustrate each answer with a number line.
   a) The price of a book decreased from $15.00 to $12.00.
   b) The number of students who take the bus to school decreased from 200 to 150.

Apply

7. Find the whole amount in each case.
   a) 15% is 125 g.
   b) 9% is 45 cm.
   c) 0.8% is 12 g.

8. Write each increase as a percent. Illustrate each answer with a number line.
   a) The price of a house increased from $320 000 to $344 000.
   b) The area of forest in southwestern Yukon affected by the spruce bark beetle increased from 41 715 ha in 2003 to 99 284 ha in 2004.

   > One hectare (1 ha) is a unit of area equal to 10 000 m².

9. Write each decrease as a percent. Illustrate each answer with a number line.
   a) The price of gasoline decreased from 109.9¢/L to 104.9¢/L.
   b) The number of students in the class who listen to MP3 players decreased from 17 to 10.

10. There were about 193 000 miners in Canada in 1986. By 2001, the number of miners was 12% less. How many miners were there in 2001?

11. The world's tallest totem pole, known as the Spirit of Lekwammen, was raised on 04 August, 1994 at Victoria, BC, prior to the Commonwealth Games. The totem pole stood about 55 m tall. A local airport was concerned that seaplanes might hit it. So, in 1998 it was partially dismantled. It then stood about 12 m tall. Find the percent decrease in the height of the totem pole.

12. Olivia has 2 puppies, George and Jemma. Each puppy had a birth mass of 1.5 kg. At the end of Week 1, Jemma's mass was 15% greater than her birth mass. At the end of Week 2, Jemma's mass was 15% greater than her mass after Week 1. At the end of Week 2, George's mass was 30% greater than his birth mass.
    a) What was each puppy's mass after Week 2?
    b) Why are the masses in part a different?

13. **Assessment Focus** In 1990, the population of Calgary, Alberta, was about 693 000. The population increased by about 24% from 1990 to 2000. From 2000 to 2005, the population increased by about 11%.
    a) In 2000, about how many people lived in Calgary?
    b) In 2005, about how many people lived in Calgary?
    c) Write the increase in population from 1990 to 2005 as a percent of the 1990 population.
    d) Is your answer in part c 35%? Should the answer be 35%? Explain why or why not.

14. In 2004, the crime rate in a city was 15 194 crimes per 100 000 population. The crime rate decreased by 6% in 2005, and by 4% in 2006.
    a) What was the crime rate at the end of 2006?
    b) Is your answer to part a the same as a decrease in the crime rate of 10%? Why or why not?

15. On average, a girl reaches 90% of her final height when she is 11 years old, and 98% of her final height when she is 17 years old.
    a) Anna is 11 years old. She is 150 cm tall. Estimate her height when she is 20 years old.
    b) Raji is 17 years old. She is 176 cm tall. Estimate her height when she is 30 years old.
    What assumptions do you make?

16. On average, a boy reaches 90% of his final height when he is 13 years old, and 98% of his final height when he is 18 years old. Use these data or the data in question 15 to estimate your height when you are 21 years old.
    Explain any assumptions you make.
    Show your work.

17. After a price reduction of 20%, the sale price of an item is $16. A student says, "So, the original price must have been 120% of the sale price."
    Is this statement correct?
    Justify your answer.

18. Take It Further A photocopier is used to reduce a square.
When the photocopier is set at 80%, the side length of the copy is 80% of its length on the original square. Suppose the side length of the square is 10 cm. It is copied at 80%. The image square is then copied again at 70%.

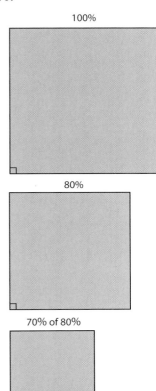

a) What is the side length on the final image square?
b) What is the percent decrease in the side length of the square?
c) What is the area of the final image square?
d) What is the percent decrease in the area of the square?

19. Take It Further A box was $\frac{3}{4}$ full of marbles. The box fell on the floor. Thirty marbles fell out. This was 20% of the marbles in the box. How many marbles would a full box contain?

20. Take It Further Shen dug a 5-m by 15-m garden along one side of his rectangular lawn.
He says that this has reduced the area of his lawn by 25%. What are the dimensions of the remaining lawn? Use a diagram to show your answer. Describe the strategy you used to solve the problem. What assumptions do you make?

Reflect

What is the difference between a percent increase and a percent decrease? Include examples in your explanation.

5.3 Solving Percent Problems   **255**

5.4 Sales Tax and Discount

Focus Investigate the use of percent in consumer math.

A sales tax is charged by the federal government and by most provincial governments. In 2007, the federal tax, the goods and services tax (GST), was 6%.

The provincial sales tax (PST) is set by each provincial or territorial government.

Some provinces have introduced a harmonized sales tax (HST), which combines both the PST and GST.

Province or Territory	Provincial Sales Tax	Goods and Services Tax
Northwest Territories (NT)	no PST	6%
Nunavut (NU)	no PST	6%
Yukon (YT)	no PST	6%
British Columbia (BC)	7%	6%
Alberta (AB)	no PST	6%
Saskatchewan (SK)	5%	6%
Manitoba (MB)	7%	6%
Ontario (ON)	8%	6%
Québec (QC)	7.5%	6%
Newfoundland and Labrador (NL)	14% HST	
Nova Scotia (NS)	14% HST	
New Brunswick (NB)	14% HST	
Prince Edward Island (PEI)	10%	6%

Investigate

Work with a partner.
Celine wants to purchase a tennis racquet in Winnipeg, Manitoba.
The racquet sells for $129.99.
To ensure she has enough money, Celine wants to calculate the final price of the racquet, including all taxes. She asks 3 friends for help.

Helen says: Calculate the PST and GST separately, then add each tax to the price of the racquet.

Zack says: Find the total sales tax as a percent, calculate the tax, then add the tax to the price of the racquet.

William says: Find 113% of the cost of the racquet to find the final price.

Use each method to calculate the final price of the racquet.

Compare your results with those of another pair of classmates.
What do you notice? Why do you think this happens?
Which of the three methods would you use? Justify your choice.

Connect

The selling price of an item is often the same throughout Canada. But the amount you pay depends on the province or territory where you buy the item.

Example 1

How much would you pay for this DVD in each place?
a) Nunavut   **b)** Saskatchewan   **c)** Nova Scotia

▶ A Solution

a) In Nunavut, there is no PST.
The GST is 6%.
6% of $25.99 = 0.06 × $25.99
≐ $1.56
So, the price you pay is: $25.99 + $1.56 = $27.55

b) In Saskatchewan, the PST is 5% and the GST is 6%.
PST: 5% of $25.99 = 0.05 × $25.99
≐ $1.30
GST: 6% of $25.99 = 0.06 × $25.99
≐ $1.56
So, the price you pay is: $25.99 + $1.30 + $1.56 = $28.85

c) In Nova Scotia, the HST is 14%.
HST: 14% of $25.99 = 0.14 × $25.99
≐ $3.64
So, the price you pay is: $25.99 + $3.64 = $29.63

When an item is on sale for 20% off, we say that there is a **discount** of 20%.
A discount of 20% on an item means that you pay:
100% − 20% = 80% of the regular price

5.4 Sales Tax and Discount   257

Example 2

A video game in Vancouver is discounted by 30%.
Its regular price is $27.99.
a) Calculate the sale price of the video game before taxes.
b) Calculate the sale price of the video game including taxes.

▶ *A Solution*

a) To find the amount of the discount, calculate 30% of $27.99.
$30\% = \frac{30}{100} = 0.3$

So, 30% of $27.99 = 0.3 \times $27.99
$\doteq 8.40
The amount of the discount is $8.40.
So, the sale price of the video game is: $27.99 − $8.40 = $19.59

b) In Vancouver, the PST is 7% and the GST is 6%.
Calculate the taxes.
The PST is 7%.
7% of $19.59 = 0.07 × $19.59
$\doteq 1.37
The GST is 6%.
6% of $19.59 = 0.06 × $19.59
$\doteq 1.18
So, the sale price, including taxes, is: $19.59 + $1.37 + $1.18 = $22.14

▶ *Example 2*
Another Solution

a) Find the sale price in one step.
The sale price of the video game is: 100% − 30%, or 70% of $27.99
70% of $27.99 = 0.7 × $27.99
= $19.59
So, the sale price of the video game is $19.59.

b) Find the sale price, including taxes, in one step.
The total sales tax is: 7% + 6% = 13%
$19.59 is 100% of the sale price.
So, the sale price including taxes is 100% +13%, or 113% of $19.59.
113% of $19.59 = 1.13 × $19.59
$\doteq 22.14
So, the sale price, including taxes, is $22.14.

258 UNIT 5: Percent, Ratio, and Rate

Example 3

The cost price of a backpack is $10.50.
This is 30% of the selling price.
a) What is the selling price of the backpack?
b) What does a customer pay for the backpack in Regina, Saskatchewan?

▶ A Solution

a) 30% of the selling price is $10.50.
So, 1% of the selling price is: $\frac{\$10.50}{30}$
and 100% of the selling price is: $\frac{\$10.50}{30} \times 100 = 35$
The selling price of the backpack is $35.00.

b) The PST in Saskatchewan is 5%.
The GST is 6%.
So, the total sales tax is: 5% + 6% = 11%
So, the price you pay is 100% + 11%, or 111% of $35.00.
111% of $35.00 = 1.11 × $35.00
= $38.85
A customer in Regina pays $38.85 for the backpack.

Discuss the ideas

1. In *Example 1b*, the PST was calculated before the GST. Suppose these calculations were reversed. Do you think you would get the same answer? Explain.
2. In *Example 2*, the discount was calculated as $8.397. Why was the answer given as $8.40?
3. Suppose you bought a taxable item for $50. How much would you pay for it in your province or territory? What strategy would you use to find out?

Practice

Use a calculator when you need to.

Check

4. Suppose you are in Flin Flon, Manitoba. Find the PST on each item.
   a) a pair of sunglasses that cost $15.00
   b) a sunscreen that costs $8.99
   c) a laser mouse that costs $21.99

5. Suppose you are in Fort Simpson, Northwest Territories. Find the GST on each item.
   a) a digital camera that costs $89.97
   b) a cordless phone that costs $24.97
   c) a soccer ball that costs $17.99

6. Suppose you are in Victoria, British Columbia. Find the sales taxes on each item.
   a) a package of light bulbs that costs $7.47
   b) an inflatable raft that costs $32.99
   c) a diving mask that costs $27.98

7. Suppose you are in Moose Jaw, Saskatchewan. For each item below:
   a) Calculate the PST and GST.
   b) Calculate the selling price including taxes.

   i)  $25.99

   ii)  $152.45

Apply

8. Suppose you are in Iqaluit, Nunavut. For each item below:
   a) Calculate the discount.
   b) Calculate the sale price before taxes.
   c) Calculate the sale price including taxes.

   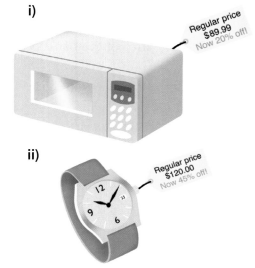

   i) Regular price $89.99 Now 20% off!

   ii) Regular price $120.00 Now 45% off!

9. A new house was purchased for $304 000. After 3 years, its market value had increased by 28%. What was the market value of the house after 3 years?

10. **Assessment Focus** A video store offers these choices.
    Choice A: 30% off each DVD with regular price $25.00
    Choice B: Buy two DVDs for $40.00.
    Which is the better deal for the customer? Justify your answer.

11. In a sale in Red Deer, Alberta, the price of a blow dryer is marked down.

a) What is the percent decrease?
b) Calculate the sale price including taxes.

12. At the end of the summer, a gift store in Vancouver reduced the price of a souvenir T-shirt. The regular price of the T-shirt was $30. The T-shirt was reduced by 25%. The manager then put this sign in the window:

Rico told his mother that the T-shirt was now half its regular price.
Was Rico correct? Justify your answer.

13. During a 20%-off sale, the sale price of an MP3 alarm clock radio was $35.96. What was the regular price of the radio?

14. The regular price of a pair of snowshoes in Brandon, Manitoba, is $129.99. The price of the snowshoes is marked down by 13%. Abbott says that the sale price including taxes will be $129.99 because the discount and taxes cancel each other out. Is Abbott's reasoning correct? Justify your answer.

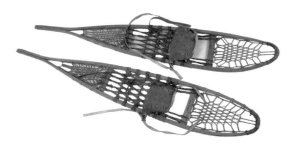

15. The price of a hair straightener in Fredericton, New Brunswick, is reduced by $28.38. This is a discount of 33%.
a) What is the regular price of the hair straightener?
b) What is the sale price of the hair straightener including taxes?

16. Anika wants to buy a snowboarding helmet. The original price is $75.00. It is on sale for 30% off. Anika will pay 13% sales tax. A customer behind Anika in the line suggested that it would be cheaper if the 30% discount was subtracted before calculating the sales tax. Another customer said it would be cheaper if the 13% sales tax was added before the discount was subtracted. Who is correct? Show how you found out.

17. Take It Further Two identical lacrosse sticks are on sale at two stores in Dawson City, Yukon. At Strictly Sports, the stick is on sale for 15% off its regular price of $45.99. At Sport City, the stick is on sale for 20% off its regular price of $49.99. Which store offers the better deal? How much would you save? Show your work.

18. Take It Further For a promotion, a store offers to pay the sales taxes on any item you buy. You are actually paying taxes, but they are calculated on a lower price.
Suppose you buy an item for $100. The store pays the 14% sales tax.
a) What is the true sale price of the item?
b) How much tax are you really paying?

19. Take It Further A pair of shoes in a clearance store went through a series of reductions. The regular price was $125. The shoes were first reduced by 20%. Three weeks later, the shoes were reduced by a further 20%. Later in the year, the shoes were advertised for sale at $\frac{3}{4}$ off the current price. Sean wants to buy the shoes. He has to pay 11% sales tax.
a) Sean has $40.00.
   Can Sean buy the shoes?
   How did you find out?
b) If your answer to part a is yes, how much change does Sean get?

20. Take It Further A skateboard in a store in Charlottetown, PEI, costs $39.99. It is on sale for 30% off. The taxes are 16%. What is the sale price of the skateboard including taxes? A student says, "$39.99 – 30% discount + 16% taxes is the same as calculating $39.99 – 14%."
Is the student's reasoning correct? Explain.

Reflect

Describe two different methods to calculate the sales tax on an item.
Which method do you prefer?
Which method is more efficient?
Include an example using each method.

Mid-Unit Review

LESSON

5.1

1. Write each percent as a fraction and as a decimal.
   a) 60%   b) 9.75%   c) 97.5%

2. Use a hundredths chart to represent 1%. Shade a hundredths chart to show each percent.
   a) 0.12%   b) $\frac{4}{5}$%   c) 0.65%   d) $\frac{1}{10}$%

3. Write each decimal as a fraction and as a percent.
   a) 0.18   b) 0.006   c) 0.875   d) 0.0075

5.2

4. Write each percent as a decimal. Draw a diagram to illustrate each percent.
   a) 145%   b) 350%   c) 0.44%   d) 0.2%

5. Jon reported his mark on a test as 112%. Is this mark possible? Give an example to support your answer.

6. A local school raised $5687.50 for the Terry Fox Foundation. Hua raised 0.8% of this total. How much money did Hua raise?

5.3

7. Find the number in each case.
   a) 15% of a number is 3.
   b) 160% of a number is 80.

8. Meryl sold 8% of the tickets to the school play. Meryl sold 56 tickets. How many tickets were sold in all?

9. Write this decrease as a percent. Illustrate this percent on a number line.
   The attendance fell from 9850 to 8274.

5.4

10. The regular price of an ink cartridge is $32. It is on sale in Kelowna, BC, for 20% off. Calculate the sale price including taxes.

11. During a 30%-off sale in Portage la Prairie, Manitoba, the sale price of a memory stick was $27.99.
    a) What was the regular price of the memory stick?
    b) What was the regular price of the memory stick including taxes?
    c) What is the sale price of the memory stick including taxes?
    d) How much do you save by buying the memory stick on sale?

12. Jessica gets a 15% employee discount on everything she buys from Fashion City, in Whitehorse, Yukon. The regular price of a pair of jeans in this store is $69. They have been reduced by 30%. How much will Jessica pay for the jeans including taxes?

5.5 Exploring Ratios

Focus Use models and diagrams to investigate ratios.

There are different ways to compare numbers.
Look at these advertisements.

How are the numbers in each advertisement compared?
Which advertisement is most effective? Why do you think so?

Investigate

Work with a partner.

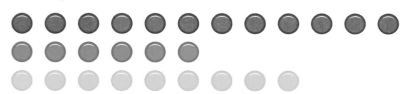

Compare the number of blue counters to the number of yellow counters.
How many different ways can you compare the counters?
Write each way you find.

Share your list with another pair of classmates.
Add any new comparisons to your list.
Talk about the different ways you compared the counters.

Connect

Here is a collection of sports balls.

➤ We can use a **two-term ratio** to compare one part of the collection to the whole collection.
There are 7 basketballs compared to 20 balls.
The ratio of basketballs to all the balls is 7 to 20, which is written as 7:20.

This is a **part-to-whole** ratio.

We can write a part-to-whole ratio as a fraction.
The ratio of basketballs to all the balls is $\frac{7}{20}$.

A part-to-whole ratio can also be written as a percent: $\frac{7}{20} = \frac{35}{100} = 35\%$
So, 35% of the balls are basketballs.

➤ We can use a two-term ratio to compare one part of the collection to another part of the collection.
There are 5 golf balls compared to 8 tennis balls.
The ratio of golf balls to tennis balls is written as 5 to 8, or 5:8.
We cannot write this ratio in fraction form because the ratio is not comparing one part to the whole.

This is a **part-to-part** ratio.

➤ We can use a **three-term ratio** to compare the three types of balls.
There are 5 golf balls to 8 tennis balls to 7 basketballs.
We can write this as the ratio:
5 to 8 to 7, or 5:8:7

5.5 Exploring Ratios

Example

At a class party, there are 16 boys, 15 girls, and 4 adults.
Show each ratio as many different ways as you can.
a) boys to girls
b) boys to girls to adults
c) adults to total number of people at the party

▶ **A Solution**
a) There are 16 boys to 15 girls.
   This is a part-to-part ratio.
   So, the ratio of boys to girls is: 16 to 15, or 16:15
b) There are 16 boys, 15 girls, and 4 adults.
   So, the ratio of boys to girls to adults is: 16 to 15 to 4, or 16:15:4
c) The total number of people is 16 + 15 + 4 = 35.
   There are 4 adults to 35 people.
   This is a part-to-whole ratio.
   So, the ratio of adults to total number of people is:
   4 to 35; 4:35, $\frac{4}{35}$, or about 11.4%

1. What is the difference between a part-to-whole ratio and a part-to-part ratio?
2. In the *Example*, explain how the part-to-whole ratio of 4:35 can be written as about 11.4%.
3. In the *Example*, why were the ratios of boys to girls and boys to girls to adults *not* written in fraction form?

Practice

Check

4. Write each part-to-whole ratio as a fraction.
   a) 5:8   b) 12:16   c) 4:9   d) 24:25

5. Write each part-to-whole ratio as a percent.
   a) 19:20   b) 12:15   c) 3:8   d) 5:6

6. Look at the candy-covered chocolates below. Explain what each ratio means.
   a) 3:5
   b) 7:5
   c) 5:15
   d) 3:5:7
   e) 3:12

7. Look at the golf balls below. Write each ratio in two different ways.

  a) orange golf balls to the total number of golf balls
  b) white golf balls to the total number of golf balls
  c) yellow golf balls to pink golf balls
  d) yellow golf balls to white golf balls to orange golf balls

Apply

8. The ratio of T-shirts to shorts in Frank's closet is 5:2.
  a) Write the ratio of T-shirts to the total number of garments.
  b) Write the ratio in part a as a percent.

9. a) Write a part-to-part ratio to compare the items in each sentence.
    i) A student had 9 green counters and 7 red counters on his desk.
    ii) In a dance team, there were 8 girls and 3 boys.
    iii) A recipe called for 3 cups of flour, 1 cup of sugar, and 2 cups of milk.
  b) Write a part-to-whole ratio for the items in each sentence in part a. Express each ratio as many ways as you can.

10. a) What is the ratio of boys to girls in your class?
  b) What is the ratio of girls to boys?
  c) What is the ratio of boys to the total number of students in your class? Write the ratio as a percent.
  d) Suppose two boys leave the room. What is the ratio in part c now?

11. A box contains 8 red, 5 green, 2 orange, 3 purple, 1 blue, and 6 yellow candies.
  a) Write each ratio.
    i) red:purple
    ii) green:blue
    iii) purple:blue:green
    iv) orange and yellow:total candies
  b) Suppose 3 red, 2 green, and 4 yellow candies were eaten. Write the new ratios for part a.

12. Suppose you were asked to tutor another student.
  a) How would you explain $\frac{2}{7}$ as a ratio?
  b) What real-life example could you use to help?

13. a) Draw two different diagrams to show the ratio 3:5.
  b) Draw a diagram to show the ratio 7:1.
  c) Draw a diagram to show the ratio 5:2:4.
  d) Why can you draw 2 diagrams in part a, but not in parts b and c?

14. **Assessment Focus** Patrick plans to make macaroni salad. The recipe calls for:
    3 cups of cooked macaroni
    3 cups of sliced oranges
    2 cups of chopped apple
    1 cup of chopped celery
    2 cups of mayonnaise
    a) What is the total amount of ingredients?
    b) What is each ratio?
        i) oranges to apples
        ii) mayonnaise to macaroni
        iii) apples to mayonnaise to celery
    c) What is the ratio of apples and oranges to the total amount of ingredients? Write this ratio as a fraction and as a percent.
    d) Patrick uses 2 cups of oranges instead of 3. What are the new ratios in parts b and c?
    e) Write your own ratio problem about this salad. Solve your problem.

15. Look at the words below.
    - ratio
    - discount
    - decimal
    - problem
    - percent
    - increase
    - taxes
    - number

    Which words represent the ratio 2:5? Explain what the ratio means each time.

16. **Take It Further** Maria shares some cranberries with Jeff. Maria says, "Two for you, three for me, two for you, three for me …" Tonya watches. At the end, she says, "So Jeff got $\frac{2}{3}$ of the cranberries." Do you agree with Tonya? Give reasons for your answer.

17. **Take It Further**
    a) Create four different ratios using these shapes.

    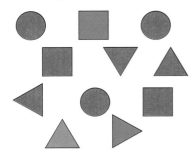

    b) How can you change one shape to create ratios 2:5 and 7:3? Explain.

18. **Take It Further** Choose a vowel-to-consonant ratio. Find 3 words that represent this ratio.

Reflect

Give 3 examples from your classroom that can be represented by the ratio 1:1.

5.6 Equivalent Ratios

Focus Write equivalent ratios.

Investigate

Work with a partner.
Which cards have the same ratio of pepperoni pieces to pizzas?

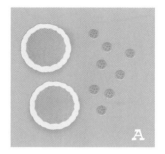
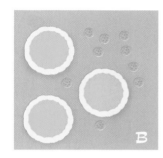
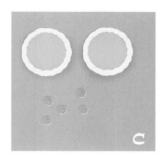
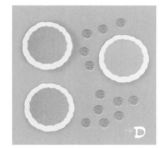
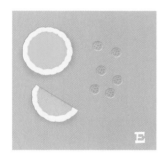
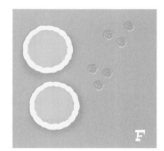
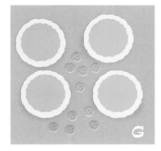
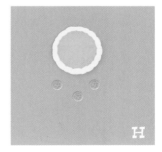

Share your answers with another pair of classmates.
What strategies did you use to identify the same ratios?
Why do you think your answers are correct?
What patterns do you see?

5.6 Equivalent Ratios   **269**

Connect

The ratio of triangles to squares is 4:3.
That is, for every 4 triangles, there are 3 squares.

The ratio of triangles to squares is 8:6.
That is, for every 8 triangles, there are 6 squares.

The ratios 8:6 and 4:3 are called **equivalent ratios**.
Equivalent ratios are equal.
8:6 = 4:3

An equivalent ratio can be formed by multiplying or dividing the terms of a ratio by the same number.
➤ We can show this with the terms of the ratios in rows.

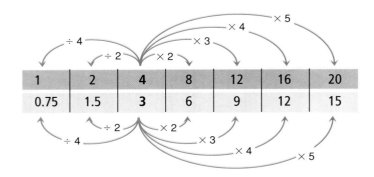

Note that:

$3 \div 2$ is $\frac{3}{2} = 1.5$

and

$3 \div 4$ is $\frac{3}{4} = 0.75$

270   UNIT 5: Percent, Ratio, and Rate

➤ We can show this with the terms of the ratios in columns.

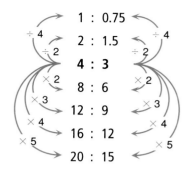

The equivalent ratios are:
1:0.75; 2:1.5; 4:3, 8:6; 12:9; 16:12; 20:15

➤ These ratios can be plotted as points on a grid.
Show the equivalent ratios in a table.

1st term	1	2	4	8	12	16	20
2nd term	0.75	1.5	3	6	9	12	15

Plot the points.
The points lie on a straight line.

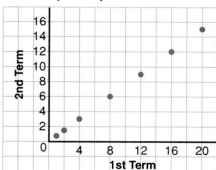

➤ When we divide the terms in a ratio by their *greatest common factor*, we write the ratio in *simplest form*.

A ratio is in simplest form when its terms have no common factors.

To write 24:16 in simplest form:
  24:16   The greatest common factor is 8.
$= \frac{24}{8} : \frac{16}{8}$ Divide by 8.
= 3:2

So, 24:16 and 3:2 are equivalent ratios.
The ratio 3:2 is in simplest form.

5.6 Equivalent Ratios

Example 1

Write 3 ratios equivalent to 2:5.

▶ A Solution

2:5
Multiply each term by 2.
$(2 \times 2):(5 \times 2)$
$= 4:10$
Multiply each term by 3.
$(2 \times 3):(5 \times 3)$
$= 6:15$
Multiply each term by 4.
$(2 \times 4):(5 \times 4)$
$= 8:20$
Three equivalent ratios are 4:10, 6:15, and 8:20.

Example 2

Write 3 ratios equivalent to 36:6.

▶ A Solution

36:6
Divide each term by 2.
$\frac{36}{2}:\frac{6}{2}$

$= 18:3$
Divide each term by 3.
$\frac{36}{3}:\frac{6}{3}$

$= 12:2$
Divide each term by 6.
$\frac{36}{6}:\frac{6}{6}$

$= 6:1$
Three equivalent ratios are 18:3, 12:2, and 6:1.

Example 3

Construction kits come in different sizes. The regular kit contains 120 long rods, 80 short rods, and 40 connectors. List 3 other kits that could be created with the same ratio of rods and connectors.

▶ **A Solution**

For the regular kit, the ratio of long rods to short rods to connectors is 120:80:40.
For each new kit, find an equivalent ratio.
For Kit A, divide each term by 2.
$120:80:40 = \frac{120}{2}:\frac{80}{2}:\frac{40}{2} = 60:40:20$
Kit A has 60 long rods, 40 short rods, and 20 connectors.

For Kit B, divide each term by 4.
$120:80:40 = \frac{120}{4}:\frac{80}{4}:\frac{40}{4}$
$= 30:20:10$
Kit B has 30 long rods, 20 short rods, and 10 connectors.

There are many more possible kits. These are just 3 of them.

For Kit C, multiply each term by 2.
$120:80:40 = (120 \times 2):(80 \times 2):(40 \times 2)$
$= 240:160:80$
Kit C has 240 long rods, 160 short rods, and 80 connectors.

Discuss the ideas

1. On pages 270 and 271, you saw equivalent ratios with terms written horizontally and vertically. Which representation is easier to follow? Justify your answer.
2. a) How do you know when a ratio is in simplest form?
   b) Which of the ratios in *Example 1* and *Example 2* is in simplest form? How do you know?
3. In *Example 1* and *Example 2*, which other equivalent ratios could be written for each ratio?
4. Are these ratios equivalent? Explain.
   3:2   2:3

Practice

Check

5. Write 3 ratios equivalent to each ratio. Use tables to show your work.
   a) 1:2   b) 2:3   c) 1:4

6. Write 3 ratios equivalent to each ratio. Use tables to show your work.
   a) 3:4   b) 14:4   c) 24:25

7. Write 3 ratios equivalent to each ratio. Use tables to show your work.
   a) 1:3:6   b) 12:5:7   c) 24:4:8

8. Write each ratio in simplest form.
   a) 5:15   b) 6:9
   c) 3:12:18   d) 110:70:15

9. Write a ratio, in simplest form, to compare the items in each sentence.
   a) In a class, there are 32 chairs and 8 tables.
   b) In a parking lot, there were 4 American cars and 12 Japanese cars.
   c) A paint mixture is made up of 6 L of blue paint, 2 L of yellow paint, and 1 L of white paint.
   d) A stamp collection contains 12 Canadians stamps, 24 American stamps, and 9 Asian stamps.

10. Find the missing number in each pair of equivalent ratios.
   a) 2:7 and ☐:28
   b) 5:12 and 25:☐
   c) ☐:24 and 5:3
   d) 3:☐:11 and 30:70:110

Apply

11. a) Find pairs of equivalent ratios:
   2:3:4   9:12:15
   8:5:4   1:2:3
   3:2:1   16:10:8
   3:6:9   6:9:12
   5:8:4   3:4:5
   b) Tell how you know they are equivalent.

12. In a class library, 3 out of 4 books are non-fiction. The rest are fiction.
   a) How many non-fiction books could there be? How many fiction books could there be?
   b) How many different answers can you find for part a? Which answers are reasonable? Explain.

13. The official Canadian flag has a length-to-width ratio of 2:1.

Doreen has a sheet of paper that measures 30 cm by 20 cm. What are the length and width of the largest Canadian flag Doreen can draw? Sketch a picture of the flag.

14. Assessment Focus
  a) Draw a set of counters to represent each ratio.
   i) red:blue = 5:6
   ii) blue:green = 3:4
   iii) red:blue:green = 10:12:16
   How many different ways can you do this? Record each way you find.
  b) Draw one set of counters that satisfies all 3 ratios. How many different ways can you do this?

15. Take It Further Are these ratios equivalent? How do you know?
  a) 16:30 and 28:42
  b) 27:63 and 49:21
  c) 56:104:88 and 42:78:66
  d) 20:70:50 and 30:105:75

16. Take It Further There are 32 students in a Grade 8 class.
The ratio of girls to boys is 5:3.
  a) How many boys are in the class?
  b) How many girls are in the class? How did you find out?

17. Take It Further Find the missing number in each pair of equivalent ratios.
  a) 10:35 and ☐:42
  b) 36:78 and ☐:182
  c) ☐:15 and 68:85
  d) 49:☐:63 and 84:36:108

18. Take It Further A quality control inspector finds that 8 out of 9 batteries from the production line meet or exceed customer requirements.
  a) Write each ratio.
   i) number of batteries that passed to number of batteries that failed
   ii) number of batteries that passed to the number of batteries tested
   iii) number of batteries that failed to the number of batteries tested
  b) After some changes to the production line, the number of batteries that did not meet or exceed customer requirements decreased by one-half. Rewrite the ratios in part a to reflect these changes.

Reflect

Choose a ratio. Use pictures, numbers, or words to show how to find two equivalent ratios.

Explaining Your Thinking

Explaining how you solved a problem helps you and others understand your thinking.

Solve this problem.

A restaurant has square tables. Each table seats 4 people. For large parties and banquets, the tables are put together in rows. How many people can be seated when 6 tables are put together? When 20 tables are put together?

Recall the problem-solving strategies you know.

Strategies

- **Make a table.**
- **Use a model.**
- **Draw a diagram.**
- **Solve a simpler problem.**
- **Work backward.**
- **Guess and test.**
- **Make an organized list.**
- **Use a pattern.**
- **Draw a graph.**
- **Use logical reasoning.**

Strategies for Success

When you have found the solution to a problem, write a few sentences to explain how you solved the problem. These sentences should help someone else understand how you solved the problem.

Here is one way to describe your thinking:
- Describe the problem.
- Describe the strategies you used—even the ones you tried that did not lead you to a solution.
- Describe the steps you took.
- Describe how you know your answer is correct.

Solve these problems.
For each problem, write a few sentences to describe your thinking.

1. There are 400 students at a school.
   Is the following statement true? Explain.
   There will always be at least 2 students in the school whose birthdays fall on the same day of the year.

2. Camden has a custard recipe that needs:
   6 eggs, 1 cup of sugar, 750 mL of milk, and 5 mL of vanilla
   He has 4 eggs. Camden adjusts the recipe to use the 4 eggs.
   How much of each other ingredient will Camden need?

3. Lo Choi wants to buy a dozen doughnuts. She has a coupon.
   This week, the doughnuts are on sale for $3.99 a dozen.
   If Lo Choi uses the coupon, each doughnut is $0.35.
   Should Lo Choi use the coupon?
   Justify your answer.

Strategies for Success: Explaining Your Thinking

Triple Play

HOW TO PLAY

Your teacher will give you a copy of *Triple Play* game cards.

1. Cut out the cards, then shuffle them. Place the cards face down in a pile.

2. Player A turns over 1 card. He says the ratio in simplest form, then says 2 equivalent ratios. One point is awarded for the ratio in simplest form. One point is awarded for each equivalent ratio.

3. Player B takes a turn.

4. Players continue to take turns. The player with the higher score after 6 rounds wins.

Triple Play Cards

10:15	28:35	24:8	12:16
8:14	35:28	15:21	27:33
40:35	9:21	14:18	24:18
7:14	33:45	15:6	10:25
7:35	42:30	6:18	24:15
56:32	27:81	22:99	21:6
12:42	30:18	27:15	18:30
21:36	33:21	14:49	13:91

YOU WILL NEED
A set of *Triple Play* game cards; paper; pencil

NUMBER OF PLAYERS
2

GOAL OF THE GAME
To get the greater number of points

TAKE IT FURTHER

Create a set of cards with the ratios in simplest form. Combine the cards you created with the game cards. Shuffle the cards, then deal 8 cards to each player. The goal of the game is to collect pairs of cards that show equivalent ratios. Students pick and discard a card each round. The player with more equivalent ratios after 8 rounds wins.

278   UNIT 5: Percent, Ratio, and Rate

5.7 Comparing Ratios

Focus: Use different strategies to compare ratios.

Investigate

Work with a partner.

Recipe A for punch calls for 2 cans of concentrate and 3 cans of water.

Recipe B for punch calls for 3 cans of concentrate and 4 cans of water.

In which recipe is the punch stronger? Or, are the drinks the same strength? Explain how you know.

Reflect & Share

Compare your answer with that of another pair of classmates. Compare strategies.
If your answers are the same, which strategy do you prefer? Would there be a situation when the other strategy would be better? Explain.
If your answers are different, find out which answer is correct.

Connect

Erica makes her coffee with 2 scoops of coffee to 5 cups of water.

Jim makes his coffee with 3 scoops of coffee to 7 cups of water.

Here are two strategies to find out which coffee is stronger.

➤ Draw a picture.
Find how much water is used for 1 scoop of coffee.

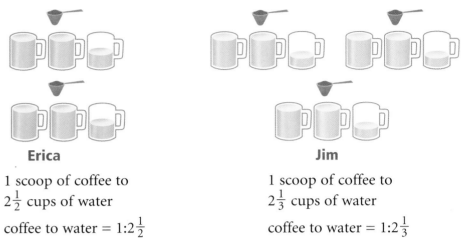

Erica

1 scoop of coffee to
$2\frac{1}{2}$ cups of water

coffee to water = $1:2\frac{1}{2}$

Jim

1 scoop of coffee to
$2\frac{1}{3}$ cups of water

coffee to water = $1:2\frac{1}{3}$

Since $2\frac{1}{3}$ is less than $2\frac{1}{2}$, Jim uses less water to 1 scoop of coffee.
So, Jim's coffee is stronger.

➤ Use equivalent ratios.
Find how much coffee is used for the same amount of water.
Write equivalent ratios with the same second term.
Then compare the first terms.

Erica

Coffee		Water
2	:	5
4	:	10
6	:	15
8	:	20
10	:	25
12	:	30
14	:	35

Since 2:5 = 14:35,
Erica uses 14 scoops of
coffee to 35 cups of water.

Jim

Coffee		Water
3	:	7
6	:	14
9	:	21
12	:	28
15	:	**35**

Since 3:7 = 15:35,
Jim uses 15 scoops of
coffee to 35 cups of water.

Jim uses more coffee for the same amount of water. So, Jim's coffee is stronger.

Another way to compare ratios is to write equivalent ratios, with 1 as the second term.

Example 1

The recommended seeding on a package of grass seed is 200 g per 9 m². Carey spread 150 g over 6.5 m². Is this more than, equal to, or less than the recommended seeding?
How do you know?

▶ A Solution

Write each ratio with second term 1.
For each ratio, divide each term by the second term.
Then, use a calculator to write each fraction as a decimal.

Package

Grass seed:Area
200:9 = $\frac{200}{9} : \frac{9}{9}$ Divide by 9.

  = $22.\overline{2}$:1

The package recommended $22.\overline{2}$ g of grass seed per 1 m².

Carey

Grass seed:Area
150:6.5 = $\frac{150}{6.5} : \frac{6.5}{6.5}$ Divide by 6.5.

  ≐ 23.1:1

Carey used about 23.1 g of grass seed per 1 m².

Since 23.1 > $22.\overline{2}$, Carey spread more seed than the package recommended.

Example 1 involved part-to-part ratios.
Some problems are solved using part-to-whole ratios.

5.7 Comparing Ratios   **281**

Example 2

a) Write each part-to-part ratio as a part-to-whole ratio.
   i) 2:3   ii) 4:3
b) Write each part-to-whole ratio on part a in fraction form. Which part-to-whole ratio is greater?

▶ **A Solution**

a) i) 2:3
      = 2:(2 + 3)
      = 2:5
   ii) 4:3
      = 4:(4 + 3)
      = 4:7

b) Write each part-to-whole ratio in fraction form.
   i) 2:5
      = $\frac{2}{5}$
   ii) 4:7
      = $\frac{4}{7}$

   Since $\frac{2}{5} < \frac{1}{2}$ and $\frac{4}{7} > \frac{1}{2}$, then $\frac{4}{7} > \frac{2}{5}$
   So, 4:7 is greater than 2:5.

We can also write ratios as percents to compare them.

Example 3

A contractor brought 2 shades of yellow paint for his clients to see. Shade 1 is made by mixing 5 cans of yellow paint with 3 cans of white paint. Shade 2 is made by mixing 7 cans of yellow paint with 4 cans of white paint.
The clients want the lighter shade.
Which shade should they choose?
What assumptions do you make?

▶ *A Solution*

Write the ratio of cans of yellow paint to cans of white paint for each shade.

Shade 1   **Shade 2**
5:3   7:4

The lighter shade will have less yellow paint.
Assume all the cans are the same size.
Write part-to-whole ratios for the number of cans
of yellow paint to the total number of cans.

Shade 1   **Shade 2**
$5:(3 + 5) = 5:8$ $7:(4 + 7) = 7:11$ Write each ratio as a fraction.
$\quad\quad = \frac{5}{8}$ $\quad\quad = \frac{7}{11}$ Write each fraction as a percent.
$\quad\quad = 62.5\%$ $\quad\quad = 63.\overline{63}\%$

Shade 1 has 62.5% yellow paint.
Shade 2 has $63.\overline{63}$% yellow paint.

Since 62.5% < $63.\overline{63}$%, Shade 1 is the lighter shade.
The clients should choose Shade 1.

Discuss the ideas

1. You have seen these strategies to compare ratios.
   - Draw a picture.
   - Use equivalent ratios.
   - Make a term 1.
   - Use percents.
   a) Which strategy do you prefer?
   b) Could you always use that strategy? Explain.

2. a) How could you use ratios with second term 1
      to solve the coffee problem in *Connect*?
   b) Which strategy is most efficient for the coffee problem?
      Justify your answer.
      - Draw a picture.
      - Use equivalent ratios.
      - Make the second term 1.

3. In *Example 3*, explain how the fraction $\frac{7}{11}$ was written as the
   percent $63.\overline{63}$%.

5.7 Comparing Ratios

Practice

Check

4. Write each ratio with first term 1.
   a) 3:12 b) 5:40 c) 8:56
   d) 9:81 e) 33:99 f) 22:132

5. Write each ratio with second term 1.
   a) 16:4 b) 55:11 c) 144:12
   d) 120:24 e) 91:13 f) 96:8

6. The principal is deciding which shade of blue to have the classrooms painted. One shade of blue requires 3 cans of white paint mixed with 4 cans of blue paint. Another shade of blue requires 5 cans of white paint mixed with 7 cans of blue paint.
   a) Which mixture will give the darker shade of blue? Explain.
   b) Which mixture will require more white paint?

7. In a hockey skills competition, Olga scored on 3 of 5 breakaways. Tara scored on 5 of 7 breakaways. Whose performance was better?
   a) To find out, write each ratio as a fraction.
   b) How can you use common denominators to help you solve the problem? Explain.

Apply

8. A chicken farmer in Manitoba compares the numbers of brown eggs and white eggs laid in 2 henhouses. The chickens in Henhouse A lay 6 brown eggs for every 10 white eggs. The chickens in Henhouse B lay 3 brown eggs for every 9 white eggs. Which henhouse produces more white eggs? What assumptions do you make?

9. The concentrate and water in each picture are mixed.

Which mixture is stronger: A or B? Draw a picture to show your answer.

10. In a basketball game, Alison made 6 of 13 free shots. Nadhu made 5 of 9 free shots. Who played better? Explain. Use two different methods to compare the ratios.

11. Two different groups at a summer camp have pizza parties. The Calgary Cougars order 2 pizzas for every 3 campers. The Alberta Antelopes order 3 pizzas for every 5 campers.
   a) Which group gets more pizza per person? How do you know?
   b) Could you use percent to find out? Why or why not?

12. Rick is comparing two recipes for oil and vinegar salad dressing. Recipe A calls for 150 mL of vinegar and 250 mL of oil. Recipe B calls for 225 mL of vinegar and 400 mL of oil. Which salad dressing will have a stronger vinegar taste? How did you find out?

13. **Assessment Focus** The ratio of fiction to non-fiction books in Ms. Arbuckle's class library is 7:5. The ratio of fiction to non-fiction books in Mr. Albright's class library is 4:3. Each classroom has 30 non-fiction books.
    a) Which room has more fiction books? How many more?
    b) What percent of the books in each class is non-fiction?

14. Look at the sets of cans of concentrate and water.

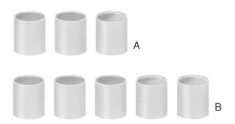

a) What is the ratio of concentrate to water in A and in B?
b) Explain how you could add concentrate or water to make both ratios the same. Draw a picture to show your answer.

15. Drew has 3 shades of paint.

Shade	Red (drops)	Yellow (drops)
A	4	12
B	3	15
C	2	3

a) Write the ratio of red to yellow paint in each shade. Draw a picture to represent each ratio.
b) For each ratio in part a, write an equivalent ratio that uses 1 drop of red paint.
c) Which shade will have the most red? How do you know?
d) Which shade will have the most yellow? How do you know?

16. Two cages contain white mice and brown mice. The ratio of white mice to brown mice in Cage A is 5:6. The ratio of white mice to brown mice in Cage B is 7:5. Which cage contains more brown mice?
    Marcel says, "Since $\frac{6}{11}$ is greater than $\frac{1}{2}$, and $\frac{5}{12}$ is less than $\frac{1}{2}$, Cage A contains more brown mice." Is Marcel's reasoning correct? Explain.

17. Take It Further All aircraft have a glide ratio. A glide ratio compares the distance moved forward to the loss in altitude. For example, an aircraft with a glide ratio of 7:1 will move forward 7 m for every 1 m of altitude lost. Two hang-gliders start at the same altitude. Glider A has a glide ratio of 14:3. Glider B has a glide ratio of 15:4. When the gliders reach the ground, which one will have covered the greater horizontal distance? How did you find out?

18. Take It Further Two boxes contain pictures of hockey and basketball players. The boxes contain the same number of pictures.
In one box, the ratio of hockey players to basketball players is 4:3. In the other box, the ratio is 3:2.
a) What could the total number of pictures be?
b) Which box contains more pictures of hockey players? Draw a picture to show your answer.

19. Take It Further Valerie uses a powdered iced tea mix. She always uses 1 more scoop of mix than cups of water. Valerie uses 5 scoops of mix for 4 cups of water. To make more iced tea, she uses 7 scoops of mix for 6 cups of water. To make less iced tea, she uses 3 scoops of mix for 2 cups of water.
a) Will all these mixtures have the same strength?
b) If your answer to part a is yes, explain your reasoning. If your answer to part a is no, which mixture will be the strongest? Justify your answer.

Reflect

Look back at your answer to *Discuss the Ideas*, question 1a.
Now that you have completed *Practice*, is your answer to this question unchanged? Explain.

5.8 Solving Ratio Problems

Focus Set up a proportion to solve a problem.

We can use ratios when we change a recipe.

Investigate

Work with a partner.
Here is a recipe for an apple pie that serves 6 people.
- 500 mL flour
- 200 mL margarine
- 500 g sliced apples
- 125 g sugar

Toby has only 350 g of sliced apples.
How much of each other ingredient does
Toby need to make the pie?
How many people will Toby's pie serve? Explain.

Compare your answers with those of another pair of classmates.
What strategies did you use to solve the problems?

Connect

In a recycle drive last week at Island Middle School, Mr. Bozyk's
Grade 8 class collected bottles and recycled some of them.
The ratio of bottles recycled to bottles collected was 3:4.
This week, his class collected 24 bottles.
Mr. Bozyk told the students that the ratio of bottles recycled to
bottles collected is the same as the ratio the preceding week.
How can the students find how many bottles are recycled this week?

Last week's ratio of bottles recycled to bottles collected is 3:4.
Let r represent the number of bottles recycled this week.
This week's ratio of bottles recycled to bottles collected is r:24.
These two ratios are equivalent.
So, r:24 = 3:4
A statement that two ratios are equal is a **proportion**.

To find the value of r, write a ratio equivalent to 3:4, with second term 24.
Since $4 \times 6 = 24$, multiply each term by 6.
$3:4 = (3 \times 6):(4 \times 6)$
$= 18:24$
So, $r:24 = 18:24$
So, $r = 18$
18 bottles were recycled.

> When we solve problems this way, we use proportional reasoning.

Example 1

Find the value of each variable.
a) $5:x = 40:56$
b) $49:35 = 14:n$

▶ **A Solution**

a) $5:x = 40:56$
Since $5 < 40$, we divide to find x.
Think: What do we divide 40 by to get 5?

$\overset{\div 8}{\frown}$
$5:x = 40:56$
Divide 56 by the same number to get x.
$x = 56 \div 8$
$ = 7$
So, $x = 7$

b) $49:35 = 14:n$
Since $14 < 49$, we divide to get n.
Think: What do we divide 49 by to get 14?
$49:35 = 14:n$
$\underset{\div ?}{\smile}$
We cannot find a factor, so we simplify the ratio first.
$49:35 = \frac{49}{7}:\frac{35}{7}$
$ = 7:5$
Now, $7:5 = 14:n$
Think: What do we multiply 7 by to get 14?
$7:5 = 14:n$
$\underset{\times 2}{\smile}$
Multiply 5 by the same number to get n.
$n = 5 \times 2$
$ = 10$
So, $n = 10$

We can use proportions to solve ratio problems.

Example 2

This is a photo of a father and his daughter.
In the photo, the father's height is 8 cm and
the daughter's height is 6 cm.
The father's actual height is 1.8 m.
What is the actual height of his daughter?

▶ **A Solution**

In the photo, the ratio of the height of the father to
the height of the daughter is 8:6.

Let h represent the actual height of the daughter,
in centimetres.
Then, the ratio of the actual height of the father to
the actual height of the daughter is 1.8:h.
1.8 m = 180 cm
So, the ratio is 180:h.
The two ratios are equivalent.
Then, 180:h = 8:6

To find the value of h, write a ratio equivalent to 8:6, with first term 180.
Since 8 does not divide into 180 exactly, simplify the ratio 8:6 first.
$8:6 = \frac{8}{2}:\frac{6}{2}$
$ = 4:3$
Write a ratio equivalent to 4:3, with first term 180.
Since 4 × 45 = 180, multiply each term by 45.
4:3 = (4 × 45):(3 × 45)
$ = 180:135$
So, 180:h = 180:135
So, $h = 135$
The daughter's actual height is 135 cm, or 1.35 m.

For some proportions, a suitable equivalent ratio cannot be written by multiplying or dividing each term by the same number.

Example 3

A bike is in fourth gear.
When the pedals turn 3 times,
the rear wheel turns 7 times.
When the pedals turn twice,
how many times does the rear wheel turn?

▶ **A Solution**

The ratio of pedal turns to rear wheel turns is 3:7.
When the pedal turns twice, let the number of
rear wheel turns be n.
The ratio of pedal turns to rear wheel turns is 2:n.
These ratios are equal.
Then 3:7 = 2:n
We cannot easily write a ratio equivalent to 3 to 7 with first term 2.
We look for a multiple of the first terms; that is,
a multiple of 2 and 3.
The least multiple is 6.
We multiply each ratio to get the first term 6.
$(3 \times 2):(7 \times 2) = (2 \times 3):(n \times 3)$
$\qquad 6:14 = 6:3n$

> When we multiply each ratio in a proportion by a different number, the new ratios are still equivalent.

The first terms are equal, so the second terms must also be equal.
That is, $3n = 14$
$\qquad \frac{3n}{3} = \frac{14}{3}$
$\qquad n = \frac{14}{3}$, or $4\frac{2}{3}$

When the pedals turn twice, the rear wheel turns $4\frac{2}{3}$ times.

Discuss the ideas

1. In *Example 1*, how could we check that our answers are correct?
2. In *Example 2*, we first converted all measurements to centimetres. Could you have solved the problem without converting to centimetres? Justify your answer.
3. In *Example 3*:
   a) We multiplied the terms of one ratio by 2 and the terms of the other ratio by 3. Explain why this did not change the proportion.
   b) Could we have multiplied each ratio by numbers other than 2 and 3? If your answer is yes, give an example of these numbers.

Practice

Check

4. Find the value of each variable.
   a) $t:18 = 6:3$ b) $v:60 = 3:10$
   c) $x:15 = 2:3$ d) $s:28 = 9:4$
   e) $6:c = 2:11$ f) $39:b = 3:2$

5. Find the value of each variable.
   a) $5:t = 15:36$ b) $45:72 = 5:n$
   c) $120:70 = 12:k$ d) $81:27 = 9:m$
   e) $27:63 = p:7$ f) $8:s = 64:80$

6. Find the value of each variable.
   a) $1:6 = a:54$ b) $3:8 = e:40$
   c) $2:15 = f:75$ d) $42:36 = g:6$
   e) $3:7 = 30:p$ f) $26:65 = 2:r$

7. Find the value of each variable.
   a) $18:a = 14:21$ b) $35:b = 15:12$
   c) $m:18 = 18:27$ d) $88:33 = h:6$
   e) $6:8 = j:44$ f) $15:42 = 20:w$

8. In the NHL, the ratio of shots taken to goals scored by an all-star player is 9:2. The player has a 50-goal season. How many shots did he take?

Apply

9. An ad said that 4 out of 5 dentists recommend a certain chewing gum for their patients. Suppose 185 dentists were interviewed. Find the number of dentists who recommend this gum.

10. These suitcases have the same length-to-width ratio. Calculate the width of suitcase B.

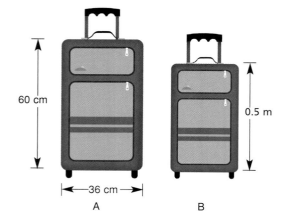

11. Assessment Focus In rectangle MNPQ, the ratio of the length of MN to the length of MP is 4:5.
  a) Does this ratio tell you how long MN is? Explain.
  b) Suppose MN is 12 cm long. How long is MP? Use a diagram to illustrate your answer. Show your work.

12. The scale on a map of British Columbia is 1:5 000 000. This means that 1 cm on the map represents 5 000 000 cm actual distance. The map distance between Kelowna and Salmon Arm is 2.1 cm. What is the actual distance between these towns?

13. The scale on a map of Saskatchewan is 1 cm represents 50 km. The actual straight line distance between Regina and Saskatoon is about 257 km. What is the map distance between these 2 cities?

14. Jacob drew this picture of his bedroom.

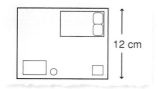

The ratio of an actual dimension to the dimension in the drawing is 60:1.
  a) The actual length of his bedroom is 9 m. What is the length of the bedroom in Jacob's drawing?
  b) The width of Jacob's bedroom in the drawing is 12 cm. What is the actual width of his bedroom?

15. On a blueprint of a new house, a particular room has length 11 cm. The actual room has length 6.6 m and width 4.8 m. What is the width of the room on the blueprint? Show how you found out.

16. Fatima plants 5 tree seedlings for every 3 that Shamar plants. Shamar plants 6 trees in 1 min.
  a) How many trees does Fatima plant in 1 min?
  b) Did you write a proportion to solve this problem? If so, how else could you have solved it?

17. Take It Further Bruno wants to mix concrete to lay a walkway. The instructions call for 2 parts cement, 3 parts sand, and 4 parts gravel. Bruno has 4 m³ of sand.
How much cement and gravel should Bruno mix with this sand?

18. Take It Further At the movie theatre, 65 student tickets were sold for one performance.
  a) The ratio of adult tickets sold to student tickets sold was 3:5. How many adult tickets were sold?
  b) The ratio of adult tickets sold to child tickets sold was 3:2. How many child tickets were sold?
  c) One adult ticket cost $13. One student ticket cost $7.50. One child ticket cost $4.50. How much money did the performance make?

19. Take It Further Bella's grandfather is confined to a wheelchair. He is coming to visit her. Bella wants to build a wheelchair ramp. Her research tells her that there should be 3.5 m of ramp for every 30 cm of elevation. The distance from the ground to the front doorstep of Bella's house is 9 cm. What should the length of the ramp be?

20. Take It Further Forty-five students take piano lessons. The ratio of the numbers of students who take piano lessons to violin lessons is 15:8. The ratio of the numbers of students who take violin lessons to clarinet lessons is 8:9.
  a) How many students take violin lessons?
  b) How many students take clarinet lessons?

21. Take It Further Suppose you want to find the height of a flagpole. You know your height. You can measure the length of the shadow of the flagpole. Your friend can measure your shadow. How can you use ratios to find the height of the flagpole? Explain. Include a sketch in your explanation.

Reflect

Can you use a proportion to solve any ratio problem?
If your answer is yes, when might you not use a proportion to solve a ratio problem?
If your answer is no, give an example of a ratio problem that you *could* not use a proportion to solve.

5.9 Exploring Rates

Focus Use models and diagrams to investigate rates.

Canadian speed skater Jeremy Wotherspoon, of Red Deer, Alberta, set the world record for the 500 m at the 2004 World Cup in Italy. He skated at an average speed of 14.44 m/s. The white-tailed deer can run at speeds of up to 30 km/h.

Who is faster?
How can you find out?

Investigate

Work with a partner.

You will need a stopwatch.
Take your pulse. Your partner is the timekeeper.
Place your index and middle fingers on the side of your neck, under your jawbone.
Count the number of beats in 20 s.
Reverse roles.
Count the number of beats in 30 s.

➤ Who has the faster heart rate?
  How do you know?
➤ Estimate how many times each person's heart would beat in 1 h.
  What assumptions do you make?
  Are these assumptions reasonable?

Reflect & Share

Compare your results with those of another pair of classmates.
How can you decide who has the fastest heart rate?

294 UNIT 5: Percent, Ratio, and Rate

Connect

When we compare two things with different units, we have a **rate**.
Here are some rates:
- We need 5 sandwiches for every 2 people.
- Oranges are on sale for $1.49 for 12.
- Gina earns $4.75 per hour for baby-sitting.
- There are 500 sheets on one roll of paper towels.

The last two rates are **unit rates**.
Each rate compares a quantity to 1 unit.

Jamal skipped rope 80 times in 1 min.
We say that Jamal's rate of skipping is 80 skips per minute.
We write this as 80 skips/min.

To find unit rates, we can use diagrams, tables, and graphs.

Example 1

A printing press prints 120 sheets in 3 min.
a) Express the printing as a unit rate.
b) How many sheets are printed in 1 h?

▶ A Solution

a) Draw a diagram.
The press prints 120 sheets in 3 min.
So, in 1 min, the press prints:
120 sheets ÷ 3 = 40 sheets
The unit rate of printing is 40 sheets/min.

b) In 1 min, the press prints 40 sheets.
One hour is 60 min.
So, in 60 min, the press prints:
60 × 40 sheets = 2400 sheets
The press prints 2400 sheets in 1 h.

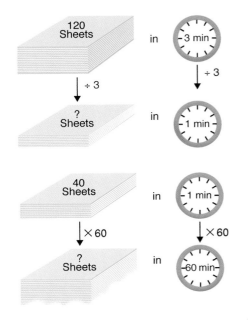

5.9 Exploring Rates

Example 2

Use the data in *Example 1*.
How long will it take to print 1000 sheets?

▶ **A Solution**

In 1 min, the press prints 40 sheets.
So, in 5 min, the press prints: 5 × 40 = 200 sheets
Make a table. Every 5 min, 200 more sheets are printed.
Extend the table until you get 1000 sheets.

Time (min)	5	10	15	20	25
Sheets printed	200	400	600	800	1000

The press takes 25 min to print 1000 sheets.

▶ **Example 2
Another Solution**

The press prints 40 sheets in 1 min.
Think: What do we multiply
40 by to get 1000?
Use division: 1000 ÷ 40 = 25
So, 40 × 25 = 1000
We multiply the time by the same number.
1 min × 25 = 25 min
The press takes 25 min to print 1000 sheets.

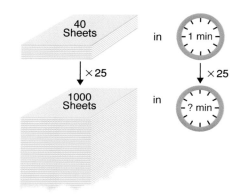

The rate at which a car travels is its average speed.
When a car travels at an average speed of 80 km/h, it travels:
  80 km in 1 h
  160 km in 2 h
  240 km in 3 h
  320 km in 4 h
  400 km in 5 h … and so on
We can show this motion on a graph.
An average speed of 80 km/h is a unit rate.

At any given time, the car may be travelling at a different speed.

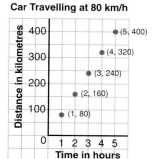

Example 3

a) A human walks at an average speed of 5 km/h.
   What is this speed in metres per second?
b) A squirrel can run at a top speed of about 5 m/s.
   What is this speed in kilometres per hour?

▶ *A Solution*

a) In 1 h, a human walks 5 km.
   There are 1000 m in 1 km.
   So, to convert kilometres to metres, multiply by 1000.
   5 km is 5 × 1000 m or 5000 m.
   There are 60 min in 1 h.
   So, in 1 min, a human walks: $\frac{5000}{60} = 83.\overline{3}$ m
   There are 60 s in 1 min.
   So, in 1 s, a human walks: $\frac{83.\overline{3}}{60}$ m $= 1.3\overline{8}$ m
   A human walks at an average speed of about 1.4 m/s.

b) In 1 s, a squirrel can run about 5 m.
   There are 60 s in 1 min.
   So, in 1 min, a squirrel can run about: 5 m × 60 = 300 m
   There are 60 min in 1 h.
   So, in 1 h, a squirrel can run about: 300 m × 60 = 18 000 m
   There are 1000 m in 1 km.
   So, to convert metres to kilometres, divide by 1000.
   18 000 m is $\frac{18\,000}{1000}$ km or 18 km.
   So, in 1 h a squirrel can run about 18 km.
   A squirrel can run at a top speed of about 18 km/h.

Discuss the ideas

1. What is the difference between a rate and a unit rate?
2. How is a ratio like a rate?
3. Look at the *Examples* in *Connect*.
   a) Can a rate be expressed as a fraction? Justify your answer.
   b) Can a rate be expressed as a percent? Justify your answer.
   c) Can a rate be expressed as a decimal? Justify your answer.

Practice

Check

4. Express each unit rate using symbols.
  a) Morag typed 60 words in 1 min.
  b) Peter swam 25 m in 1 min.
  c) Abdul read 20 pages in 1 h.

5. Express as a unit rate.
  a) June cycled 30 km in 2 h.
  b) A caribou travelled 12 km in 30 min.
  c) A plane flew 150 km in 15 min.

6. Express as a unit rate.
  a) Elsie delivered 220 flyers in 4 h.
  b) Winston iced 90 cupcakes in 1.5 h.
  c) The temperature rose 18°C in 4 h.

7. Which sentences use ratios? Which sentences use rates? How do you know?
  a) The car was travelling at 60 km/h.
  b) Katie earns $8/h at her part-time job.
  c) The punch has 3 cups of cranberry juice and 4 cups of ginger ale.
  d) The soccer team won 25 games and lost 15 games.

Apply

8. Find the unit price for each item.
  a) Milk costs $4.50 for 4 L.
  b) Corn costs $3.00 for 12 cobs.
  c) 24 cans of iced tea cost $9.99.

9. Lindsay has scored 15 goals in 10 lacrosse games this season.
  a) What is her unit rate of scoring?
  b) How many goals might Lindsay score in 35 games? What assumptions do you make?

10. Before running in a 100-m race, Gaalen's heart rate was 70 beats/min. Which do you think is more likely after the race: 60 beats/min or 120 beats/min? Explain.

11. Ribbon costs $1.44 for 3 m.
  a) What is the cost per metre?
  b) How much would 5 m of ribbon cost?
  c) How much ribbon could you buy for $12?

12. Monique worked as a cleaner at the Calgary Stampede and Exhibition. She was paid $84 for an 8-h day.
  a) What was Monique's hourly rate of pay?
  b) How much would Monique earn for 35 h of work?

13. A 400-g package of Swiss cheese costs $4.80.
  a) What is the cost per 100 g?
  b) How much would 250 g cost?
  c) How much would 1 kg cost?
  d) How much Swiss cheese could you buy with $18?

14. Write each speed in metres per second.
    a) The sailfish is the fastest swimmer, reaching a speed of up to 109 km/h.
    b) A bald eagle can fly at a top speed of about 50 km/h.

15. Write each speed in kilometres per hour.
    a) A pitcher throws a baseball. The ball crosses home plate at a speed of 40 m/s.
    b) A cockroach can travel at a speed of 1.5 m/s.

16. The graph shows how a cyclist travelled in 3 h.
    a) How far did the cyclist travel in 1 h?
    b) What is the average speed of the cyclist? How do you know?

    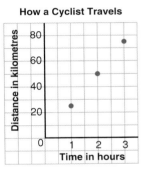
    How a Cyclist Travels

17. James and Lucinda came from England to Canada on holiday. The rate of exchange for their money was $2.50 Can to £1.
    a) How many Canadian dollars would James get for £20?
    b) What is the value in English pounds of a gift Lucinda bought for $30 Can?

18. **Assessment Focus** Petra works on an assembly line. She can paint the eyes on 225 dolls in 1 h.
    a) How many dolls can she paint in 15 min?
    b) How many dolls can she paint in 30 s?
    What assumptions do you make?

19. **Take It Further** When a person runs a long-distance race, she thinks of the time she takes to run 1 km (min/km), rather than the distance run in 1 min (km/min). On a training run, Judy took 3 h 20 min to run 25 km. What was Judy's rate in minutes per kilometre?

20. **Take It Further** Leo trained for the marathon. On Day 1, he took 70 min to run 10 km. On Day 10, he took 2 h 40 min to run 20 km. On Day 20, he took 4 h 15 min to run 30 km.
    a) What was Leo's running rate, in minutes per kilometre, for each day?
       i) Day 1   ii) Day 10   iii) Day 20
    b) What do you think Leo's running rate, in minutes per kilometre, might be for the 44 km of the marathon? How long do you think it will take him? Why do you think so?

Reflect

Explain how ratios and rates are different.
Describe a situation where you might use a ratio.
Describe a situation where you might use a rate.

5.10 Comparing Rates

Focus Use unit rates to compare rates.

Many grocery items come in different sized packages.

| 250 mL | 500 mL | 1 L | 2 L |
| $1.49 | $1.69 | $2.79 | $3.99 |

How can you find out which is the best buy?

Investigate

Work on your own.

Use a calculator if it helps.

12 garbage bags for $1.99   48 garbage bags for $5.29

Which box of garbage bags is the better buy?
What do you need to consider before you decide?

Compare your results with those of a classmate.
Did both of you choose the same box? If so, justify your choice.
If not, can both of you be correct? Explain.

Connect

Great Start cereal can be purchased in three different sizes and prices.

The smallest box costs the least, but that does not mean it is the best buy.
Find the unit cost for each box of cereal, then compare the unit costs.
It is difficult to calculate the cost of 1 g; so, we calculate the cost of 100 g for each box of cereal.

Box A has mass 450 g and costs $4.69.
450 g is 4.5 × 100 g; so, the cost of 100 g of Box A is:
$\frac{\$4.69}{4.5} \doteq \$1.04$

Box B has mass 600 g and costs $6.49.
The cost of 100 g of Box B is: $\frac{\$6.49}{6} \doteq \$1.08$

Box C has mass 1000 g and costs $7.89.
The cost of 100 g of Box C is: $\frac{\$7.89}{10} \doteq \$0.79$

Each unit cost can be written as a unit rate.
The "unit" for unit rate is 100 g.
The least unit rate is $0.79/100 g, for Box C.
The greatest unit rate is $1.08/100 g, for Box B.

Math Link

Your World

In a supermarket, the tag on the shelf with the bar code of an item often shows the cost of 1 g of the item.

With this information, you can compare the costs of packages of the item in different sizes.

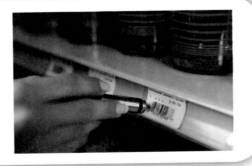

Example 1

Paper towels can be purchased in 3 different sizes and prices.

A
2 rolls $0.99

B
6 rolls $3.99

C
12 rolls $4.99

Which package is the best buy?
How do you know?

▶ **A Solution**

Find the unit cost of each package, then compare the unit costs.
Since 12 is a multiple of both 2 and 6, use 12 rolls as the unit.

Package A:
12 rolls is 6 × 2 rolls; so, the cost of 12 rolls is:
$0.99 × 6 = $5.94

Package B:
12 rolls is 2 × 6 rolls; so, the cost of 12 rolls is:
$3.99 × 2 = $7.98

Package C:
The cost of 12 rolls is $4.99.

The least unit rate is $4.99/12 rolls, for Package C.
So, the best buy is Package C.

Example 2

Mariah is looking for a part-time job. She wants to work 15 h a week.
She has been offered three positions.

Day Camp Counsellor   Cashier   Library Assistant
$7.50 per hour $25.00 for 3 h $44.00 for 5 h

a) Which job pays the most?
b) For the job in part a, how much will Mariah earn in one week?

▶ **A Solution**

a) Calculate the unit rate for each job.
The unit rate is the hourly rate of pay.
For day camp counsellor, the unit rate is $7.50/h.
For cashier, the unit rate is: $\frac{\$25.00}{3\text{ h}} \doteq \$8.33/\text{h}$

For library assistant, the unit rate is: $\frac{\$44.00}{5\text{ h}} = \$8.80/\text{h}$
The library assistant job pays the most.

b) Mariah works 15 h a week, at a rate of $8.80/h.
She will earn: $15 \times \$8.80 = \$132.00$
Mariah will earn $132.00 a week as a part-time library assistant.

Discuss the ideas

1. In *Connect*, we compared the prices of cereal by dividing to find unit rates, where 100 g was the unit.
   a) What other unit could we have used for the unit rate?
   b) What is the price for each cereal at the unit rate in part a?
2. In *Connect*, the best buy for the cereal is Box C. Why might you not buy Box C?
3. In *Connect*, why did we not use a unit of 1 kg for the cereal?
4. For the cereal, we used a unit of 100 g. For the paper towels, we used a unit of 12 rolls. Which strategy did you find easier to understand? Justify your choice.

Practice

Check

5. Write a unit rate for each statement.
   a) $399 earned in 3 weeks
   b) 680 km travelled in 8 h
   c) 12 bottles of juice for $3.49
   d) 3 cans of soup for $0.99

6. Which is the greater rate? How do you know?
   a) $24.00 in 3 h or $36.00 in 4 h
   b) $4.50 for 6 muffins or $6.00 for 1 dozen muffins
   c) $0.99 for 250 mL or $3.59 for 1 L

7. Delaney goes to the store to buy some mushroom soup. She finds that a 110-mL can costs $1.49. A 500-mL can of the same brand costs $4.29.
  a) Which is the better buy?
  b) Delaney buys the 110-mL can. Why might she have done this?
  c) Another customer bought a 500-mL can. How might you explain this?

Apply

8. Which is the better buy?
  a) 5 grapefruit for $1.99 or
     8 grapefruit for $2.99
  b) 2 L of juice for $4.49 or
     1 L of juice for $2.89
  c) 100 mL of toothpaste for $1.79 or
     150 mL of toothpaste for $2.19
  d) 500 g of yogurt for $3.49 or
     125 g of yogurt for $0.79

9. Mr. Gomez travelled 525 km in 6 h.
  a) Assume Mr. Gomez travelled the same distance each hour. What is this distance?
  b) How is the distance in part a related to the average speed?
  c) At this rate, how long will it take Mr. Gomez to travel 700 km?

10. a) Which is the greatest average speed?
     i) 60 km in 3 h
     ii) 68 km in 4 h
     iii) 70 km in 5 h
  b) Draw a graph to illustrate your answers in part a.

11. Each week, Petra earns $370 for 40 h of work as a lifeguard. Giorgos earns $315 for 35 h of work as a starter at the golf course.
  a) Which job pays more?
  b) Would you take the job in part a instead of the other job? Justify your answer.

12. In the first 9 basketball games of the season, Lashonda scored 114 points.
  a) On average, how many points does Lashonda score per game?
  b) At this rate, how many points will Lashonda score after 24 games?

13. Lakelse Lake, BC, had the most snow for one day in Canada, which was 118.1 cm. Assume the snow fell at a constant rate. How much snow fell in 1 h?

14. **Assessment Focus** Becky's dog will eat two different brands of dog food. Brand A costs $12.99 for a 3.6-kg bag. Brand B costs $39.99 for an 18.1-kg bag.
    a) Without calculating, which brand do you think is the better buy? Explain your choice.
    b) Find the unit cost of each brand of dog food.
    c) Which brand is the better buy? How does this compare with your prediction in part a?
    d) Why might Becky not buy the brand in part c?

15. The food we eat provides energy in calories. When we exercise, we burn calories. The tables show data for different foods and different exercises.

Food	Energy Provided (Calories)
Medium apple	60
Slice of white bread	70
Medium peach	50
Vanilla fudge ice cream	290
Chocolate iced doughnut	204

Activity	Calories Burned per Hour
Skipping	492
Swimming	570
Cycling	216
Aerobics	480
Walking	270

    These data are for a female with a mass of 56 kg. The data vary for men and women, and for different masses.

    a) How long would a person have to:
       i) cycle to burn the calories after eating one apple?
       ii) walk to burn the calories after eating two slices of bread?
    b) Suppose a person ate vanilla fudge ice cream and a chocolate iced doughnut.
       i) Which activity would burn the calories quickest? How long would it take?
       ii) Which 2 activities would burn the calories in about 2 h?
    c) Use the data in the tables to write your own problem. Solve your problem. Show your work.

16. A 2.5-kg bag of grass seed covers an area of 1200 m². How much seed is needed to cover a square park with side length 500 m?

17. Suppose you were asked to tutor another student.
    a) How would you explain $\frac{40}{5}$ as a rate? What real-life example could you use to help?
    b) How would you explain $\frac{1.75}{100}$ as a unit rate? What real-life example could you use to help?

18. Take It Further This table shows the city fuel consumption for 5 different cars.

Car	Fuel Consumption
Toyota Echo	26.8 L/400 km
Ford Focus	23.0 L/250 km
Honda Civic	11.25 L/150 km
Saturn Ion	33.25 L/350 km
Hyundai Accent	16.2 L/200 km

a) Which of these cars is the most economical for city driving?
b) Over a distance of 500 km, how much more fuel is needed to drive the least fuel-efficient car compared to the most fuel-efficient car? How did you find out? What assumptions do you make?

19. Take It Further A local garden nursery sells a 2-kg bag of plant food for $6.99. Hayley decides to market her own plant food. She is going to sell it in a 2.5-kg bag. What is the most Hayley should sell her plant food for to undercut the competitor? How did you find out?

20. Take It Further Population density is expressed as a rate. It compares the number of people in a population with the area of the land where they live. Population density is measured in number of people per square kilometre.
a) Find the population density for each country.
   i) Canada: 32 623 490 people in 9 984 670 km^2
   ii) China: 1 307 560 000 people in 9 562 000 km^2
   iii) Japan: 127 760 000 people in 377 727 km^2
b) How do the population densities in part a compare?

21. Troy rides his bike to school. He cycles at an average speed of 20 km/h. It takes Troy 24 min to get to school.
a) How far is it from Troy's home to school?
b) One morning, Troy is late leaving. He has 15 min to get to school. How much faster will Troy have to cycle to get to school on time? Explain.

Reflect

What is a unit rate?
Describe the types of problems you can solve using unit rates.
Write your own problem that involves a unit rate.

Unit Review

What Do I Need to Know?

✓ To calculate a *percent decrease*: divide the decrease by the original amount, then write the quotient as a percent.
Percent decrease (%) = $\frac{\text{Decrease}}{\text{Original amount}} \times 100$

✓ To calculate a *percent increase*: divide the increase by the original amount, then write the quotient as a percent.
Percent increase (%) = $\frac{\text{Increase}}{\text{Original amount}} \times 100$

✓ A *part-to-whole ratio* can be written in fraction form and as a percent.
For example:
There are 9 girls in a class of 20 students.
The ratio of girls to all the students is: 9:20, or $\frac{9}{20}$
This can be written as a percent: $\frac{9}{20} = \frac{45}{100} = 45\%$
So, 45% of the students are girls.

✓ An *equivalent ratio* can be formed by multiplying or dividing the terms of a ratio by the same number.
For example,
10:16 = (10 × 2):(16 × 2)
     = 20:32
10:16 = $\frac{10}{2} : \frac{16}{2}$
     = 5:8
5:8, 10:16, and 20:32 are equivalent ratios.

✓ A *proportion* is a statement that two ratios are equal.
For example, x:3 = 6:9

✓ A *rate* is a comparison of two quantities with different units.
For example, 500 km in 4 h is a rate.
Divide 500 km by 4 h to get the *unit rate* of $\frac{500 \text{ km}}{4}$, or 125 km/h.

Unit Review **307**

What Should I Be Able to Do?

LESSON 5.1

1. Write each decimal as a fraction and as a percent.
   a) 0.65   b) 0.0069
   c) 0.0375   d) 0.9825

2. Conner got 21 out of 24 on a science quiz. Rose got $83.\overline{3}\%$ on the quiz. Who did better?
   How did you find out?

3. Write each percent as a fraction and as a decimal.
   a) 38%   b) 93.75%   c) 0.79%   d) 0.2%

5.2

4. Write each percent as a decimal.
   a) 160%   b) 310%   c) 0.27%   d) 0.9%

5. The average attendance at a regular season home game of the Winnipeg Blue Bombers in 2006 was 26 988. The attendance at the 2006 Grey Cup game in Winnipeg was about 166% of the regular season attendance.
   a) How many people attended the 2006 Grey Cup game?
   b) Estimate to check your answer is reasonable.

5.3

6. Joline collects hockey cards. She needs 5 cards to complete a set. This is 20% of the set. How many cards are in the set? Justify your answer.

7. Wei's mark on a test was 39. This mark was 60%. What was the total possible mark?

8. Two mandrills, Amy and Joe, were born at the zoo. Each mandrill had a birth mass of 1 kg. After Month 1, Amy's mass was 25% greater than her birth mass. After Month 2, Amy's mass was 20% greater than her mass after Month 1. After Month 1, Joe's mass was 20% greater than his birth mass. After Month 2, Joe's mass was 25% greater than his mass after Month 1.
   a) Which mandrill had the greater mass after Month 2? Explain.
   b) Could you have found Amy's mass after Month 2 by finding 145% of her birth mass? Justify your answer.

LESSON

9. A ball bounced 64% of the height from which it was dropped. The bounce was 72 cm high. What is the height from which the ball was dropped?

10. A new queen size bed sheet measures 210 cm by 240 cm. The length and width each shrinks by 2% after the first wash.
   a) What are the dimensions of the sheet after washing?
   b) What is the percent decrease in the area of the sheet?

5.4

11. A portable CD player is on sale. It is advertised as, "Save $20. You pay $49.99."
   a) What is the regular price?
   b) What is the percent decrease?

12. A gym suit sells for $89.99 at Aerobics for All in Victoria, BC. In August, there is a 25% discount as a *Back to School* special. Calculate the total price including taxes.

13. The regular price of a sleeping bag is $39.95. It is on sale for 20% off. There is a 13% sales tax. The discount is usually applied before the tax is added. Suppose the tax is calculated first. Would the total cost be more or less in this case? Explain.

5.5

14. A box contains 5 cream chocolates, 2 chocolate-covered almonds, and 3 caramel chocolates.
   a) What is each ratio below? Sketch a picture for each ratio. When possible, write the ratio as a percent.
      i) almond chocolates to caramel chocolates
      ii) cream chocolates to caramel chocolates
      iii) cream chocolates to all chocolates
      iv) cream chocolates to almond chocolates to caramel chocolates
   b) Lesley ate one of each kind of chocolate. What is each new ratio for part a?

15. In Mary's closet, there are 7 T-shirts, 4 pairs of shorts, and 3 sweatshirts. Write each ratio in as many different ways as you can.
   a) T-shirts to shorts
   b) sweatshirts to shorts
   c) sweatshirts to T-shirts and shorts

16. a) Draw two different diagrams to show the ratio 5:6.
   b) Draw a diagram to show the ratio 5:3.
   c) Draw a diagram to show the ratio 4:1:2.

Unit Review

LESSON 5.6

17. a) Write each ratio below in simplest form.
   i) green squares to red squares
   ii) yellow squares to purple squares
   iii) red squares to total number of squares
   iv) purple squares to blue squares to yellow squares

b) State the colours for each ratio.
   i) 1:6   **ii)** 2:5   **iii)** 2:4:5

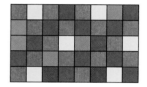

18. In Ms. Bell's class, the ratio of boys to girls is 5:4.
   a) There are 15 boys in the class. How many girls are there?
   b) Two girls leave. What is the new ratio of boys to girls?

19. Explain two different ways to get ratios equivalent to 25:10:30.

20. Find the value of the variable in each pair of equivalent ratios.
   a) 3:7 and h:70
   b) 3:8 and 15:r
   c) s:42 and 5:6
   d) 84:t:96 and 21:27:24

5.7

21. Write each ratio with second term 1.
   a) 32:4   **b)** 132:144
   c) 9:6   **d)** 22:8

22. A jug of orange juice requires 3 cans of orange concentrate and 5 cans of water.
   a) Accidentally, 4 cans of concentrate were mixed with 5 cans of water. Is the mixture stronger or weaker than it should be? Explain.
   b) Suppose 6 cans of water were mixed with 3 cans of concentrate. Is the mixture stronger or weaker than it should be? Explain.

23. The ratio of computers to students in Ms. Beveridge's class is 2:3. The ratio of computers to students in Mr. Walker's class is 3:5. Each class has the same number of students.
   a) Which room has more computers? How did you find out?
   b) Did you use percent to find out? Why or why not?

5.8

24. The numbers of bass and pike in a lake are estimated to be in the ratio of 5:3. There are approximately 300 bass in the lake. About how many pike are there?

25. The St. Croix junior hockey team won 3 out of every 4 games it played. The team played 56 games. How many games did it lose?

LESSON

26. A punch recipe calls for orange juice and pop in the ratio of 2:5. The recipe requires 1 L of pop to serve 7 people.
   a) How much orange juice is needed for 7 people?
   b) About how much orange juice and pop do you need to serve 15 people? 20 people? Justify your answers.

5.9

27. Express as a unit rate.
   a) A bus travelled 120 km in 3 h.
   b) An athlete ran 1500 m in 6 min.
   c) A security guard earned $16.00 for 2 h of work.

28. A cougar can run 312 m in 20 s. A wild horse can run 200 m in 15 s.
   a) Which animal is faster?
   b) What is the ratio of their average speeds?

29. Write each speed in metres per second.
   a) The dolphin can swim at a top speed of 60 km/h.
   b) The polar bear can run at a top speed of 40 km/h.

30. Milena worked in the gift shop of the Costume Museum of Canada in downtown Winnipeg. She was paid $57 for a 6-h shift.
   a) What was Milena's hourly rate of pay?
   b) How much would Milena earn for 25 h of work?

5.10

31. a) Find the unit cost of each item.
   i) 4 L of milk for $4.29
   ii) 2.4 kg of beef for $10.72
   iii) 454 g of margarine for $1.99
   b) For each item in part a, which unit did you choose? Justify your choice.

32. Which is the better buy? Justify each answer.
   a) 6.2 L of gas for $5.39 or 8.5 L of gas for $7.31
   b) 5 candles for $3.00 or 12 candles for $5.99
   c) 2 kg of grass seed for $1.38 or 5 kg of grass seed for $2.79

33. Kheran ran 8 laps of the track in 18 min. Jevon ran 6 laps of the track in 10 min. Who had the greater average speed? How do you know?

34. Each week, Aaron earns $186 for 24 h of work as a ticket seller. Kayla earns $225 for 30 h of work as a cashier. Which job pays more? How did you find out?

Practice Test

1. Find.
   a) 165% of 80
   b) $3\frac{1}{2}$% of 400
   c) 0.4% of 500
   d) 35% of 51

2. Find the value of the variable in each proportion.
   a) $3:5 = g:65$
   b) $h:8 = 56:64$
   c) $16:m = 20:35$
   d) $36:8 = 81:f$

3. Express as a unit rate.
   a) Ethan earned $41.25 in 8 h.
   b) Brianna completed 8 Sudoku puzzles in 96 min.
   c) A car travelled 20 km in 15 min.

4. The Grade 8 class at Wheatly Middle School sold 77 boxes of greeting cards to raise money for the under privileged. This is 22% of all the boxes of greeting cards sold. Calculate how many boxes were sold.

5. The manager of a clothing store reduces the price of an item by 25% if it has been on the rack for 4 weeks.
   The manager reduces the price a further 15% if the item has not been sold after 6 weeks.
   a) What is the sale price of the jacket after 6 weeks?
   b) Calculate the sale price including PST of 5% and GST of 6%.

6. The price of a house increased by 20% from 2004 to 2005.
The price of the house decreased by 20% from 2005 to 2006.
Will the price of the house at the beginning of 2004 be equal
to its price at the end of 2006?
If your answer is yes, show how you know.
If your answer is no, when is the house less expensive?

7. Which is the better buy?
   a) 8 batteries for $3.49 or 24 batteries for $9.29
   b) 100 g of iced tea mix for $0.29 or 500 g of iced tea mix for $1.69
   Show your work.

8. In the baseball league, the Leos play 8 games.
Their win to loss ratio is 5:3.
The Tigers play 11 games.
Their win to loss ratio is 7:4.
   a) Which team has the better record?
   b) Suppose the Leos win their next game and the Tigers lose theirs.
   Which team would have the better record? Explain.

9. Taylor is comparing the prices of 3 different sizes of Rise and Shine orange juice.

1 L for $2.79   1.84 L for $3.99   2.78 L for $6.29

Should Taylor buy the largest carton?
Justify your answer.

10. Use the number $\frac{3}{4}$.

Provide a context to explain each meaning of $\frac{3}{4}$.
   a) as a fraction   **b)** as a ratio
   c) as a quotient   **d)** as a rate

Unit Problem: What Is the Smartest, Fastest, Oldest?

Mr. Peabody has three **hypotheses** about the animal world. For a Science Fair project, you decide to analyse data to test each of Mr. Peabody's hypotheses.

> A **hypothesis** is something that seems likely to be true. It needs to be tested to be proved or disproved.

Part A

Mr. Peabody believes you can predict an animal's intelligence by looking at the size of its brain. He thinks the smartest animal has the greatest brain mass compared with its body mass. Investigate this hypothesis.

Species	Comparing Mass	
	Body Mass (g)	Brain Mass (g)
Cat	3 300	30
Monkey	7 000	100
Human	56 000	1400

1. Find each ratio of brain mass to body mass.
- How does a human compare to a cat?
- How does a human compare to a monkey?
- How does a cat compare to a monkey?

According to Mr. Peabody's hypothesis, which animal is smartest? Explain your reasoning.

Part B

Mr. Peabody believes that among animals of the same family, the lesser the average mass of a species, the greater the average running speed. Here are some data for the cat family.

Species	Average Mass (kg)	Running Rate
Cheetah	45	150 m in 5 s
Lion	250	300 m in 15 s
Leopard	50	170 m in 10 s

2. Find the average speed of each animal in metres per second, metres per minute, metres per hour, and kilometres per hour. Do the data support Mr. Peabody's hypothesis that the lesser average mass indicates a greater average speed? Explain your reasoning.

Part C

Mr. Peabody believes that an animal's heart rate can be used to predict its life expectancy. The animal with the least heart rate will have the greatest life expectancy.

Species	Heart Rate	Average Life Span
Cat	30 beats per 12 s	15 years
Monkey	48 beats per 15 s	25 years
Human	10 beats per 10 s	70 years

3. Find the heart rate of each animal in beats per minute. Which animal has the greatest heart rate? The least heart rate? Calculate the number of heartbeats in the average life span of each animal. What assumptions did you make?
Do the data support Mr. Peabody's hypothesis that a lesser heart rate indicates a longer life expectancy?
Explain your reasoning.

Part D

Review your results. Write a short letter to Mr. Peabody telling him whether you agree or disagree with each of his hypotheses, and explain why. Use math language to support your opinions.

Check List

Your work should show:
- ✓ how you calculated each ratio and rate
- ✓ your conclusions and reasoning for each hypothesis
- ✓ the correct use of math language

Reflect on Your Learning

How are ratio, rate, and percent related?
How can you use your knowledge of ratios, rates, and percents outside the classroom?
Include examples in your explanation.

UNIT 6
Linear Equations and Graphing

Suppose you are planning a ski trip to Nakiska, a ski resort 90 km west of Calgary.

What do you need to know to organize this trip?

You want to calculate the total cost for all Grade 8 students. What assumptions do you make? What other things do you need to consider?

What You'll Learn

- Model and solve problems using linear equations.
- Solve equations using models, pictures, and symbols.
- Graph and analyse two-variable linear relations.

Why It's Important

- Solving equations is a useful problem-solving tool.
- You find linear relations in everyday situations, such as rates of pay, car rentals, and dosages of medicine.

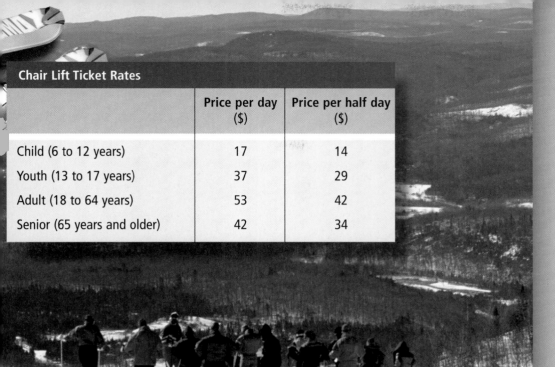

Chair Lift Ticket Rates

	Price per day ($)	Price per half day ($)
Child (6 to 12 years)	17	14
Youth (13 to 17 years)	37	29
Adult (18 to 64 years)	53	42
Senior (65 years and older)	42	34

Key Words

- distributive property
- expand
- linear relation
- ordered pair
- discrete data

Equipment Rental

Skis and Snowboards	Price per 1 day ($)	Price per day for 3 or more days ($)
Complete Package	32	30
Skis or Boards	23	22
Boots	9	8
Poles	7	7

6.1 Solving Equations Using Models

Focus   Use concrete materials to model and solve linear equations.

Recall that one yellow unit tile represents +1.
One red unit tile represents −1.
What happens when you combine
one red unit tile and one yellow unit tile?

The yellow variable tile  represents any variable, such as n or x.

Investigate

Work with a partner.

Marie received three $100 savings bonds on her first birthday.
Her grandmother promised to give her 2 savings bonds each year after that for her birthday. How old will Marie be when she has 13 savings bonds?

Let n represent Marie's age in years.
Write an equation you can use to solve for n.
Use tiles to represent the equation.
Use the tiles to solve the equation.
Sketch the tiles you used.

Reflect & Share

Compare your equation and its solution with those of another pair of classmates.
Did you write the same equation?
If your answer is yes, find another way to write the equation.
If your answer is no, are both equations correct?
How do you know?
Share your strategies for solving the equation using tiles.

Connect

We can use a balance-scales model to solve an equation.

To keep the scales balanced, we must do the same to both sides.
For example,
we add the same mass:

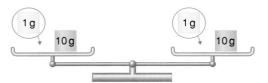

The scales remain balanced.

or remove the same mass:

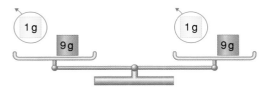

The scales remain balanced.

Example 1

Herman is in the last round of the Yellowknife Elementary School Spell-A-Thon.
A contestant receives 3 points for every word spelled correctly.
Herman has 42 points. How many words has he spelled correctly?

▶ **A Solution**

Let h represent the number of words Herman has spelled correctly.
Then, the number of points is 3 times h, or $3h$.

Since Herman has 42 points, the algebraic equation is $3h = 42$.
Use a balance-scales model to represent this equation.

On the left side, show masses to represent $3h$.
On the right side, show a mass to represent 42.

Since there are 3 identical unknown masses in the left pan,
replace 42 g in the right pan with 3 equal masses.
Each mass is 14 g.

So, each unknown mass is 14 g.
$h = 14$

Herman has spelled 14 words correctly.

Check: 14 words worth 3 points each is $14 \times 3 = 42$ points.
   The solution is correct.

Another strategy to solve an equation is to use algebra tiles.
We rearrange the tiles to end up with variable tiles on one side and
unit tiles on the other side.

Example 2

Jodee is also a contestant in the Yellowknife Elementary School Spell-A-Thon.
A contestant receives 3 points for every word spelled correctly.
Because of a technical penalty, Jodee loses 5 points. She now has 19 points.
How many words has Jodee spelled correctly?

▶ *A Solution*

Let j represent the number of words Jodee has spelled correctly.
Then, the number of points she receives is $3j$.
When the penalty is considered, the number of points is $3j - 5$.
So, the equation is: $3j - 5 = 19$

On the left side, place tiles to represent $3j - 5$.
On the right side, place tiles to represent 19.

To isolate the *j*-tiles on the left side, add 5 positive unit tiles to make zero pairs.
To keep the balance, add 5 positive unit tiles to the right side, too.

There are 3 *j*-tiles. So, arrange the unit tiles into 3 equal groups.

The solution is $j = 8$.
Jodee has spelled 8 words correctly.

We can verify the solution by replacing each positive variable tile with
8 positive unit tiles. Then:

becomes:

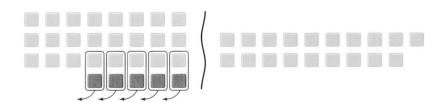

Since there are now 19 positive unit tiles on each side, the solution is correct.

Many different types of equations can be modelled and solved with algebra tiles.

The opposite of x is $-x$.
So, the red variable tile ▬ represents $-x$.
As with the unit tiles, a positive variable tile and
a negative variable tile combine to model 0.

To make some equations easier to solve,
it helps to make the variables positive.

Example 3

Use algebra tiles to solve: $8 = -6 - 2x$
Verify the solution.

▶ **A Solution**

Place tiles to model the equation.

To make the variable tiles positive,
add 2 positive variable tiles to each side.

These tiles remain.

To isolate the variable tiles, add 8 negative unit tiles to each side.
Remove zero pairs.

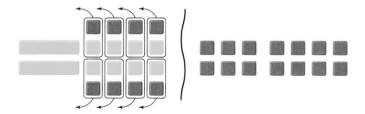

Arrange the tiles remaining on each side into 2 equal groups.

The solution is $x = -7$.

We can verify the solution.
The original equation has negative variable tiles.
If $x = -7$, then $-x = -(-7) = 7$
So, replace each variable tile in the original equation with 7 positive unit tiles.

Then:

becomes:

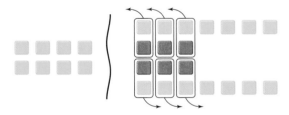

Remove zero pairs.

Since there are now 8 positive unit tiles on each side, the solution is correct.

6.1 Solving Equations Using Models

Discuss the ideas

1. In *Example 1*, we used h to represent the number of words Herman spelled correctly. Why do you think we used h? Could we have used a different letter? Explain.
2. In *Example 2*, we solved the equation with tiles to represent $3j - 5$ on the left side and 19 on the right side. Could we have solved the equation with 19 on the left side and $3j - 5$ on the right side? Would the solution have been different? Justify your answer.
3. In *Example 3*, how would you solve the equation by leaving the variable tiles on the right side of the equation?
4. Could we have solved the equation in *Example 3* using a balance-scales model? Why or why not?

Practice

Check

Use a model to solve each equation.

5. Draw pictures to represent the steps you took to solve each equation.
    a) $4s = 16$ b) $5t = -15$
    c) $18 = 6a$ d) $-18 = 3b$

6. Draw pictures to represent the steps you took to solve each equation.
    a) $3x + 2 = 8$ b) $4s - 3 = 9$
    c) $10 = 6c + 4$ d) $-4 = 5m + 6$

7. Three more than six times a number is 21. Let n represent the number.
    a) Write an equation you can use to solve for n.
    b) Represent the equation with tiles. Use the tiles to solve the equation. Sketch the tiles you used.
    c) Verify the solution.

Apply

8. Three less than six times a number is 21. Let n represent the number.
    a) Write an equation you can use to solve for n.
    b) Represent the equation with tiles. Use the tiles to solve the equation. Sketch the tiles you used.
    c) Verify the solution.

9. Maeve wants her friend to guess how many cards she has in her hand. She says that if the number of cards in her hand is tripled, and 4 are added, then she has 22 cards.
    a) Choose a variable. Write an equation for this situation.
    b) Use a balance-scales model to solve the equation.
    c) Verify the solution. Show how you did this.

10. Curtis is practising modelling equations. He is trying to model the equation: $4x - 2 = 18$
   Curtis begins by using algebra tiles.
   a) Check Curtis' work. Is this the correct model? Explain.

   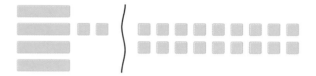

   b) If your answer to part a is yes, use the tiles to try to solve the equation. If your answer to part a is no, describe the error, correct it, then use algebra tiles to solve the equation.

11. Use a model to solve each equation. Verify the solution.
    a) $-2x = -6$
    b) $-15 = 3x$
    c) $-24 = -4x$
    d) $9x = -27$

12. **Assessment Focus** Breanna and 3 friends need $29 to buy a game. Breanna has $5. Each friend will contribute an equal amount of money. Breanna wants to know how much money each friend should contribute. She uses a to represent this amount in dollars.
    Breanna is trying to model and solve the equation: $3a + 5 = 29$

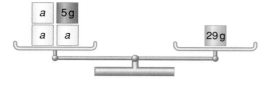

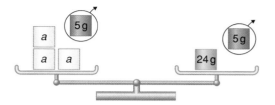

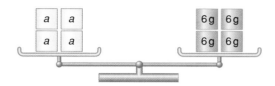

Breanna's solution is $a = 6$.
But, when she checks it, she notices that 3 times $6 plus her $5 is $18 + $5, which is $23.
The money is not enough to buy the game.
a) Find Breanna's mistake.
b) Sketch the balance scales to model the correct solution.
c) Verify the solution.

13. Use a model to solve each equation. Verify the solution.
   a) $-2x + 3 = 13$
   b) $-2x - 3 = -13$
   c) $2x - 3 = -13$
   d) $2x + 3 = -13$

14. Roger brings 4 cakes for dessert to the community potluck feast and powwow. His son brings 2 individual servings of dessert. Altogether, there will be 34 people at the feast. Each person has 1 serving of dessert.
   a) Choose a variable to represent the number of pieces into which each cake must be cut.
   Write an equation to describe this situation.
   b) Use a model to solve the equation.
   c) Verify the solution.
   Show your work.

15. Take It Further A pattern rule for a number pattern is represented by $5 - 8n$, where n is the term number. What is the term number for each term value?
   a) -3
   b) -35
   c) -155

16. Take It Further
   a) Write an equation you could solve with balance scales and with algebra tiles.
   b) Write an equation you could solve with algebra tiles but not with balance scales.
   c) Is there an equation that could be solved with balance scales but not with algebra tiles?
   Justify your answer each time.

17. Take It Further
   a) Write an equation in words to describe these balance scales.

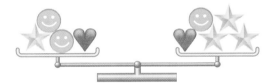

   b) i) The mass of a star is 11 g. What is the mass of a smiley face? What is the mass of a heart? Justify your answers.
   ii) Compare the strategy you used in part a with that of another classmate. If the strategies are different, is one more efficient than the other?
   c) Verify your answers. Write to explain your thinking.

Reflect

You have learned two models to solve an equation.
Are there situations where you prefer one model over the other?
Give an example.

6.2 Solving Equations Using Algebra

Focus Solve a linear equation symbolically.

Investigate

Work with a partner to solve this problem.

Asuka invites some friends over to celebrate her spelling bee award.
Her mom has some veggie burgers, but she doesn't have enough.
Asuka's mom tells her that she needs twice as many veggie burgers, plus 5 more.
Her mom leaves to run errands, and Asuka has forgotten to ask how many veggie burgers her mom has.
There will be 33 people at the party—Asuka and 32 friends.
Asuka uses this diagram to model the situation:

What does *n* represent?
Write an equation for the balance-scales model.
How many different ways can you solve the equation?
Show each way.

Compare the equation you wrote with that of another pair of classmates.
If the equations are different, is each equation correct?
How can you find out?
Discuss the strategies you used to solve the equation.
If you did not solve the equation using algebra, work together to do that now.

Connect

We can solve an equation using algebra.
To help us visualize the equation, we think about a model,
such as balance scales or algebra tiles.

To solve an equation, we need to isolate the variable on one side of the equation.
To do this, we get rid of the numbers on that side of the equation.

When we solve an equation using algebra, we must also preserve the equality.
Whatever we do to one side of the equation, we must do to the other side, too.

Example 1

Fabian charges $3 for each bag of leaves he rakes, and $5 for mowing the lawn.
On Sunday, Fabian mowed 1 lawn and raked leaves. He earned $14.
How many bags of leaves did Fabian rake?
a) Write an equation to represent this problem.
b) Use algebra tiles to solve the equation.
   Use symbols to record each step.
c) Verify the solution using algebra.

▶ *A Solution*

a) Let b represent the number of bags of leaves Fabian raked.
   An equation is: $3b + 5 = 14$

b) Model the equation with tiles.

$3b + 5 = 14$

Add 5 negative unit tiles to each side to isolate the variable.
Remove zero pairs.

$3b + 5 - 5 = 14 - 5$

328 UNIT 6: Linear Equations and Graphing

These tiles remain.

  $3b = 9$

Divide each side into 3 equal groups.

 $\frac{3b}{3} = \frac{9}{3}$

$b = 3$

Fabian raked 3 bags of leaves.

c) To verify the solution, substitute $b = 3$ into $3b + 5 = 14$.
Left side $= 3b + 5$ Right side $= 14$
$= 3(3) + 5$
$= 9 + 5$
$= 14$
Since the left side equals the right side, $b = 3$ is correct.
Fabian raked 3 bags of leaves.

It can be inconvenient to model an equation using algebra tiles when large numbers are involved, or when the solution is a fraction or a decimal.

Example 2

a) Use algebra to solve the equation:
$16t - 69 = -13$
b) Verify the solution.

▶ **A Solution**

a) $16t - 69 = -13$
To isolate the variable term, add 69 to each side.
$16t - 69 + 69 = -13 + 69$
$16t = 56$

To isolate the variable, divide each side by 16.
$$\frac{16t}{16} = \frac{56}{16}$$
At this stage in the solution, we can continue in 2 ways.

As a fraction:
$t = \frac{56}{16}$
Divide by the common factor 8.
$t = \frac{56 \div 8}{16 \div 8}$
$t = \frac{7}{2}$

As a decimal:
$t = \frac{56}{16}$
Use a calculator.
$t = 3.5$

b) To verify the solution, substitute $t = \frac{7}{2}$ into:

$16t - 69 = -13$
Left side $= 16t - 69$
$= 16\left(\frac{7}{2}\right) - 69$
$= {}^8\cancel{16}\left(\frac{7}{\cancel{2}}\right) - 69$
$= 56 - 69$
$= -13$
Right side $= -13$

Since the left side equals the right side, $t = \frac{7}{2}$ is correct.

To verify the solution, substitute $t = 3.5$ into:

$16t - 69 = -13$
Left side $= 16t - 69$
$= 16(3.5) - 69$
$= 56 - 69$
$= -13$
Right side $= -13$

Since the left side equals the right side, $t = 3.5$ is correct.

Discuss the ideas

1. When is it easier to solve an equation using algebra instead of using algebra tiles or a balance-scales model?
2. When is it easier to verify a solution using substitution instead of algebra tiles or a balance-scales model?
3. When you solve an equation using algebra, why do you add or subtract the same number on each side? Why do you divide each side by the coefficient of the variable term?
4. Suppose the solution to an equation is a fraction or a decimal. Which would you prefer to use to verify the solution? Give reasons for your answer.

Practice

Check

5. Model each equation. Then solve it using concrete materials. Use algebra to record each step you take. Verify each solution.
   a) $2x - 1 = 7$
   b) $11 = 4a - 1$
   c) $5 + 2m = 9$
   d) $1 = 10 - 3x$
   e) $13 - 2x = 5$
   f) $3x - 6 = 12$

6. Use algebra to solve each equation. Verify the solution.
   a) $4x = -16$
   b) $12 = -3x$
   c) $-21 = 7x$
   d) $6x = -30$

7. Check each student's work. Rewrite a correct and complete algebraic solution where necessary.
   a) Student A:
   $$-3x + 15 = 30$$
   $$-3x + 15 - 15 = 30 + 15 - 15$$
   $$-3x = 30$$
   $$\frac{-3x}{-3} = \frac{30}{-3}$$
   $$x = -10$$
   b) Student B:
   $$7 = 1 + 2n$$
   $$7 - 1 = 1 - 1 + 2n$$
   $$8 = 2n$$
   $$\frac{8}{2} = \frac{2n}{2}$$
   $$4 = n$$
   $$n = 4$$
   c) Student C:
   $$3 + 2t = 4$$
   $$3 + 2t - 3 = 4 - 3$$
   $$2t = 1$$
   $$t = 2$$
   d) Student D:
   $$-5 = -8 + 5f$$
   $$-5 + 8 = -8 + 8 + 5f$$
   $$3 = 5f$$
   $$\frac{3}{5} = \frac{5f}{5}$$
   $$\frac{3}{5} = f$$
   $$f = \frac{3}{5}$$

Apply

8. Solve each equation. Verify the solution.
   a) $2x + 5 = -7$
   b) $-3x + 11 = 2$
   c) $-9 = 5 + 7x$
   d) $18 = -4x + 2$

9. Navid now has $72 in her savings account. Each week she will save $24. After how many weeks will Navid have a total savings of $288?
   a) Write an equation you can use to solve the problem.
   b) Solve the equation. When will Navid have $288 in her savings account?
   c) Verify the solution.

10. Assessment Focus

The Grade 8 students had an end-of-the-year dance.
The disc jockey charged $85 for setting up the equipment, plus $2 for each student who attended the dance.
The disc jockey was paid $197.
How many students attended the dance?
a) Write an equation you can use to solve the problem.
b) Solve the equation.
c) Check your answer and explain how you know it is correct.

11. Solve each equation. Verify your solution.
a) $-8x + 11 = 59$ b) $11c + 21 = -34$
c) $23 = -5b + 3$ d) $-45 = 6a - 15$
e) $52 = 25 - 9f$ f) $-13 + 4d = 31$

12. Solve each equation. Verify your solution.
a) $3n + 7 = 8$ b) $6x + 6 = 15$
c) $-23 = 5p - 27$ d) $5p + 6 = 7$
e) $8e - 9 = -3$ f) $-17 + 10g = -9$

13. The high temperature today is 7°C higher than twice the high temperature yesterday. The high temperature today is −3°C. What was the high temperature yesterday?
a) Write an equation you can use to solve the problem.
b) Solve the equation. Verify the solution.

14. Take It Further
Use this information.
Boat rental: $300
Fishing rod rental: $20
a) Write a problem that can be solved using an equation.
b) Write the equation, then solve the problem.
c) How could you have solved the problem without writing an equation? Explain.

15. Take It Further
Use this information:
Water is pumped out of a flooded basement at a rate of 15 L/min.
a) Write a problem that can be solved using an equation.
b) Write, then solve, the equation.

Reflect

Which types of equations do you prefer to solve using algebra?
Explain why you may not want to use algebra tiles or a balance-scales model.

6.3 Solving Equations Involving Fractions

Focus Solve an equation that involves a fraction.

Which number could you multiply $\frac{3}{7}$ by to get the product 3?
Which number could you multiply $\frac{5}{6}$ by to get the product 5?
How did you find out?

Investigate

Work with a partner.

A fair comes to Behchoko, NWT.
Nicole has some tickets for the midway.
She shares the tickets with 2 friends so that
they have 9 tickets each.
How many tickets did Nicole begin with?

Let t represent the number of tickets Nicole began with.
Write an equation you can use to solve for t.
Use any strategy to solve the problem.

Compare the equation you wrote with that of another pair of classmates.

If the equations are different, are both equations correct? How do you know?

If you did not write an equation that involved division, work together to do so now.
What strategy would you use to solve this equation?
How could you check your answer?

Equations that involve fractions cannot be easily modelled
with algebra tiles or balance scales.
We can write and solve these equations using algebra.

Example 1

Grandpa has enough gift certificates to give the same number
to each of his 4 grandchildren.
After Grandpa gives them the gift certificates, each grandchild has 5 gift certificates.
How many gift certificates does Grandpa have?
a) Write an equation to represent this problem.
b) Solve the equation.
c) Verify the solution.

▶ **A Solution**

a) Let n represent the number of gift certificates Grandpa has.
He shares them equally among 4 grandchildren.
Each grandchild will receive $\frac{n}{4}$ gift certificates.
Each grandchild will then have 5 gift certificates.
One possible equation is: $\frac{n}{4} = 5$

b) Solve the equation using algebra.
$\frac{n}{4} = 5$ To isolate the variable, multiply each side by 4.
$\frac{n}{4} \times 4 = 5 \times 4$
$n = 20$
Grandpa has 20 gift certificates.

c) To verify the solution,
substitute $n = 20$ into $\frac{n}{4} = 5$.

Left side $= \frac{n}{4}$ Right side $= 5$
$= \frac{20}{4}$
$= 5$

Since the left side equals the right side, $n = 20$ is correct.
Grandpa has 20 gift certificates.

Example 2

The school's student council sold T-shirts for charity.
The council bought the T-shirts in boxes of 40.
The student council added $6 to the cost of each T-shirt.
Each T-shirt sold for $26.
What did the student council pay for 1 box of T-shirts?

a) Write an equation to represent this problem.
   Solve the equation.
b) Verify the solution.

▶ **A Solution**

a) Let c dollars represent what the student council paid for 1 box of T-shirts.
   The cost of each T-shirt the student council bought was: $\frac{c}{40}$
   $6 was added to the cost of each T-shirt: $\frac{c}{40} + 6$
   Each T-shirt sold for $26.
   So, an equation is: $\frac{c}{40} + 6 = 26$

   Solve the equation using algebra.
   $\frac{c}{40} + 6 = 26$ To isolate the variable term, subtract 6 from each side.
   $\frac{c}{40} + 6 - 6 = 26 - 6$
   $\frac{c}{40} = 20$ To isolate the variable, multiply each side by 40.
   $\frac{c}{40} \times 40 = 20 \times 40$
   $c = 800$
   The student council paid $800 for 1 box of T-shirts.

b) To verify the solution,
   substitute $c = 800$ into $\frac{c}{40} + 6 = 26$.
   Left side $= \frac{c}{40} + 6$ Right side $= 26$
   $= \frac{800}{40} + 6$
   $= 20 + 6$
   $= 26$
   Since the left side equals the right side, $c = 800$ is correct.

Discuss the ideas

1. Why would it be difficult to model and solve equations like those in *Examples 1* and *2* with algebra tiles or balance scales?
2. In *Example 2*, why did we subtract 6 from each side before we multiplied by 40?

Practice

Check

3. Solve each equation. Verify the solution.
   a) $\frac{t}{5} = 6$ b) $\frac{a}{7} = 8$
   c) $\frac{b}{6} = 3$ d) $\frac{c}{3} = 9$

4. Solve each equation. Verify the solution.
   a) $\frac{d}{-4} = 5$ b) $\frac{f}{8} = -5$
   c) $\frac{k}{9} = -4$ d) $\frac{m}{-5} = -7$

5. One-quarter of the golf balls in the bag are yellow.
   There are 8 yellow golf balls.
   How many golf balls are in the bag?
   a) Write an equation you can use to solve the problem.
   b) Solve the equation.
   c) Verify the solution.

6. For each sentence, write an equation. Solve the equation to find the number.
   a) A number divided by 6 is 9.
   b) A number divided by -4 is -3.
   c) A number divided by -5 is 7.

7. Solve each equation. Verify the solution.
   a) $\frac{n}{4} + 3 = 10$ b) $\frac{m}{3} - 2 = 9$
   c) $13 + \frac{x}{2} = 25$ d) $-9 + \frac{s}{2} = 2$

Apply

8. Solve each equation. Verify the solution.
   a) $\frac{p}{-3} + 9 = 3$ b) $\frac{t}{-6} + 12 = 18$
   c) $-24 + \frac{w}{5} = -29$ d) $-17 + \frac{e}{-7} = -8$

9. For each sentence, write an equation. Solve the equation to find the number.
   a) Add 1 to a number divided by -3 and the sum is 6.
   b) Subtract a number divided by 9 from 3 and the difference is 0.
   c) Add 4 to a number divided by -2 and the sum is -3.

10. One-half of the team's supply of baseballs was taken from the dressing room to the dugout. During the game, 11 baseballs were caught by fans. At the end of the game, there were 12 baseballs left in the dugout. What was the team's original supply of baseballs?
    a) Write an equation you can use to solve the problem.
    b) Solve the equation.
    c) Verify the solution.

11. **Assessment Focus** Five students in Mrs. Lamert's tutorial class after school are solving equations. She brought a bag of treats. Mrs. Lamert explained that if the 5 students shared the bag of treats equally, then gave one treat each to the teacher, each student would still have 9 treats. How many treats were in the bag? Here is the equation Jerry suggested: $\frac{n}{5} - 1 = 9$
    a) Is Jerry's equation correct? Explain why or why not.
    b) If your answer to part a is yes, solve the equation using algebra.
       If your answer to part a is no, correct the equation, then solve the equation using algebra.
    c) Verify the solution.

12. One-third of the Grade 8 students went to the track-and-field meet. Five track coaches went too. There were 41 people on the bus, not including the driver. How many students are in Grade 8?
    a) Write an equation you can use to solve the problem.
    b) Solve the equation.
    c) Verify the solution.

13. Check each student's work below. Rewrite a correct and complete algebraic solution where necessary.
    a) Student A:
    $$\frac{h}{-9} = 4$$
    $$\frac{h}{-9} \times (-9) = 4 \times (-9)$$
    $$\frac{-9h}{-9} = -36$$
    $$h = -36$$

    b) Student B:
    $$\frac{t}{8} - 2 = 4$$
    $$8 \times \frac{t}{8} - 2 = 4 \times 8$$
    $$\frac{8t}{8} - 2 = 32$$
    $$t - 2 = 32$$
    $$t - 2 + 2 = 32 + 2$$
    $$t = 34$$

    c) Student C:
    $$\frac{r}{-4} + 3 = 13$$
    $$\frac{r}{-4} + 3 - 3 = 13 - 3$$
    $$\frac{r}{-4} = 10$$
    $$4 \times \frac{r}{-4} = 10 \times 4$$
    $$r = 40$$

14. **Take It Further** Jonah used the equation $3 + \frac{n}{7} = 18$ to solve a word problem.
    a) What might the word problem be?
    b) Solve the problem.
    c) Verify the solution.

Reflect

How did your knowledge of multiplying fractions help you in this lesson? Include examples in your explanation.

6.4 The Distributive Property

Focus Relate the distributive property to algebra.

How does this diagram model the product of 4 × 37?
What is the product?

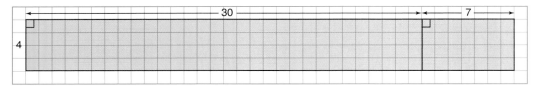

Investigate

Work with a partner.

Gina's Trattoria Ristorante has only round tables.
Each table seats 5 people.
Some tables are in the lounge and some are on the patio.

➤ There are 20 tables in the lounge and 8 tables on the patio.
Draw a diagram to show how to calculate the total number of people who can be seated.

➤ Gina removes some tables from the lounge.
Use a variable to represent the number of tables that remain.
Write an algebraic expression for the number of people the restaurant can now seat.

Compare your diagram and expression with those of another pair of classmates.
If your diagrams and expressions are different, are both of them correct? How can you check?
If your expressions are the same, work together to write a different expression for the number of people the restaurant can now seat.

Connect

➤ A charity sells pots of flowers for $10 each to raise money.
8 people pay with a $10 bill.
5 people pay with a $10 cheque.
The total amount of money collected, in dollars, is:

- Add the total number of bills and cheques, then multiply by 10:
  $10 \times (8 + 5) = 130$

Or

- Multiply the number of bills by 10 and the number of cheques by 10, then add:
  $10 \times 8 + 10 \times 5 = 130$

We write: $10 \times (8 + 5) = 10 \times 8 + 10 \times 5$

We can use the same strategy to write two equivalent expressions for any numbers of bills and cheques.

➤ Suppose you are selling event tickets for $20 each.
Some people pay with a $20 bill and some pay with a $20 cheque.
Suppose b is the number of $20 bills you receive,
and c is the number of $20 cheques you receive.
The total amount of money you collect, in dollars, is:

- Add the total number of bills and cheques, then multiply by 20:
  $20 \times (b + c)$

Or

- Multiply the number of bills by 20 and the number of cheques by 20, then add:
  $20b + 20c$

6.4 The Distributive Property

We write: $20(b + c) = 20b + 20c$ **Multiplication has been *distributed* over addition.**

➤ We can also model the distributive property with algebra tiles.

For example:

To model $4(x + 5)$, you need 4 groups of 1 positive variable tile and 5 positive unit tiles.

To model $4x + 20$, you need the same tiles, but their arrangement is grouped differently.

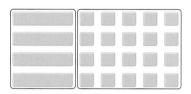

We can see that $4(x + 5) = 4x + 20$ because the two diagrams show the same numbers of tiles.

When an expression for the distributive property uses only variables, we can illustrate the property with this diagram.

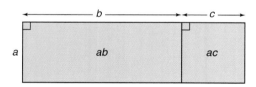

$a(b + c) = ab + ac$
That is, the product $a(b + c)$ is equal to the sum $ab + ac$.

Example 1

Use the distributive property to write each expression as a sum of terms. Sketch a diagram in each case.
a) $7(c + 2)$ b) $42(a + b)$

▶ **A Solution**

a) $7(c + 2) = 7(c) + 7(2)$
$= 7c + 14$

b) $42(a + b) = 42(a) + 42(b)$
$= 42a + 42b$

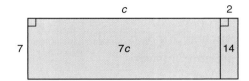
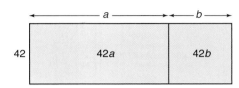

In *Example 1*, when we use the distributive property, we **expand**.

We can also use the distributive property with any integers.

Example 2

Expand.
a) $-3(x + 5)$
b) $-4(-5 + a)$

▶ **A Solution**

a) $-3(x + 5) = -3(x) + (-3)(5)$
$= -3x - 15$

b) $-4(-5 + a) = -4(-5) + (-4)(a)$
$= 20 - 4a$

Subtraction can be thought of as "adding the opposite."
For example, $3 - 4 = 3 + (-4)$

Example 3

Expand.
a) $6(x - 3)$
b) $5(8 - c)$

▶ **A Solution**

Rewrite each expression using addition.

a) $6(x - 3) = 6[x + (-3)]$
$= 6(x) + 6(-3)$
$= 6x - 18$

b) $5(8 - c) = 5[8 + (-c)]$
$= 5(8) + 5(-c)$
$= 40 - 5c$

Discuss the ideas

1. Look at the algebra tile diagrams on page 340. Why are the tiles grouped in different ways?
2. Can you draw a diagram to model each product in *Example 2*? Justify your answer.
3. Do you think the distributive property can be applied when there is a sum of 3 terms, such as $2(a + b + c)$? Draw a diagram to illustrate your answer.

Practice

Check

4. Evaluate each pair of expressions. What do you notice?
   a) i) $7(3 + 8)$
      ii) $7 \times 3 + 7 \times 8$
   b) i) $5(7 - 2)$
      ii) $5 \times 7 + 5 \times (-2)$
   c) i) $-2(9 - 4)$
      ii) $(-2) \times 9 + (-2) \times (-4)$

5. Use algebra tiles to show that $5(x + 2)$ and $5x + 10$ are equivalent. Draw a diagram to record your work. Explain your diagram in words.

6. Draw a rectangle to show that $7(4 + s)$ and $28 + 7s$ are equivalent. Explain your diagram in words.

7. Expand.
   a) $2(x + 10)$ b) $5(a + 1)$
   c) $10(f + 2)$ d) $6(12 + g)$
   e) $8(8 + y)$ f) $5(s + 6)$
   g) $3(9 + p)$ h) $4(11 + r)$
   i) $7(g + 15)$ j) $9(7 + h)$

Apply

8. Expand.
   a) $3(x - 7)$ b) $4(a - 3)$
   c) $9(h - 5)$ d) $7(8 - f)$
   e) $5(1 - s)$ f) $6(p - 2)$
   g) $8(11 - t)$ h) $2(15 - v)$
   i) $10(b - 8)$ j) $11(c - 4)$

9. Write two formulas for the perimeter, P, of a rectangle. Explain how the formulas illustrate the distributive property.

10. Explain how you know $hb = bh$. Use an example to justify your answer.

11. Which expression is equal to $9(6 - t)$? How do you know?
    a) $54 - 9t$
    b) $96 - 9t$
    c) $54 - t$

342   UNIT 6: Linear Equations and Graphing

12. Expand.
  a) $-6(c + 4)$ b) $-8(a - 5)$
  c) $10(f - 7)$ d) $3(-8 - g)$
  e) $-8(8 - y)$ f) $-2(-s + 5)$
  g) $-5(-t - 8)$ h) $-9(9 - w)$

13. **Assessment Focus** Which pairs of expressions are equivalent? Explain your reasoning.
  a) $2x + 20$ and $2(x + 20)$
  b) $3x + 7$ and $10x$
  c) $6 + 2t$ and $2(t + 3)$
  d) $9 + x$ and $x + 9$

14. There are 15 players on the Grade 8 baseball team. Each player needs a baseball cap and a team jersey. A team jersey costs $25. A baseball cap costs $14.
  a) Write 2 different expressions to find the cost of supplying the team with caps and jerseys.
  b) Evaluate each expression. Which expression did you find easier to evaluate? Explain.

15. Five friends go to the movies. They each pay $9 to get in, and $8 for a popcorn and drink combo.
  a) Write 2 different expressions to find the total cost of the outing.
  b) Evaluate each expression. Which expression was easier to evaluate? Justify your choice.

16. Match each expression in Column 1 with an equivalent expression in Column 2.
  Column 1   **Column 2**
  a) $6(t - 6)$ i) $6t + 36$
  b) $-6(t - 6)$ ii) $-6t + 36$
  c) $-6(t + 6)$ iii) $-6t - 36$
  d) $6(6 + t)$ iv) $6t - 36$

17. **Take It Further**
Harvey won some money on a scratch-and-win ticket. Then, he won a $2 bonus. When he arrived at the counter, he noticed that he had also won a "triple your winnings" ticket. As Harvey was cashing in his prize, the cashier told him he was the 100th customer, so his total winnings were automatically doubled. Write two algebraic expressions to describe Harvey's winnings.

18. **Take It Further**
  a) Expand.
    i) $7(5 + y - 2)$
    ii) $-3(-t + 8 - 3)$
    iii) $-8(-9 + s + 5)$
    iv) $12(-10 - p + 7)$
  b) Choose an expression in part a. How many different ways can you expand the expression? Show your work.

19. **Take It Further** Expand.
  a) $2(7 + b + c)$ b) $11(-6 + e - f)$
  c) $-(-r + s - 8)$ d) $-10(-6 - v - w)$
  e) $5(j - 15 - k)$ f) $-4(-g + 12 - h)$

Reflect

How did your knowledge of operations with integers help you in this lesson?

6.5 Solving Equations Involving the Distributive Property

Focus Apply the distributive property to solve an equation.

The distributive property is needed to solve some algebraic equations.

Investigate

Work with a partner.

➤ Alison thought of her favourite number.
She subtracted 2.
Then Alison multiplied the difference by 5.
The product was 60.
What is Alison's favourite number?

Use any strategy to solve the problem.

➤ Write a similar number problem.
Trade problems with another pair of classmates.
Write an algebraic equation, then solve it.

Compare your strategy and answer with the same pair of classmates.
Are the equations the same?
How can you check the solution is correct?
How did you solve your classmates' problem?

344   UNIT 6: Linear Equations and Graphing

Connect

These examples show how we can use the distributive property to solve some algebraic equations.

Example 1

John and Lorraine are landscaping their yard. They are buying pyramidal cedars that cost $12 each. John and Lorraine need 11 cedars to shade their patio on two adjacent sides. They would like to purchase as many more cedars as they can for the far end of their lot. John and Lorraine have $336 to buy cedars. How many more cedars can they buy?

a) Write an equation that models this problem.
b) Solve the equation.
c) Verify the solution.

▶ **A Solution**

a) Let e represent how many more cedars John and Lorraine can buy.
Then they will buy a total of $(e + 11)$ cedars. Since the cedars are $12 each, an equation is: $12(e + 11) = 336$

b) $12(e + 11) = 336$ Use the distributive property to remove the brackets.
$12(e) + 12(11) = 336$
$12e + 132 = 336$
$12e + 132 - 132 = 336 - 132$
$12e = 204$
$\frac{12e}{12} = \frac{204}{12}$ Use a calculator.
$e = 17$

John and Lorraine can buy 17 more cedars.

c) To verify the solution, substitute $e = 17$ into $12(e + 11) = 336$.
Left side $= 12(e + 11)$ Right side $= 336$
$= 12(17 + 11)$
$= 12(28)$
$= 336$

Since the left side equals the right side, $e = 17$ is correct.

Example 2

Solve: $14 = 3(x + 4)$
Verify the solution.

▶ **A Solution**

$14 = 3(x + 4)$
Expand.
$$14 = 3(x + 4)$$
$$14 = 3(x) + (3)(4)$$
$$14 = 3x + 12$$
$$14 - 12 = 3x + 12 - 12$$
$$2 = 3x$$
$$\frac{2}{3} = \frac{3x}{3}$$
$$\frac{2}{3} = x$$
$$x = \frac{2}{3}$$

To verify the solution, substitute $x = \frac{2}{3}$ into $14 = 3(x + 4)$.

Left side = 14   Right side = $3(x + 4)$
$= 3(\frac{2}{3} + 4)$
$= 3(\frac{2}{3} + \frac{12}{3})$
$= 3(\frac{14}{3})$
$= 14$

Since the left side equals the right side, $x = \frac{2}{3}$ is correct.

Discuss the ideas

1. How could you solve the equation in *Example 1* without using the distributive property?
   What is the first step in the solution?
2. Why would it be more difficult to use this method to solve the equation in *Example 2*?
3. For the solution in *Example 2*, could you use the decimal form of the fraction to verify the solution?
   Justify your answer.

Practice

Check

4. Solve each equation using the distributive property. Verify the solution.
 a) $3(x + 5) = 36$
 b) $4(p - 6) = 36$
 c) $5(y + 2) = 25$
 d) $10(a + 8) = 30$

5. Solve each equation. Verify the solution.
 a) $-2(a + 4) = 18$
 b) $-3(r - 5) = -27$
 c) $7(-y + 2) = 28$
 d) $-6(c - 9) = -42$

6. Marc has some hockey cards. His friend gives him 3 more cards. Marc says that if he now doubles the number of cards he has, he will have 20 cards. How many cards did Marc start with?
 a) Choose a variable to represent the number of cards Marc started with. Write an equation to model this problem.
 b) Solve the equation using the distributive property.
 c) Verify the solution. Explain your thinking in words.

7. A student wrote this equation to solve the problem in question 6:
$2n + 3 = 20$
How would you explain to the student why this is incorrect?

Apply

8. The perimeter of a rectangle is 26 cm. The rectangle has length 8 cm. What is the width of the rectangle?
 a) Write an equation that can be solved using the distributive property.
 b) Solve the equation.
 c) Verify the solution.

9. **Assessment Focus** The price of a souvenir T-shirt was reduced by $5. Jason bought 6 T-shirts for his friends. The total cost of the T-shirts, before taxes, was $90. What was the price of a T-shirt before it was reduced?
 a) Write an equation to model this problem.
 b) Solve the equation.
 c) Verify the solution. Explain how you know it is correct.

10. Chuck and 7 friends went to Red Deer's Westerner Days fair. The cost of admission was $6 per person. They each bought an unlimited midway ride ticket. The total cost of admission and rides for Chuck and his friends was $264. What was the price of an unlimited midway ride ticket?
 a) Write an equation to model this problem.
 b) Solve the equation. Verify the solution.

6.5 Solving Equations Involving the Distributive Property

11. Inge chose an integer. She added 9, then multiplied the sum by −5. The product was 15. Which integer did Inge choose?
  a) Write an equation you can use to solve the problem.
  b) Solve the equation.
  c) Verify the solution.

12. Mario chose an integer. He subtracted 7, then multiplied the difference by −4. The product was 36.
  Which integer did Mario choose?
  a) Write an equation you can use to solve the problem.
  b) Solve the equation.
  c) Verify the solution.

13. Kirsten used the distributive property to solve this equation: $8(-x + 3) = 8$
  a) Check Kirsten's work.
    Is her solution correct?
    $8(-x + 3) = 8$
    $8(-x) + 8(3) = 8$
    $-8x + 24 = 8$
    $-8x + 24 - 24 = 8 - 24$
    $-8x = -16$
    $\frac{-8x}{-8} = \frac{-16}{8}$
    $x = -2$
  b) If your answer is yes, verify the solution. If your answer is no, describe the error, then correct it.

14. Solve each equation using the distributive property. Verify the solution.
  a) $-10 = 5(t - 2)$ b) $7 = 2(p - 3)$
  c) $4(r + 5) = 23$ d) $-3(s + 6) = 18$

15. Take It Further
Amanda's office has 40 employees. The employees want to have a catered dinner. They have found a company that will provide what they need for $25 per person. Amanda knows that some people will bring a guest. The company has budgeted $1500 for this event. How many guests can they invite? Assume the price of $25 includes all taxes.
  a) Write an equation for this problem.
  b) Solve the equation.
  c) Verify the solution.

16. Take It Further
Glenn used the equation $7(n - 2) = 42$ to solve a word problem.
  a) Create a word problem that can be solved using this equation.
  b) Solve the problem.
    Verify the solution.

17. Take It Further
Solve each equation using the distributive property.
Verify the solution.
  a) $7(2 + p - 5) = 14$
  b) $8(x - 9 + 7) = -13$
  c) $-2(10 - s + 1) = -21$

Reflect

How do you know when to use the distributive property to help you solve an equation? Include examples in your explanation.

Make the Number

In this game, an ace is 1, a jack is 11, a queen is 12, and a king is 13.

HOW TO PLAY

1. Shuffle the cards.
   Place 4 cards face up. These are the "*seed*" cards.
   Place the remaining cards face down in a pile.
   Turn over the top card. This is the "*target*" card.

2. Each player chooses a "*secret integer*" between −9 and 9.
   Give the integer a name, such as x.

3. Use your secret integer, the "seed" cards, and any operations or brackets to write an expression equal to the "target" number.
   You may use up to 5 integers in your expression.

   You score 5 points for each "seed" card number used, and 2 additional points if you use a "seed" card number more than once.

4. When all players have written their equations, share the equations.
   Each player solves everyone else's equations.
   If a player writes an incorrect equation so the solution is not the secret integer, that player loses all the points scored for writing the equation.

5. Play continues. The player with the most points after 3 rounds wins.

YOU WILL NEED
A standard deck of cards with the jokers removed; paper and pencil

NUMBER OF PLAYERS
2 to 4

GOAL OF THE GAME
To get the most points

Example
"Seed" cards: 4♥, Q♠, 7♥, 7♣
"Target" card: 2♦
"Secret Integer": 5
Expression: $12(5 − 4) − 7 − 4 + 1$
Equation: $12(x − 4) − 7 − 4 + 1 = 2$
Score: 3×5 points (3 "seed" cards used) +
       2 points (one "seed" card used twice)
       = 17 points

Mid-Unit Review

LESSON

6.1

1. Use a model to solve each equation. Verify the solution.
  a) $4x = -36$
  b) $-7x = 63$
  c) $4x + 7 = 19$
  d) $-3x + 5 = 17$

2. Alice has some granola bars in her backpack. If she triples the number of granola bars then adds 4, she will get 13. How many granola bars does Alice have?
  a) Choose a variable. Write an equation for this situation.
  b) Use a model to solve the equation.
  c) Verify the solution. Show how you did this.

6.2

3. Solve each equation. Verify the solution.
  a) $4x + 9 = -27$
  b) $-5x + 8 = 23$
  c) $3x - 4 = -3$
  d) $10 = 6x + 5$

4. The school's sports teams held a banquet. The teams were charged $125 for the rental of the hall, plus $12 for each meal served. The total bill was $545. How many people attended the banquet?
  a) Write an equation you could use to solve the problem.
  b) Solve the equation. Verify the solution.

6.3

5. Solve each equation. Verify the solution.
  a) $\frac{n}{4} = -8$
  b) $\frac{m}{3} - 2 = 3$
  c) $\frac{b}{-3} = 6$
  d) $\frac{f}{-8} + 8 = 12$

6. For each sentence, write an equation. Solve the equation to find the number.
  a) A number divided by –7 is 4.
  b) A number divided by –9 is –3.
  c) Add 5 to a number divided by –2 and the sum is 0.

6.4

7. Draw a rectangle to show that: $6(3 + a) = 18 + 6a$

8. Expand.
  a) $3(x + 11)$
  b) $5(12 + y)$
  c) $-7(a - 4)$
  d) $-12(-t + 6)$

6.5

9. Use the distributive property to solve each equation. Verify the solution.
  a) $3(x + 2) = 21$
  b) $4(p - 3) = 16$
  c) $-5(r + 4) = -15$
  d) $6(-s - 3) = 24$

10. Jon is playing a game. He starts with some points. On his first turn, Jon wins 6 points. On his second turn, Jon's points are doubled. He then has 26 points. How many points did Jon start with?
  a) Write an equation to model this problem.
  b) Solve the equation. Verify the solution.

6.6 Creating a Table of Values

Focus Create a table of values from the equation of a linear relation.

How many different ways can you describe the relation shown in this graph?

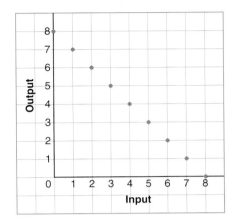

Investigate

Work with a partner.

At the country fair, Mischa sells hot dogs for $3 each, and drinks for $2 each. A meal consists of hot dogs and one drink.

The number of hot dogs in a meal, h, is related to the total cost of the meal in dollars. The relation is: h is related to $3h + 2$.

➤ Copy and complete the a table of values for the relation.

➤ How can you use the table of values to find:
 • the cost of a meal when a person orders 9 hot dogs?
 • the number of hot dogs ordered when a meal costs $35?

Input h	Output $3h + 2$
1	
2	

Compare your answers with those of another pair of classmates. When you know the total cost of a meal, how can you determine the number of hot dogs ordered?
What helped you solve this problem?
What else can you find out using the table or the relation?
Work together to write, then answer, 3 questions about this relation.

Connect

We can represent a relation in different ways.
For example, consider this relation: x is related to $20 - 3x$

➤ We can create a table of values.

When $x = 1$,　　　　　　When $x = 2$,
$20 - 3x = 20 - 3(1)$　　$20 - 3x = 20 - 3(2)$
　　　　　$= 20 - 3$　　　　　　　　$= 20 - 6$
　　　　　$= 17$　　　　　　　　　　$= 14$

A table of values is:

Input x	Output $20 - 3x$
1	17
2	14
3	11
4	8
5	5
6	2
7	−1

When the input increases by 1, the output decreases by 3.

When the change in the input is constant and the change in the output is constant, the relation is a **linear relation**.

➤ We can draw a graph.
Use the data in the table.
The input is plotted on the horizontal axis.
The output is plotted on the vertical axis.
On the horizontal axis, the scale is
1 square represents 1 unit.
On the vertical axis, the scale is
1 square represents 2 units.

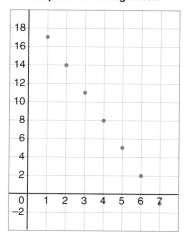

Graph of $20 - 3x$ against x

352　　UNIT 6: Linear Equations and Graphing

➤ We can write an equation for the relation.
We introduce a second variable, y.
Then, an efficient way to write the relation x is related to $20 - 3x$ is:
$y = 20 - 3x$
We say the equation of the linear relation is $y = 20 - 3x$.

We write the table of values as:

x	1	2	3	4	5	6	7
y	17	14	11	8	5	2	−1

A related pair of x and y values is called an **ordered pair**.
Some ordered pairs for this relation are:
$(1, 17), (2, 14), (3, 11), (4, 8), (5, 5), (6, 2), (7, -1), (x, y)$

Example 1

Saskatoon Pizza charges $11 for a medium cheese pizza, plus $2 for each topping. An equation for this relation is $c = 11 + 2n$, where n represents the number of toppings and c represents the cost of the pizza in dollars.

a) Use the equation to create a table of values.
b) Use the equation to find the cost of a pizza with 5 toppings. Check the answer.
c) Use the equation to find how many toppings are on a pizza that costs $27.

▶ A Solution

a) Since it is possible to order a pizza with no toppings, start the table of values with $n = 0$.
When $n = 0$, the cost is:
$c = 11 + 2n$
$ = 11 + 2(0)$
$ = 11 + 0$
$ = 11$
A pizza with no toppings costs $11.

6.6 Creating a Table of Values   353

When $n = 1$,
$$\begin{aligned} c &= 11 + 2n \\ &= 11 + 2(1) \\ &= 11 + 2 \\ &= 13 \end{aligned}$$

When $n = 2$,
$$\begin{aligned} c &= 11 + 2n \\ &= 11 + 2(2) \\ &= 11 + 4 \\ &= 15 \end{aligned}$$

When $n = 3$,
$$\begin{aligned} c &= 11 + 2n \\ &= 11 + 2(3) \\ &= 11 + 6 \\ &= 17 \end{aligned}$$

A table of values is:

n	*c*
0	11
1	13
2	15
3	17

b) To find the cost of a pizza with 5 toppings, substitute $n = 5$.

$$\begin{aligned} c &= 11 + 2n \\ &= 11 + 2(5) \\ &= 11 + 10 \\ &= 21 \end{aligned}$$

A five-topping pizza will cost \$21.
To check the answer, extend the table.
In the second column, the value of c increases by 2 each time.
So, when $n = 4, c = 19$; and when $n = 5, c = 21$.

c) To find out how many toppings are on a pizza that costs \$27, substitute $c = 27$.

$$27 = 11 + 2n$$

Solve for n.

$$\begin{aligned} 27 - 11 &= 11 + 2n - 11 \\ 16 &= 2n \\ \frac{16}{2} &= \frac{2n}{2} \\ 8 &= n \end{aligned}$$

There are 8 toppings on a pizza that costs \$27.

To check the answer, calculate the cost of a pizza with 8 toppings.
For 8 toppings, the cost is \$11 plus $8 \times \$2$, which is $\$11 + \$16 = \$27$.
Since this matches the given cost, the number of toppings is correct.

Example 2

The equation of a linear relation is: $y = -5x - 3$
Some ordered pairs in the relation are:
$(0, -3), (1, -8), (2, -13), (3, \ ), (4, -23), (\ , -28)$
Find the missing numbers in the ordered pairs.

▶ **A Solution**

The first missing number is in the ordered pair $(3, \ )$.
The missing number is the value of y when $x = 3$.
Substitute $x = 3$ in the equation $y = -5x - 3$.
$y = -5(3) - 3$
$ = -15 - 3$
$ = -18$
The ordered pair is $(3, -18)$.
The second missing number is in the ordered pair $(\ , -28)$.
The missing number is the value of x when $y = -28$.
Substitute $y = -28$ in the equation $y = -5x - 3$.
$ -28 = -5x - 3$
Solve for x.
$-28 + 3 = -5x - 3 + 3$
$ -25 = -5x$
$ \frac{-25}{-5} = \frac{-5x}{-5}$
$ 5 = x$
The ordered pair is $(5, -28)$.

▶ **Example 2**
Another Solution

To find the missing number in $(3, \ )$:
There is a pattern in the x-values.
So, list the y-values in order, and
look for a pattern in the y-values.
$-3, -8, -13, ?, -23, -28$
The numbers decrease by 5 each time.
So, the first missing number is: $-13 - 5 = -18$

To find the missing number in $(\ , -28)$:
The first ordered pair is $(0, -3)$.
-28 is in the 6th ordered pair.
So the missing number is 5.

Discuss the ideas

1. Look at the tables of values in *Connect*.
   Give some examples of pairs of numbers that will never appear in these tables.
2. Why do you think the numbers (4, 8) are called an ordered pair, and not simply a pair?
3. To check that a solution to an equation is correct, you can either extend the table of values or substitute in the equation.
   a) When would it be easier to extend the table of values?
   b) When would it be easier to substitute?

Practice

Check

4. Copy and complete each table of values.

   a) $y = x + 1$

x	y
1	
2	
3	
4	
5	

   b) $y = x + 3$

x	y
1	
2	
3	
4	
5	

   c) $y = 2x$

x	y
1	
2	
3	
4	
5	

5. Make a table of values for each relation.
   a) $y = 2x + 1$
   b) $y = 2x - 1$
   c) $y = -2x + 1$

6. The equation of a linear relation is:
   $y = 9x - 7$
   Some ordered pairs in the relation are:
   (0, −7), (1, 2), (2,   ), (3, 20),
   (4,   ), (, 38)
   Find the missing numbers in the ordered pairs.

7. Melanie earns $7 per hour when she baby-sits. An equation for this relation is $w = 7h$, where h represents the number of hours and w represents Melanie's wage in dollars.
   a) Use the equation to create a table of values.
   b) In one week, Melanie earned $105. How many hours did she baby-sit?
   c) In one month, Melanie baby-sat for 24 h. How much did she earn from baby-sitting in that month?

Apply

8. Copy and complete each table of values.

a) $y = x + 2$

x	y
−3	
−2	
−1	
0	
1	
2	
3	

b) $y = x − 3$

x	y
−3	
−2	
−1	
0	
1	
2	
3	

c) $y = x + 4$

x	y
−3	
−2	
−1	
0	
1	
2	
3	

9. Make a table of values for each relation.
 a) $y = −2x + 3$
 b) $y = −5x − 4$
 c) $y = 8x − 3$

10. The equation of a linear relation is:
$y = −3x + 5$
Some ordered pairs in the relation are:
(−3, 14), (−1, 8), (1,), (3, −4), (5,), (, −16)
Find the missing numbers in the ordered pairs.

11. The equation of a linear relation is:
$y = −2x + 7$
Find the missing number in each ordered pair.
 a) (−8,)   b) (12,)
 c) (, 31)   d) (, −23)

12. **Assessment Focus** Herbie has a mass of 100 kg.
His personal trainer sets a goal for him to lose 2 kg per month until he reaches his goal mass. An equation for this relation is $m = 100 − 2n$, where n represents the number of months and m represents Herbie's mass in kilograms.
 a) Use the equation to create a table of values.
 b) At some time, Herbie should have a mass of 60 kg. How many months will he have trained?
 c) By his birthday, Herbie had trained for 7 months. What was his mass then?

6.6 Creating a Table of Values

13. Candice is experimenting with divisibility rules. She can't remember the rule for "divisibility by 9", so she makes a table of values to study the multiples of 9.

She uses the equation $m = 9t$ to find multiples of 9.

  a) In the equation $m = 9t$, what does m represent?
  What does t represent?

  b) Make a table of values for this equation.

  c) What patterns do you see in the table?
  Use these patterns to write a rule for divisibility by 9.

  d) Is 126 divisible by 9? How do you know?

  e) What is the 17th multiple of 9? How do you know?

14. **Take It Further**
These ordered pairs are in the same linear relation:
(−2, −6), (0, 2), (2, 10), (4, 18)
The ordered pairs below are also in this relation. Find the missing number in each ordered pair.
Describe the strategy you used.

  a) (−4,)   b) (, −26)
  c) (3,)   d) (, −2)

15. **Take It Further**
These ordered pairs are in the same linear relation: (0, −8), (−4, −28), (4, 12)
The ordered pairs below are also in this relation. Find the missing number in each ordered pair.
Describe the strategy you used.

  a) (−2,)   b) (, −48)
  c) (6,)   d) (, −3)

Math Link

Science

Pressure is force per unit area.
Pressure is measured in pascals (Pa).
A formula for pressure is:
$$\text{Pressure} = \frac{\text{Force}}{\text{Area}}$$
When we know the pressure in pascals and the area in square metres, we can use this formula to find the force newtons (N).

Reflect

You have learned 2 ways to find the missing number in an ordered pair.
Which strategy do you prefer?
When might you use one strategy rather than the other?

6.7 Graphing Linear Relations

Focus Construct a graph from the equation of a linear relation and describe the graph.

Look at the coordinate grid.
Point A has coordinates (4, 2).
What are the coordinates of Point B? Point C? Point D?

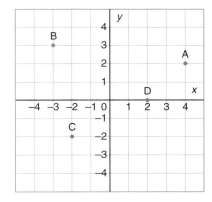

Investigate

Work with a partner.

Benny is designing banners for a school event.
On a banner, the school motto is 50 cm long
and the school logo is 20 cm long.

In each design, the banner has the school logo at each end, and the school motto in the middle. The motto can be repeated any number of times.

An equation that relates the length of the banner to the number of mottos is
$\ell = 40 + 50n$, where ℓ is the length of the banner in centimetres with n mottos.

➤ Explain each term in the equation of the relation.
  What does each term represent?
➤ Use the equation to make a table of values for the relation.
➤ Graph the relation.
➤ What can you find out from the table of values? From the graph?

Compare your table of values and graph with those of another pair of classmates.
Should you draw a line through the points? Justify your decision.
Is the relation linear? How can you tell?

Connect

Sylvia works at a garden nursery. She is paid $6 for every tray of tomatoes she plants. Let n represent the number of trays Sylvia plants. Let p represent her pay in dollars. An equation that relates Sylvia's pay to the number of trays she plants is: $p = 6n$

Substitute values for n to find corresponding values of p.
When $n = 0$, $p = 6(0)$
$ = 0$
When $n = 1$, $p = 6(1)$
$ = 6$

Here is a table of values.

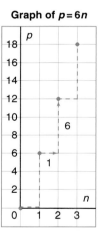

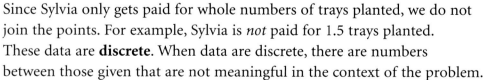

To graph the relation, plot n along the horizontal axis and p along the vertical axis.
Label the axes and write the equation of the relation on the graph.
The points lie on a straight line, so the relation is linear.
Since Sylvia only gets paid for whole numbers of trays planted, we do not join the points. For example, Sylvia is *not* paid for 1.5 trays planted. These data are **discrete**. When data are discrete, there are numbers between those given that are not meaningful in the context of the problem.

The graph shows that for every tray Sylvia plants, her pay increases by $6. As the number of trays increases, so does her pay.

Example 1

A Grade 8 class is going on a field trip. The bus seats 24 students. An equation that relates the number of boys on the bus to the number of girls is $b = 24 - g$, where g represents the number of girls and b represents the number of boys.

a) Create a table of values for the relation.
b) Graph the relation.
c) Describe the relationship between the variables in the graph.

▶ **A Solution**

a) Substitute values for g to find corresponding values of b.
When $g = 0, b = 24 - 0$ When $g = 1, b = 24 - 1$
         $= 24$ $= 23$

A table of values is:

g	b
0	24
1	23
2	22
3	21
4	20
...	...
24	0

b)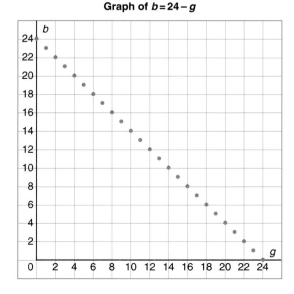

c) The variables represent the number of boys and the number of girls. As the number of girls increases by 1, the number of boys decreases by 1. The graph begins and ends at 24 on each axis. It is not possible to have more than either 24 boys or 24 girls on the bus.

Example 2

The equation of a linear relation is: $y = -4x + 1$

a) Create a table of values for the relation for integer values of x from -4 to 4.
b) Graph the relation.
c) Describe the relationship between the variables in the graph.

▶ **A Solution**

a) When $x = -4$,
$y = -4x + 1$
$= -4(-4) + 1$
$= 16 + 1$
$= 17$

When $x = -3$,
$y = -4x + 1$
$= -4(-3) + 1$
$= 12 + 1$
$= 13$

When $x = -2$,
$y = -4x + 1$
$= -4(-2) + 1$
$= 8 + 1$
$= 9$

A table of values is:

x	y
−4	17
−3	13
−2	9
−1	5
0	1
1	−3
2	−7
3	−11
4	−15

Each $+1$ increase in x corresponds to a -4 change in y.

b)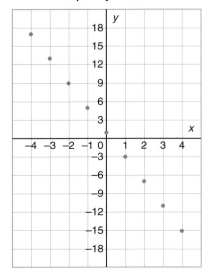

Graph of $y = -4x + 1$

c) The variables are x and y. When x increases by 1, y decreases by 4. The points lie on a line that goes down to the right.

Discuss the ideas

1. In *Example 1*, is it possible to have negative values for *b* and *g*? Justify your answer.
2. In *Example 1*, the relation is linear and the points lie on a line. Why did we not draw a line through the points?
3. In *Example 1*, the equation for the relation could have been written as $g = 24 - b$.
   How would the table of values change?
   How would the graph change?

Practice

Check

You will need grid paper.

4. Each graph below is a graph of a linear relation. Describe the relationship between the variables in each graph.
   a) $y = 4x - 1$ b) $y = -3x + 9$

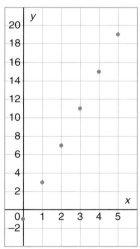
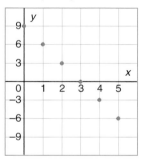

5. Graph each relation for integer values of *x* from 0 to 5.
   a) $y = 2x$ b) $y = 3x$
   c) $y = 4x$ d) $y = 5x$
   e) $y = -2x$ f) $y = -3x$
   g) $y = -4x$ h) $y = -5x$

6. Graph each relation for integer values of *x* from 0 to 5.
   a) $y = 2x + 1$ b) $y = 2x - 1$
   c) $y = -2x + 1$ d) $y = -2x - 1$
   e) $y = 3x + 1$ f) $y = 3x - 1$
   g) $y = -3x + 1$ h) $y = -3x - 1$

Apply

7. Here is a graph of the linear relation $y = 8x + 3$.

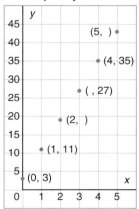

Each point on the graph is labelled with an ordered pair.
Some numbers in the ordered pairs are missing. Find the missing numbers. Explain how you did this.

6.7 Graphing Linear Relations   363

8. Here is a graph of the linear relation $y = -6x - 5$.

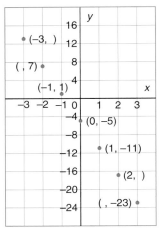
Graph of $y = -6x - 5$

Each point on the graph is labelled with an ordered pair.
Some numbers in the ordered pairs are missing. Find the missing numbers. Explain how you did this.

9. Use the data from *Example 1*, page 361.
An equation for the linear relation is: $c = 11 + 2n$,
where n is the number of toppings on the pizza, and c is the total cost of the pizza in dollars. Here is a table of values.

n	0	1	2	3	4	5	6	7	8
c	11	13	15	17	19	21	23	25	27

a) Construct a graph for the data.
b) Describe the relationship between the variables in the graph.
c) Find the ordered pair on the graph that shows the cost of a pizza with 6 toppings.

10. Use the data from Lesson 6.6 *Practice* question 12, page 357.
An equation for the linear relation is: $m = 100 - 2n$,
where n is the number of months that Herbie trains and m is his mass at any time in kilograms.
Here is a table of values.

n	0	2	4	6	8	10
m	100	96	92	88	84	80

a) Construct a graph for the data.
b) Describe the relationship between the variables in the graph.
c) Find the ordered pair on the graph that indicates Herbie's mass after 7 months. Explain how you did this.

11. **Assessment Focus** Regina plans a marshmallow roast. She will buy 8 marshmallows for each person who attends, and 12 extra marshmallows in case someone shows up unexpectedly. Let n represent the number of people who attend. Let m represent the number of marshmallows Regina must buy. An equation that relates the number of marshmallows to the number of people is: $m = 8n + 12$
a) Create a table of values for the relation.
b) Graph the relation.
c) Describe the relationship between the variables in the graph.
d) Is the relation linear? How do you know?

12. Graph each relation for integer values of x from -4 to 4.
a) $y = 8x + 2$ b) $y = -8x - 2$
c) $y = -7x + 4$ d) $y = 5x - 4$

13. Peter's Promoting is organizing a concert. The cost of the venue and the rock band is $15 000. Each concert ticket sells for $300. Peter's profit is the money he makes from selling tickets minus the cost. Let n represent the number of tickets sold. Let p represent Peter's profit. An equation that relates the profit to the number of tickets sold is:
$p = 300n - 15\ 000$
a) Create a table of values for the relation. Use these values of n: 10, 20, 30, 40, 50, 60, 70, 80
b) Graph the relation. What do negative values of p represent?
c) Describe the relationship between the variables in the graph.
d) How can you use the graph to find the profit when 75 tickets are sold?

14. Take It Further A computer repair company charges $60 to make a house call, plus an additional $40 for each hour spent repairing the computer. An equation that relates the total cost to the time in hours for a house call is $C = 60 + 40n$, where n represents the time in hours, and C represents the total cost of the house call in dollars.
a) Graph the relation.
b) Describe the relationship between the variables in the graph.
c) Does the point $(-1, 20)$ lie on the graph? What does this point represent? Does this point make sense in the context of the problem? Explain.

15. Take It Further
a) Graph each relation. Describe the relationship between the variables in the graph.
  i) $y = -9x + 4$ ii) $y = 6x - 3$
  iii) $y = -7x - 2$ iv) $y = 4x + 11$
  v) $y = 7x + 5$ vi) $y = 3x - 8$
  vii) $y = -9x - 6$ viii) $y = -8x + 7$
b) Which graphs go up to the right? Which graphs go down to the right?
c) How can you use the equation of a linear relation to tell if its graph goes up to the right or down to the right?

Reflect

You now know these ways to represent a relation:
• table of values   • equation   • graph
Which way do you think tells you the most about the relation?

Using Spreadsheets to Graph Linear Relations

Focus Use technology to construct a graph from the equation of a linear relation.

Spreadsheet software can be used to create a table of values from the equation of a linear relation.
The software can then be used to graph the data in the table of values.

Chris and her family belong to an outdoor club.
They are able to rent ATVs at reasonable rates.
It costs $55 to rent a "Rambler,"
plus $3 for each hour rented.
It costs $60 to rent a "Northern,"
plus $2 for each hour rented.
Chris wants to rent an ATV for a few hours over the weekend.
She writes these equations to help her decide which ATV to rent.

Let h represent the number of hours.
Let c represent the total cost of the rental in dollars.
An equation for the total cost to rent a "Rambler" is:
$c = 55 + 3h$
An equation for the total cost to rent a "Northern" is:
$c = 60 + 2h$

To create a table of values

➤ Open a spreadsheet program.
  Make a table of values for the "Rambler" relation using spreadsheet operations.

  This is what you might see:

A	B	C
h	c	
0	55	
1	58	
2	61	
3	64	
4	67	
5	70	
6	73	
7	76	
8	79	
9	82	
10	85	

366 UNIT 6: Linear Equations and Graphing

To graph the relation

➤ Highlight the data.
Click the graph/chart icon.
Select *XY (Scatter)*.
Label the graph and the axes.
Your graph may look similar to this:

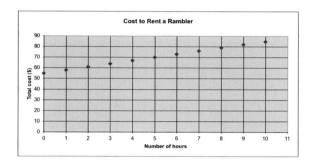

The points lie on a line so the graph represents a linear relation.
When the input increases by 1, the output increases by 3.
The graph goes up to the right.
This is because the total cost increases for every hour the "Rambler" is rented.

Check

1. Create a table of values for the "Northern" ATV.
   Use the table of values to graph the relation.
   Describe the relationship between the two variables in the graph.

2. Suppose Chris likes both ATVs equally.
   What conclusions can you make from the tables or the graphs to help Chris save money?

Choosing a Strategy

Have you ever solved a problem, looked back, and realized you could have solved the problem a different way? There are many different strategies for solving problems.

Try to solve this problem in at least two different ways.

The intramural dodgeball league at your school has 10 teams. Each team must play every other team exactly once. How many games need to be scheduled?

Try these Steps

What information are you given in the problem?
Is there any information that is not needed?
What are you asked to find?
Is an estimate okay or do you need an exact answer?

What strategies might work for this problem?

Try the strategy you think will work best.
If you have trouble solving the problem, try a different strategy.
You might have to try 3 or 4 strategies.

Have you answered the question?
Does your answer seem reasonable?
How do you know you have found all the answers?

Strategies

- **Make a table.**
- **Use a model.**
- **Draw a diagram.**
- **Solve a simpler problem.**
- **Work backward.**
- **Guess and test.**
- **Make an organized list.**
- **Use a pattern.**
- **Draw a graph.**
- **Use logical reasoning.**

Strategies for Success

Use at least two different strategies to solve each problem.

1. A rectangular field has length 550 m and width 210 m.
   Fence posts are placed 10 m apart along the perimeter of the field, with one post in each corner.
   How many fence posts are needed?

2. A basketball tournament starts at 10:00 a.m. The winners play every 1.5 h and the losers are eliminated. The winning team finishes their last game at 4:00 p.m. How many teams are in the tournament?

3. Marsha and Ivan have money to spend at the Raven Mad Days Celebration in Yellowknife.
   If Marsha gives Ivan $5, each person will have the same amount.
   If, instead, Ivan gives Marsha $5, Marsha will have twice as much as Ivan.
   How much money does each person have?

4. Briony wants to print copies of her new brochure.
   The local print shop charges 15¢ a copy for the first 25 copies,
   12¢ a copy for the next 50 copies, and 8¢ a copy for any additional copies.
   How much would Briony pay for each number of copies?
   a) 60 copies   b) 240 copies

5. How many different necklaces can you make with:
   a) one red bead, one yellow bead, and one green bead?
   b) two red beads, one yellow bead, and one green bead?
   Justify your answers.

Unit Review

What Do I Need to Know?

✓ **Distributive Property**

The product of a number and the sum of two numbers can be written as a sum of two products:
$a(b + c) = ab + ac$

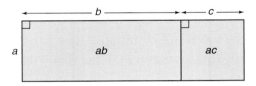

The distributive property can be used to solve some algebraic equations.

✓ We can represent a linear relation in different ways:
- as a two-variable equation
- as a table of values
- as a graph

$y = -2x + 5$

x	y
−3	11
−2	9
−1	7
0	5
1	3
2	1
3	−1

Graph of $y = -2x + 5$

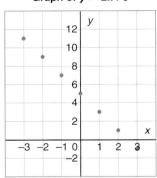

A related pair of x and y values is an *ordered pair*.

When data are *discrete*, the numbers between those given do not make sense in the context of the problem.

When the graph of a relation is a straight line, the relation is *linear*.

370   UNIT 6: Linear Equations and Graphing

What Should I Be Able to Do?

LESSON
6.1

1. Tracey put one coin in the candy machine, and 7 candies were dispensed.
   When she put 2 coins in the machine, 14 candies were dispensed.
   How many coins would Tracey have to put in to have 56 candies dispensed?
   a) Choose a variable to represent the number of coins Tracey would have to put in. Write an equation to describe the situation.
   b) Use a model to solve the equation.
   c) Verify the solution.

2. Use a model to solve each equation. Verify the solution.
   a) $-6x = -30$
   b) $11x = -22$
   c) $4x + 1 = 17$
   d) $-4x + 1 = 17$
   e) $11 = 3x - 7$
   f) $-11 = 3x + 7$

6.2

3. Troy is taking care of Mr. Green's property. He is paid $8 for mowing the lawn, and $3 for each garden that he weeds. On Saturday, Mr. Green paid Troy $29.
   How many gardens did Troy weed on Saturday?
   a) Write an equation to represent this problem.
   b) Solve the equation.
   c) Verify the solution.

4. Solve each equation. Verify the solution.
   a) $-6x + 8 = -16$
   b) $11x - 5 = 28$
   c) $3x + 5 = 7$
   d) $-4x - 9 = 19$
   e) $13 = 3x + 12$
   f) $7x - 19 = -5$

5. The high temperature today is 6°C higher than three times the high temperature yesterday. The high temperature today is 3°C. What was the high temperature yesterday?
   a) Write an equation you can use to solve the problem.
   b) Solve the equation.
   c) Verify the solution.

6.3

6. Solve each equation. Verify the solution.
   a) $\frac{p}{-4} + 11 = 8$
   b) $\frac{t}{10} + 12 = 3$
   c) $-21 + \frac{w}{6} = -30$
   d) $-12 + \frac{e}{-8} = -1$

7. Check this student's work. Rewrite a correct and complete algebraic solution if necessary.
   $$\frac{h}{-2} = -7$$
   $$\frac{h}{-2} \times 2 = -7 \times 2$$
   $$\frac{2h}{-2} = -14$$
   $$h = -14$$

LESSON

8. One-fifth of the fish in the lake are pickerel. There are 52 pickerel in the lake. How many fish are in the lake?
   a) Write an equation you can use to solve the problem.
   b) Solve the equation.
   c) Verify the solution.

6.4

9. Draw a rectangle or algebra tiles to show that $3(x + 4)$ and $3x + 12$ are equivalent.
   Explain your diagram in words.

10. Expand.
    a) $6(x + 9)$
    b) $3(11 - 4c)$
    c) $5(-7s + 5)$
    d) $-4(3a - 2)$

11. Which expression is equal to $-5(-t + 4)$? How do you know?
    a) $-5t - 20$
    b) $5t - 20$
    c) $5t + 20$

6.5

12. Solve each equation using the distributive property.
    Verify the solution.
    a) $7(x + 2) = 35$
    b) $5(b - 6) = 30$
    c) $-3(p - 9) = -24$
    d) $5(s - 3) = 9$

13. Martina chose an integer.
    She subtracted 7, then multiplied the difference by -4. The product was 36. Which integer did Martina choose?
    a) Write an equation you can use to solve the problem.
    b) Solve the problem.
    c) Verify the solution.

14. Chas used the distributive property to solve this equation: $-2(c - 5) = 28$
    a) Check Chas' work.
       Is his solution correct?
       $$-2(c - 5) = 28$$
       $$-2(c) - 2(-5) = 28$$
       $$-2c - 10 = 28$$
       $$-2c - 10 + 10 = 28 + 10$$
       $$-2c = 38$$
       $$\frac{-2c}{-2} = \frac{38}{-2}$$
       $$c = -19$$
    b) If your answer to part a is yes, verify the solution.
       If your answer to part a is no, describe the error, then correct it.

6.6

15. Copy and complete each table of values.
    a) $y = x - 8$
    b) $y = -x + 5$

x	-3	-2	-1	0	1	2	3
y							

x	-3	-2	-1	0	1	2	3
y							

16. Lauree is making friendship bracelets. She needs 6 strands of yarn for each bracelet.
    An equation for this relation is $s = 6n$, where n represents the number of bracelets, and s represents the number of strands of yarn needed.

LESSON

a) Use the equation to create a table of values.
b) Suppose Lauree makes 7 bracelets. How many strands of yarn does she need?
c) Suppose Lauree has 66 strands of yarn.
   How many bracelets can she make?
d) Yarn comes in packages of 20 strands. How many packages of yarn will Lauree need to make 18 bracelets? Explain your answer.

6.7
17. Use the data from question 16.
   a) Construct a graph for the data.
   b) Describe the relationship between the variables in the graph.
   c) Find the ordered pair on the graph that shows how many bracelets can be made with 54 strands of yarn.

6.6
18. The equation of a linear relation is:
    $y = -7x + 4$
    Find the missing number in each ordered pair.
    a) $(-2, \ )$ b) $(\ , -17)$
    c) $(8, \ )$ d) $(\ , 4)$

19. Francis sells memberships to a local health club. He is paid $200 per week, plus $40 for each membership he sells. An equation for this relation is $p = 200 + 40n$, where n represents the number of memberships Francis sells, and p represents his pay in dollars.
    a) Use the equation to create a table of values.
    b) One week, Francis sold 9 memberships. What was his pay for that week?
    c) One week, Francis was paid $480. How many memberships did he sell that week?

6.7
20. Use the data from question 19.
    a) Construct a graph for the data.
    b) Describe the relationship between the variables in the graph.
    c) Find the ordered pair on the graph that shows Francis' pay when he sells 5 memberships.

21. Graph each relation for integer values of x from -3 to 3.
    a) $y = -6x$ b) $y = -4x + 3$
    c) $y = 7x - 1$ d) $y = 9x + 8$

22. Here is a graph of the linear relation: $y = -x + 7$. Each point on the graph is labelled with an ordered pair. Some numbers in the ordered pairs are missing. Find the missing numbers. Explain how you did this.

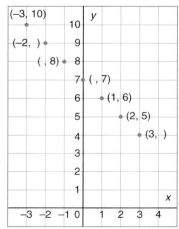

Graph of $y = -x + 7$

Practice Test

1. a) Use a model to solve this equation:
   $-6s + 5 = -7$
   Draw a picture to show the steps you took to solve the equation.
   b) Verify the solution. Show how you did this.

2. Blair is using algebra tiles to model and solve equations.
   Here is Blair's work for one question.

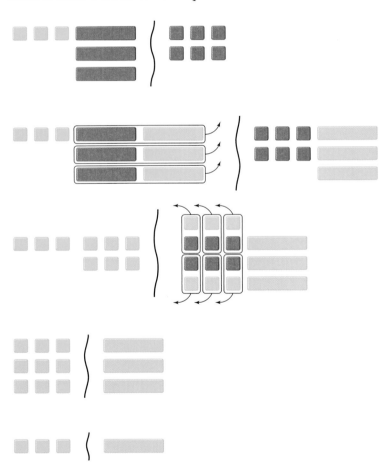

   a) Use algebra to record each step Blair took.
   b) Verify the solution in two different ways.

3. Use grid paper.
   a) Draw a picture to show that $4(x + 3)$ and $4x + 12$ are equivalent.
   b) Could you draw a picture to show that $-4(x + 3)$ and $-4x - 12$ are equivalent? Why or why not?

4. Solve each equation. Verify the solution.
   a) $5(x - 3) = 45$
   b) $\frac{n}{7} + 7 = 4$
   c) $\frac{p}{-6} = -7$
   d) $12x - 23 = 73$

5. A school soccer team rented a bus for the day. The bus cost $200 for the day, plus $14 for each person on the bus. The total cost of the bus rental was $424. How many people were on the bus?
   a) Write an equation you can use to solve the problem.
   b) Solve the equation.
   c) Check your answer. Write to explain how you know it is correct.

6. Create a table of values for each relation.
   a) $y = -6x + 1$
   b) $y = 7x - 4$

7. A chocolate bar is divided into 10 equal pieces. Both Andy and Greg want some. Let a represent the number of pieces Andy gets. Let g represent the number of pieces Greg gets. An equation that relates the number of pieces Andy gets to the number of pieces Greg gets is: $g = 10 - a$
   a) Create a table of values for the relation.
   b) Graph the relation.
   c) Describe the relationship between the variables in the graph.
   d) Which ordered pair shows the fairest way to divide the chocolate bar? Justify your answer. Identify the point on the graph.

8. The equation of a linear relation is: $y = -5x + 2$
   Find the missing number in each ordered pair. Explain how you did this.
   a) (, 2)
   b) (−4,)
   c) (, −13)
   d) (8,)

Unit Problem: Planning a Ski Trip

In 1988, the alpine events of the Winter Olympic Games were held at Nakiska Mountain, just west of Calgary. The Grade 8 students are planning a ski trip to Nakiska. An organizing committee has been formed to research the trip.

Part 1

Two local bus companies, Company A and Company B, offer packages for school trips.
The amount each company charges is given by these equations:
Company A: $C = \$300 + \$55n$
Company B: $C = \$100 + \$75n$
where n is the number of people on the bus and C is the total cost in dollars.
Suppose 85 students and 10 adults go on the trip.
Which company should the committee choose?
What strategy did you use to find out?
Justify your answer.

Part 2

The daily cost per person for ski rental is $23.
On one day, the total cost of ski rental was $851.
Write an equation you can use to find out how many students rented skis that day.
Solve the equation. Verify the solution.

Part 3

When you are skiing, you must be aware of extreme temperatures. Nakiska Mountain has an elevation of about 2250 m.
An equation for calculating the temperature at this elevation is:
$T = c - 15$, where c is the temperature in degrees Celsius at sea level, and T is the temperature at an elevation of 2250 m.
a) Suppose the temperature at sea level is 0°C.
   What is the temperature at the peak of Nakiska Mountain?
b) Suppose the temperature at the peak of the mountain is −23°C.
   What is the temperature at sea level?

Part 4

The hotel in Calgary is offering a group discount.
It charges a daily rate of $1500, plus $30 per person.
An equation for this relation is $h = 1500 + 30p$, where h represents the total cost of the hotel per day in dollars, and p represents the number of people.
a) Create a table of values for the relation.
   Use $p = 0, 10, 20, 30, 40, 50, 60, 70, 80, 90, 100$.
b) Graph the relation.
c) Describe the relationship between the variables in the graph.
d) Use the graph to find the total cost of the hotel when 95 people go on the trip.
   What does each person have to pay for the hotel?
   How did you find out?

Check List

Your work should show:
- ✓ an accurate graph and all calculations in detail
- ✓ your understanding of linear equations and linear relations
- ✓ the strategies you used to solve the problems
- ✓ a clear explanation of the thinking behind your solutions

Reflect on Your Learning

Write about equations and linear relations and how variables are used in both.
Include examples in your explanation.

Units 1–6 Cumulative Review

UNIT 1

1. Find.
   a) $\sqrt{1}$ b) 16^2
   c) $\sqrt{43}$ d) the square of 11

2. Copy each square on 1-cm grid paper. Find its area, then write the side length of the square.
   a)   b)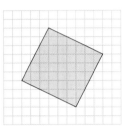

UNIT 2

3. A watch loses 3 s per hour for 24 h. Use integers to find the total number of seconds the watch lost over 24 h.

4. Evaluate.
   a) $(-15) \times (+4)$ b) $(-8) \times (-10)$
   c) $(+57) \div (-3)$ d) $\frac{-38}{+2}$

UNIT 3

5. Use an area model to find each product.
   a) $\frac{5}{9} \times \frac{3}{4}$ b) $2\frac{2}{3} \times \frac{3}{5}$
   c) $\frac{21}{5} \times \frac{4}{3}$ d) $1\frac{1}{4} \times 3\frac{5}{6}$

6. A dental hygienist takes $1\frac{1}{6}$ h to clean a patient's teeth.
   a) Estimate the number of patients the hygienist can see in $5\frac{1}{4}$ h.
   b) Calculate the number of patients the hygienist can see in $5\frac{1}{4}$ h.
   c) What assumptions do you make?

UNIT 4

7. a) Predict the object this net will form.
   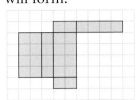
   b) Fold the net to verify your prediction.
   c) Describe the object.

8. Find the surface area of the object.
   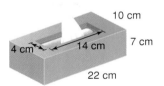

9. A triangular prism has a base that is a right triangle. The side lengths of the triangle are 5 cm, 12 cm, and 13 cm. The prism is 20 cm high.
   a) What is the volume of the prism?
   b) Sketch a net for the prism. Label the net with the dimensions of the prism.
   c) What is the surface area of the prism?

UNIT 5

10. There were 25 people in Sebastian's Tae Kwon Doe class. Then, three people dropped out.
    Write this decrease as a percent. Illustrate the percent on a number line.

11. a) Write each ratio in simplest form.

   i) red squares to blue squares
   ii) blue squares to green squares
   iii) red squares and blue squares to total number of squares
   iv) green squares to blue squares to red squares

   b) Suppose the grid is increased to a rectangle measuring 9 units by 4 units. The ratios of the colours remain the same. How many red squares will there be in the new rectangle?

12. Which is the better buy?
   a) 2 rolls of paper towel for $0.99 or 12 rolls of paper towel for $5.59
   b) 500 mL of mouthwash for $3.99 or 125 mL of mouthwash for $1.29

13. The price of a set of headphones in Prince Albert, Saskatchewan, is $34.99. This is a discount of 25%.
   a) What is the regular price of the headphones?
   b) What is the sale price of the headphones including taxes?

14. Solve each equation. Verify the solution.
   a) $-5x - 7 = 18$ **b)** $4(s - 6) = -52$
   c) $\frac{t}{-12} = -8$ **d)** $\frac{f}{8} + 9 = 4$

15. Write two expressions for the area of the shaded rectangle.

16. Rashad chose an integer. He added 11, then multiplied the sum by –2. The product was –4. Which integer did Rashad choose?
   a) Write an equation you can use to solve the problem.
   b) Solve the problem. Verify the solution.

17. The equation of a linear relation is $y = -9x + 5$. Find the missing number in each ordered pair.
   a) (2,)   **b)** (, 5)
   c) (–3,)   **d)** (, –31)

18. Cecilia rides an ice-cream bicycle. She charges $50 to go to an event, plus $3 for each ice-cream bar she sells. An equation for the relation is $p = 50 + 3n$, where n represents the number of ice-cream bars Cecilia sells, and p represents her pay in dollars.
   a) Use the equation to create a table of values.
   b) One day, Cecilia sold 45 ice-cream bars. What was her pay for that day?
   c) One day, Cecilia was paid $260. How many ice-cream bars did she sell that day?

Cumulative Review

UNIT 7
Data Analysis and Probability

Spencer, Jenny, and Simrit are working on a project on nutrition. They look at the nutrition facts label on a box of breakfast cereal.
- Which type of graph would best display the nutrition facts?
- How could you use a graph to make the cereal appear more nutritious than it really is?
- Who might want to make the cereal appear more nutritious than it is?

What You'll Learn
- Critique ways in which data are presented.
- Solve problems that involve the probability of independent events.

Why It's Important
Graphs are pictures of data. Knowing the best way to present data can help you communicate your ideas. In the media, you hear and read statements about the probability of everyday events. To make sense of these statements, you need to understand probability.

Key Words

- circle graph
- line graph
- bar graph
- double bar graph
- discrete data
- pictograph
- outcome
- event
- independent events
- probability of an event

Nutrition Facts/Valeur nutritive		
Serving 1 cup (55 g) / Portion de 1 tasse (55 g)		
Amount per serving	Cereal	With $\frac{1}{2}$ cup 2% milk Avec $\frac{1}{2}$ tasse de lait 2%
Calories / Calories	185	260
	% Daily Value / % valeur quotidienne	
Fat / Lipides 1 g		
Saturated / satures 0 g + Trans / trans 0 g	0%	8%
Cholesterol / Cholesterol 0 mg	0%	3%
Sodium / Sodium 345 mg	14%	17%
Potassium / Potassium 315 mg	9%	15%
Carbohydrate / Glucides 45 g	15%	17%
Fibre / Fibres 7 g	25%	25%
Sugars / Sucres 16 g		
Starch / Amidon 22 g		
Protein / Proteines 4 g		

"How could this information be displayed differently?"

"I know. We could draw a circle graph or a bar graph. Which one do you think would be better?"

"I don't know. I just want to know my chances of winning this game on the back of the cereal box."

7.1 Choosing an Appropriate Graph

Focus Identify the advantages and disadvantages of different types of graphs.

Investigate

Work with a partner.
All these graphs represent the same data.

Graph A

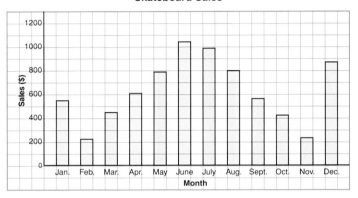

Graph B

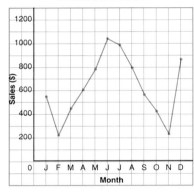

Graph C

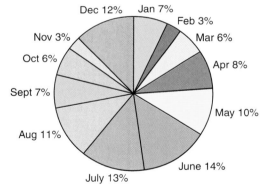

What do you know from each graph?
Which graph is most helpful in answering each question below?
- Which two months had the greatest skateboard sales?
- What is the range in skateboard sales?
- What percent of total skateboard sales occurred in May?

Compare your answers with those of another pair of classmates.
What are the advantages and disadvantages of each type of graph?
What can you find out from the bar graph or line graph that you cannot find out from the circle graph?

Connect

Each type of graph has its strengths and limitations.

Line Graphs

A **line graph** displays data that change over time.
On a line graph, line segments join adjacent data points.

A line graph:
• is easy to draw and to read
• can have a zigzag symbol on the vertical axis when the data start at a large number
• is best used to show data gathered over time
• can be used to estimate values between data points and beyond data points. (This should be done carefully as the trend may not continue.)

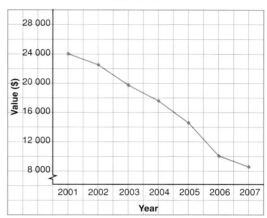

Double Bar Graphs

A **double bar graph** displays two sets of data that can be counted.
The lengths of the bars are used to represent and to compare data.

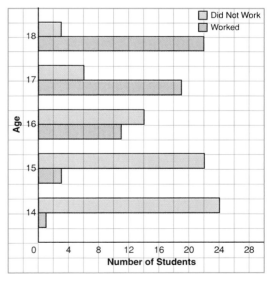

A double bar graph:
• is easy to draw and to read
• can be used to directly compare two sets of data
• can only be used to show discrete data
• may be difficult to read accurately depending on the scale used

Discrete data can be counted.

7.1 Choosing an Appropriate Graph 383

Example 1

Three students surveyed Grade 8 students in their school.
They asked: "How many times did you use a vending machine last week:
0 times, 1–3 times, 4–9 times, or 10 or more times?"
Amrit displayed the results on a circle graph.
Fred used a bar graph. Stella used a pictograph.
a) What are the strengths and limitations of each graph?
b) Which graph is appropriate? Justify your answer.

Grade 8 Students' Vending Machine Survey
- 10 or more times, 11%
- 4–9 times, 13%
- 0 times, 35%
- 1–3 times, 41%

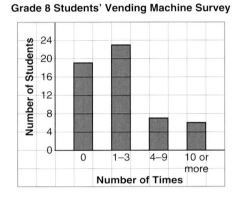

Grade 8 Students' Vending Machine Survey

Key: = 2 students

▶ **A Solution**

a)

Type of Graph	Strengths	Limitations
Circle Graph	• Shows parts of the whole. • Each response is shown as a percent of the number of students who responded. • The sizes of the sectors can be used to compare responses.	• The graph does not show the number of students who chose each response. • The total number of students cannot be calculated. • Difficult to draw accurately
Bar Graph	• The heights of the bars can be used to compare responses. • The scale on the vertical axis is: 1 grid square represents 4 students. The scale can be used to calculate the total number of students who responded. • Easy to draw	• Some people may find this graph difficult to read accurately because none of the bars end on a grid line. • The graph does not show the percent of the students who chose each response.
Pictograph	• The lengths of the rows of symbols give immediate comparison of responses. • The graph is visually appealing. • The key is 1 symbol equals 2 students. The key can be used to calculate the total number of students who responded. One-half of a symbol represents 1 student.	• There are a lot of symbols on the graph. For example, there are $11\frac{1}{2}$ symbols for 1–3 times. This makes the graph more difficult to read. • It might be difficult to draw so many vending machine symbols all of the same size. • The graph does not show the percent of the students who chose each response.

b) To decide which graph is appropriate, we need to know what the students want to display.

For example, if they want to display the fraction of Grade 8 students who did not use a vending machine, then the circle graph is appropriate. The size of that sector can be compared to the whole graph.

If the students want to display the number of students who did not use a vending machine, then the bar graph or pictograph is appropriate. The height of the bar or the key can be used to find the number of students.

Example 2

This table shows the favourite types of video games of the Grade 8 students at L'ecole Orléans.

Type	Number of Students
Action	15
Role Playing	10
Arcade	4
Strategy	7
Simulation	11
Other	3

a) Graph these data. Justify your choice of graph.
b) What are the advantages and disadvantages of the graph you drew?

▶ A Solution

a) The circle graph can display these data. The total number of students is the whole. Each sector represents the percent of students who chose each type of game.

Type	Number of Students	Fraction	Percent (%)
Action	15	$\frac{15}{50} = \frac{3}{10}$	30
Role Playing	10	$\frac{10}{50} = \frac{1}{5}$	20
Arcade	4	$\frac{4}{50} = \frac{2}{25}$	8
Strategy	7	$\frac{7}{50}$	14
Simulation	11	$\frac{11}{50}$	22
Other	3	$\frac{3}{50}$	6

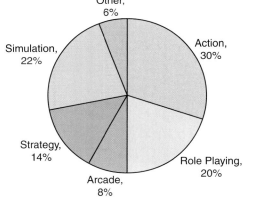

Favourite Types of Video Games of Grade 8 Students

b) **Advantages:** The circle graph shows the percent of Grade 8 students who chose each type of video game. The size of each sector can be compared to the whole and to other sectors to make conclusions; such as, action video games were most popular with these Grade 8 students.

Disadvantages: The circle graph does not show the number of students who chose each type of game and the number of students who were surveyed.
In a circle graph, the original data are lost.
A circle graph may be difficult to draw accurately because some calculations may involve approximations.

Discuss the ideas

1. Can more than one type of graph be appropriate to display a set of data?
2. In *Example 2*, which other types of graphs could you draw?

Practice

Check

3. Each graph below shows how much time Canadians spend watching TV each week.

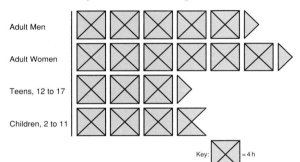

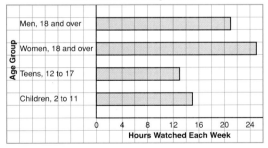

a) List 3 things you know from the bar graph.
b) List 3 things you know from the pictograph.
c) Which graph is more appropriate to display the data? Justify your choice.

4. Each graph below shows the number of times students in a Grade 8 class littered last week.

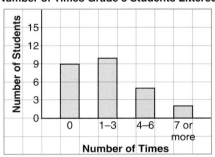

a) What are the strengths of each graph?
b) What are the limitations of each graph?
c) Which graph is more appropriate to display these data? Justify your choice.
d) Should you use a line graph or a circle graph to display these data? Why or why not?

7.1 Choosing an Appropriate Graph **387**

Apply

5. These graphs show the final grades for Mr. Sidley's Grade 8 math class.

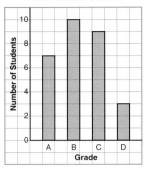
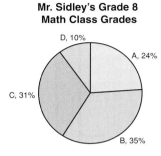

a) List 3 things you know from the bar graph.
b) List 3 things you know from the circle graph.
c) Which graph best shows the number of students who got B as a final grade? Justify your choice.
d) These graphs show the final grades for Ms. Taylor's Grade 8 math class.

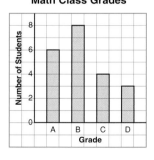
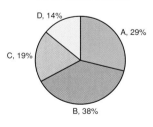

Which graphs should Mr. Sidley use to show his class has the higher grades? Explain.

e) Which class do you think did better? Why do you think so?

6. a) What data does each graph below show?

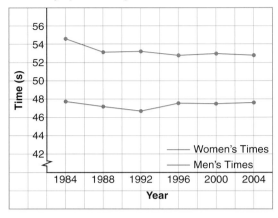
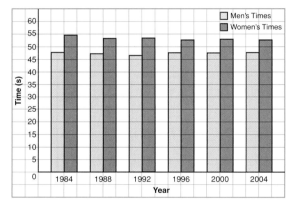

b) What is an advantage of each graph?
c) What is a disadvantage of each graph?
d) Which graph would you choose in each case? Explain your choice.
   i) You want to show how the times changed over time.
   ii) You want to show the differences between the times for each year.

7. Describe data that could be best represented by each graph below. Explain why you chose each type of data.
   a) line graph   b) bar graph
   c) double bar graph   d) pictograph
   e) circle graph

8. **Assessment Focus** Nina owns a shoe store. These tables show data about the shoe store.

Table A

Sizes of Shoes Sold in May

Size	Number of Pairs Sold
6	60
7	239
8	217
9	156
10	61
11	43
12	36

Table B

Yearly Sales

Year	Sales ($)
2000	579 000
2001	621 000
2002	598 000
2003	634 000
2004	657 000
2005	642 000
2006	675 000

a) Which data would be best represented with a line graph? Justify your choice.
b) Which type of graph would be suitable for the other table? Explain.

9. Each graph shows the mean home-game attendance of the school's soccer team over the past 5 years.

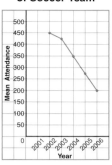
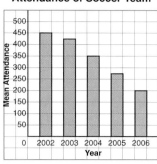

a) What are the strengths of each graph?
b) What are the limitations of each graph?
c) Which type of graph is more appropriate to display these data? Justify your choice.
d) Could you use a circle graph to display these data? Why or why not?

10. This table shows the number of people employed by the construction industry in Canada from 2002 to 2006.
a) Graph these data. Justify your choice of graph.
b) What are the advantages and disadvantages of the graph you drew?

Number of People Employed in the Construction Industry

Year	Number of People (thousands)
2002	865.2
2003	906.0
2004	951.7
2005	1019.5
2006	1069.7

11. This table shows the Canadian Aboriginal population, by province and territory, in 2001.

Canadian Aboriginal Population, 2001	
Province or Territory	Population
Newfoundland and Labrador	18 780
Prince Edward Island	1 345
Nova Scotia	17 015
New Brunswick	16 990
Quebec	79 400
Ontario	188 315
Manitoba	150 040
Saskatchewan	130 185
Alberta	156 220
British Columbia	170 025
Yukon	6 540
Northwest Territories	18 730
Nunavut	22 720

a) Graph these data. Justify your choice of graph.
b) What are the advantages and disadvantages of the graph you drew?

12. Take It Further Over a 2-month period, Dinah collected these data about her family.
 i) How many times each week her dad fell asleep while watching television
 ii) How many times her dad cooked dinner and how many times her mom cooked dinner each week
 iii) How many hours a day her brother spent playing video games, doing homework, chatting on-line, doing chores, and eating
 iv) The weekly height of a tomato plant in the garden

Dinah wants to display these data for a school project. Which type of graph would you suggest for each data set? Justify your choices.

13. Take It Further Madan measured the mass of his guinea pig every 5 months, until the pet was 25 months old. The data are shown in the table.

Age (months)	Mass (g)
5	200
10	350
15	480
20	510
25	520

a) Graph these data. Justify your choice of graph.
b) What are the advantages and disadvantages of the graph you drew?
c) Use your graph to predict the mass of the guinea pig at 8 months and at 30 months.

Reflect

Suppose you are given a data set.
How do you choose the most appropriate type of graph to illustrate these data?

Using Spreadsheets to Record and Graph Data

Focus Display data on graphs using spreadsheets.

Spreadsheets can be used to record and graph data. We can use a spreadsheet to generate different types of graphs for a set of data.

Week	Amount ($)
July 12	96
July 19	112
July 26	72
Aug. 2	120
Aug. 9	160
Aug. 16	128

Carissa had a part-time job cutting grass in the summer. She opened a bank account to save all the money she earned. This table shows Carissa's weekly deposits.

➤ Enter the data into rows and columns of a spreadsheet.
   Highlight the data. Include the column heads.
   Click the graph/chart icon. Select the *Vertical bar graph*.
   Label the graph and the axes.
   Your graph may look similar to this:

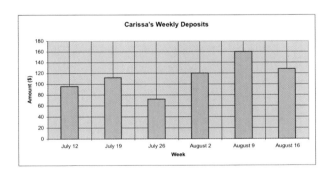

The bar for August 9 is the tallest. This shows Carissa deposited more that day than any other.

➤ Use the same data.
   Highlight the data. Include the column heads.
   Click the graph/chart icon. Select the *Line graph*.
   Label the graph and the axes.
   Your graph may look similar to this:

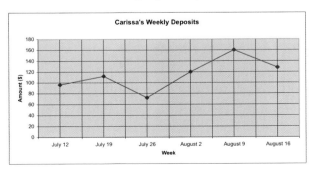

Generally, the line segments go up to the right. So, Carissa's deposits increased over time.

Technology: Using Spreadsheets to Record and Graph Data

In most spreadsheet programs, circle graphs are called pie charts.

➤ Use the same data.
Highlight the data. Do not include the column heads.
Click the graph/chart icon. Select the *Pie chart*.
Label the graph.
Your graph may look similar to this:

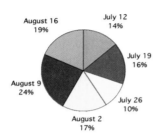

Carissa's Weekly Deposits

From the circle graph, we know that the greatest amount Carissa deposited in a week is 24% of the total amount she deposited.

What are the strengths of each graph?
What are the limitations of each graph?
Which graph is most appropriate to display these data?
Justify your choice by suggesting what could be found out from each graph.

Check

1. The table shows the places where some Canadians met their best friends.

Place	Number of People
School	5700
Work	4100
Club or organization	1400
Religious organization	700
Home/neighbourhood	4300
Through family	1200
Through friend	1100
Other	600

a) Use a spreadsheet to draw a bar graph and a circle graph.
b) Which graph represents these data better? Justify your answer.

2. a) Use a spreadsheet.
Create one graph to display these data.
b) Which types of graphs would you not use to represent these data? Explain why you would not use these graphs.

Average Hours of Television Viewed per Week by Adolescents Aged 12–17 years, Fall 2004

Province	Number of Hours
Newfoundland/Labrador	12.3
Prince Edward Island	12.3
Nova Scotia	13.8
New Brunswick	12.6
Quebec	13.5

3. a) Use a spreadsheet.
Draw two different graphs to display the data in the table.
b) Which type of graph represents these data better? Justify your choice.
c) Which type of graph would you not use to represent these data?
Explain why you would not use that graph.

Average Value of US$1

Year	Value in C$
2002	1.570
2003	1.401
2004	1.302
2005	1.212
2006	1.134

7.2 Misrepresenting Data

Focus Examine how data may be misinterpreted.

There are many ways to represent data.

Investigate

Work on your own.
What data do the two graphs below show?
How are the graphs similar? How are they different?
Explain.

Graph A

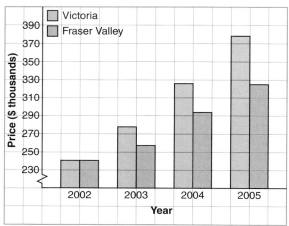

Graph B

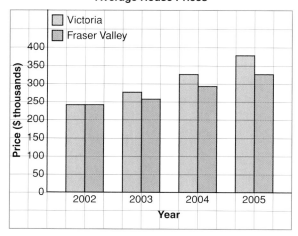

Discuss with a classmate:
- What impression does each graph create?
- Who might want to use each graph?

Connect

Different formats of a graph may lead to misinterpretation of data.

394 UNIT 7: Data Analysis and Probability

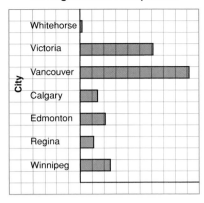

This bar graph is misleading.
It suggests that Vancouver has more than 30 times as much precipitation as Whitehorse. This graph has no measurements of the amount of precipitation.

Visually, this graph shows the same information as the first graph. However, the horizontal scale is labelled with the amount of precipitation, and the scale does not start at 0.

The scale shows that Vancouver has about 1170 mm of precipitation and Whitehorse has about 270 mm.

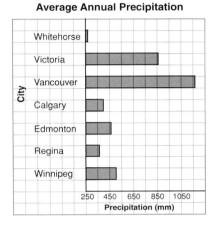

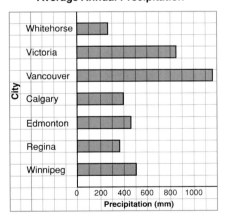

This graph accurately shows the data. The horizontal scale starts at 0.

The lengths of the bars are shown in the correct ratio. Vancouver has between 4 and 5 times as much precipitation as Whitehorse.

There are many ways in which graphs can be drawn to **misrepresent data**. Graphs like these may be found in the media to create false impressions.

▶ In this bar graph, the wider bar creates the impression that many more boys than girls scored higher than 80%. In fact, the number of girls who scored higher than 80% is greater than the number of boys.

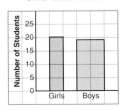

7.2 Misrepresenting Data

➤ In this pictograph, the symbols have different sizes.
The three large ice-cream cone symbols give the impression that bubble gum is the favourite flavour.
When the key is used, chocolate is the favourite flavour.

➤ In the bar graph below left, the scale on the vertical axis is 1 square represents 3 students.
The differences among the heights of the bars are easily seen.
In the bar graph below right, the scale on the vertical axis is 1 square represents 10 students.
This change in scale makes the differences among the heights of the bars less evident.

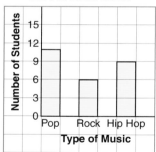 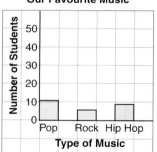

➤ A part of a graph may be treated differently to draw attention to it.
A milk company uses this circle graph to draw attention to the milk sector.
The sector for milk is not as large as the sector for water, but the special treatment makes it seem larger.

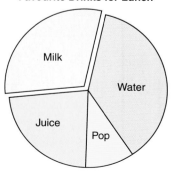

Example 1

From this line graph, Shiva made the conclusion that salaries have almost tripled in 6 years.

a) Shiva's conclusion is not consistent with the data. Explain her misinterpretation.
b) What changes should be made to the graph to accurately show how salaries have changed in 6 years?

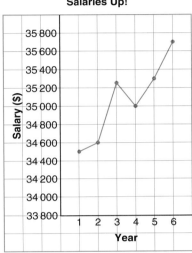

A Solution

a) The scale on the vertical axis does not start at 0. The upward trend of the graph suggests a rapid rise in salaries. This is not so.
The increase in salary is: $35\,700 - $34\,500 = $1200
So, the mean annual increase is $200.

b) To accurately display these data, the scale on the vertical axis should start at $0. Then use a scale of 1 square to represent $5000.

From this graph, it is clear that salaries have increased very little in 6 years.

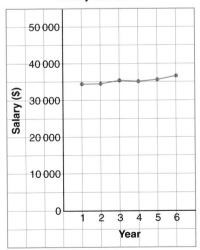

Example 2

This table shows the number of tonnes of plastic bottles processed by a local recycling company.
Draw a graph to show how the company can display these data to support each statement.
Explain how you created each impression.
a) More money is needed to advertise the importance of recycling.
b) Recycling is increasing so more staff should be hired.

Year	Amount Recycled (t)
2003	2600
2004	2750
2005	2925
2006	3130

▶ **A Solution**

a) The company could draw a line graph that suggests the amount recycled has not changed much over the years. For example, the scale on the vertical axis could be 1 grid square represents 1000 t.

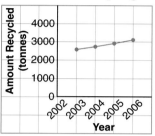

b) The company could draw a line graph that suggests the amount recycled is increasing rapidly. For example, the scale on the vertical axis could start at 2600, and each grid square could represent 50 t. This graph exaggerates the increase in recycling.

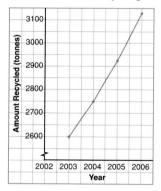

1. Look at the graph in *Example 1*. Who might have drawn the graph to create this impression?
2. How can the same data set be used to support different views on a topic?

398   UNIT 7: Data Analysis and Probability

Practice

Check

3. These graphs display the same data. Which graph is misleading? Why?

Graph A

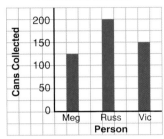

Graph B

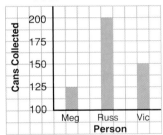

4. A Grade 8 class in Vancouver was surveyed to find out where they would like to go for an end-of-year trip. The results were graphed by two students.

Where We Would Like to Go

Graph A

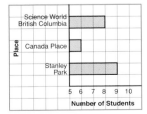

Graph B

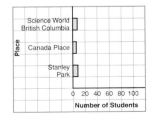

a) What impression does each graph give?

b) How is each graph misleading?
c) Where do you think the creator of Graph A would like to go? Why do you think so?
d) Where do you think the creator of Graph B would like to go? How was the graph drawn to give this impression?
e) What changes would you make to graph these data accurately?

Apply

5. Graphs A and B display the Read Books Company's profits for a four-month period.

Graph A   **Graph B**

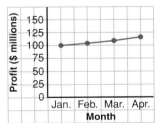

Which conclusions are incorrect?
a) The profits have tripled in 3 months.
b) The profit in March is double the profit in January.
c) The profit in March is about $10 000 000 more than the profit in January.
d) The profit in April is $16 000 000 more than the profit in January.

Explain how each incorrect conclusion may have been made.

7.2 Misrepresenting Data

6. The pictograph shows the number of students in a Grade 8 class who have different types of pets.
From this graph, Nick concluded that the most popular pet is a bird.
   a) Is Nick's conclusion correct? If yes, justify his conclusion. If not, explain his misinterpretation.
   b) If your answer to part b was no, what changes would you make to accurately display these data?
   c) Which type of pet do you think the creator of this pictograph has? Why do you think so?

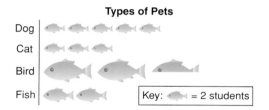

7. This graph appears to show that Paige's math mark dropped greatly in the 4th term.

   a) Do you think Paige should be very concerned about this drop? Why or why not?
   b) What changes could you make to graph these data more accurately?

8. Manufacturer A uses this graph to advertise that more than 98 out of 100 of its trucks sold in the last 10 years are still on the road.

   a) What impression does this graph give?
   b) How many trucks, out of 100, are still on the road for Manufacturer B? C? D?
   c) Do you think Manufacturer A's trucks are more dependable than the other manufacturers' trucks? Why or why not?
   d) What changes would you make to the graph to accurately display these data?

9. Four students conducted a science experiment to see who could grow the tallest plant over a given period of time. This pictograph shows the results.

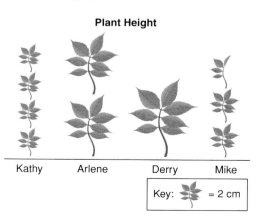

400   UNIT 7: Data Analysis and Probability

a) Whose plant grew the most?
b) How does the graph misrepresent these data?
c) How could the graph be changed to present the data accurately?
d) Do you think Kathy drew this pictograph? Why or why not?

10. **Assessment Focus** Giada surveyed her classmates to find out which sports they participated in. She drew these graphs.

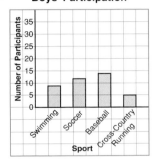

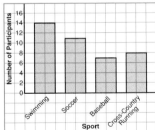

a) What impressions do these graphs give?
b) Describe how the graphs create a false impression.
c) What features of the graphs make it seem that the girls participate in sports more than the boys?
d) How could the graphs be changed to present the data accurately?
e) Suggest a different graph that could be used to accurately display these data.

11. This graph was used by a sales manager at the Dust Destroyer Vacuum Cleaner Sales Company. She said that sales had increased greatly in 5 years.

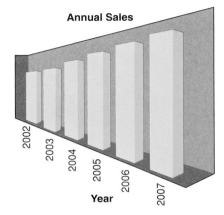

This table shows the actual data.

Annual Sales

Year	Sales ($1000s)
2002	15 450
2003	15 550
2004	16 000
2005	16 300
2006	16 600
2007	16 800

a) How is the graph misleading?
b) What changes could you make to the graph to display these data accurately?

12. Use newspapers, magazines, or the Internet. Find a graph that creates a false impression.
a) Describe how the graph creates a false impression.
b) Why might the misleading graph be used?
c) How could the graph be changed to present the data accurately?

13. Why do some graphs display data in a misleading way? Describe three ways a graph might be drawn to misrepresent data.

14. The graphs show how two students spend their allowance. From these graphs, a student concluded that Mark spends more money on movies than Tina does.
    a) From the graphs, can you tell which student spends more money on movies? Why or why not?
    b) Which type of graph could you use to display these data accurately?

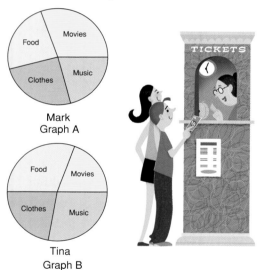

How Students Spend Their Allowance

Mark Graph A

Tina Graph B

15. **Take It Further** Draw a graph to show how the data in the table can be displayed in each way.

Board of Directors' Expenses

Quarter	Amount ($)
1st	85 000
2nd	104 000
3rd	125 000
4th	155 000

a) The directors want the expenses to look low.
b) The shareholders want to show the expenses are too high.
c) Prospective shareholders want to see the data displayed accurately.
Explain how you created each impression.

16. **Take It Further** Draw a graph to show how the data in the table below can be displayed in each way.

Submarine Sandwich Sales at Submarine Sue's

Type	Mean Daily Sales
Pizza	275
Assorted	200
Veggie	260
Meatball	250

a) The pizza submarine is by far the most popular sandwich.
b) About an equal number of each type of submarine are sold daily.
c) Very few assorted submarines are sold. Explain how you created each impression.

Reflect

When you see a graph in the media, what can you do to check whether it represents the data accurately?

Using Spreadsheets to Investigate Formatting

Focus | Recognize how formatting may lead to misinterpretation of graphs.

Spreadsheets can be used to record, then graph, data.
The format of the graph can affect its visual impact.

This table shows the quarterly sales figures for a cereal company.

Quarterly Sales	
Quarter	Sales ($1000s)
Jan. to Mar.	74 600
Apr. to June	74 820
July to Sept.	75 000
Oct. to Dec.	75 250

➤ Use a spreadsheet. Enter the data into rows and columns.
Highlight the data, then create a bar graph.
Your graph should look similar to this:

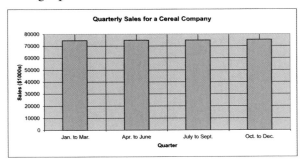

The table shows a slight increase in sales over the year.
This is difficult to see on the bar graph.

➤ Investigate the effect of changing the scale on the vertical axis.
Create a graph that exaggerates the increase in sales.
Your graph may look similar to this:

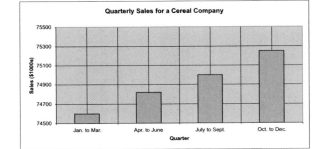

What impression does this graph give? Who might want to use this graph?

Technology: Using Spreadsheets to Investigate Formatting

These data come from the Statistics Canada Web site.

Average Annual Earnings of Canadians

Year	Women ($)	Men ($)
1998	24 900	39 700
1999	25 300	40 400
2000	25 800	41 700
2001	25 800	41 600
2002	26 200	41 600
2003	25 800	41 000
2004	26 200	41 300
2005	26 800	41 900

➤ Use a spreadsheet. Enter the data into rows and columns. Highlight the data, then create a double line graph. Your graph should look similar to the one below.

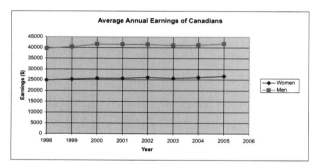

➤ Investigate the effect of not graphing both data sets on one grid. Create 2 line graphs to exaggerate the increase in women's average annual earnings compared to men's average annual earnings. Your graphs may look similar to these:

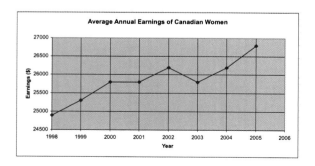
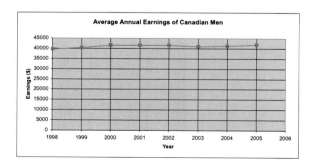

1. This table shows the NHL's top point scorers for the 2005–2006 season.
   a) Create a graph that Joe Thornton might use to negotiate a new contract. Justify your choice.
   b) Create a graph that the NY Rangers might use to negotiate a new contract with Jaromir Jagr. Justify your choice.

Top Point Scorers, 2005–2006

Player	Points
Joe Thornton, San Jose	125
Jaromir Jagr, NY Rangers	123
Alexander Ovechkin, Washington	106
Dany Heatley, Ottawa	103
Daniel Alfredsson, Ottawa	103
Sidney Crosby, Pittsburgh	102
Eric Staal, Carolina	100

2. This table shows the numbers of male and female athletes in the Winter Olympic Games 1988-2006.

Number of Athletes in Winter Olympic Games

Year	Location	Men	Women
1988	Calgary, Canada	1122	301
1992	Albertville, France	1313	488
1994	Lillehammer, Norway	1215	522
1998	Nagano, Japan	1389	787
2002	Salt Lake City, USA	1513	886
2006	Torino, Italy	1548	960

   a) Create a graph to accurately display these data. Justify your choice of graph.
   b) Create two graphs to exaggerate the increase in the number of female athletes.
   c) Create two graphs to give the impression that there are more female athletes than male athletes.

3. To collect data about students from 8 to 18 years old, Statistics Canada developed the *Census at School* Web site.
   ➤ Your teacher will give you the Web address for *Census at School*. Open the Web site.
   ➤ At the *Home Page*, on the left menu bar, click: *Data and results*
   ➤ Under *Canadian summary results*, click on any topic that interests you from the list under "*Summary tables for 2005/06.*"
   ➤ Create a graph to accurately display these data.
   ➤ Change the formatting of the graph. Explain the effect of your new graph.

Mid-Unit Review

LESSON

7.1

1. a) What data does each graph show?

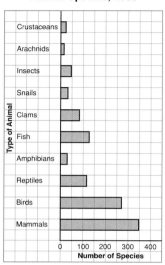

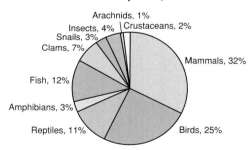

b) What are the strengths of each graph?

c) What are the limitations of each graph?

d) Which graph is more appropriate to display these data? Justify your choice.

e) Should a line graph be used to display these data? Why or why not?

7.2

2. The coach of the Ravens used these graphs to show how much better his team is than the Hawks. He said, "We have scored more points per game and we are improving faster."

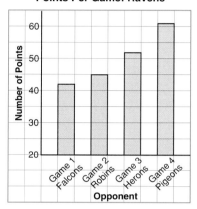

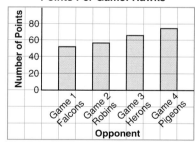

a) Are the coach's claims accurate? Explain.

b) What features of the graphs make it seem that the Ravens are the better team?

c) How could the graphs be changed to compare the teams accurately?

406 UNIT 7: Data Analysis and Probability

7.3 Probability of Independent Events

Focus Develop and apply a rule to determine the probability of two independent events.

James played a game.
He spun the pointer on this spinner and tossed the coin.
What is the probability that the pointer lands on red?
Does the spinner result affect the coin toss result?

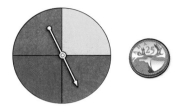

Investigate

Work with a partner.

➤ Use a tree diagram.
   List the possible outcomes of spinning the pointer
   on this spinner and tossing the two-coloured counter.
   What is the probability of each event?
   • landing on F
   • tossing red
   • landing on F and tossing red

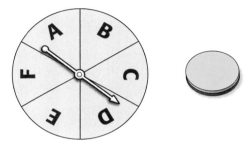

➤ Each of Kelsey and Sidney has a standard deck of 52 playing cards.
   Each student turns over a card, then the students compare suits.
   Make a table to list the possible outcomes.
   What is the probability of each event?
   • Kelsey turns over a spade.
   • Sidney turns over a heart.
   • Kelsey turns over a spade and Sidney turns over a heart.

➤ In each situation above, how does the probability of each individual event
   relate to the probability of the combined events?
   Write a rule to find the probability of two independent events.
   Use your rule to find the probability of tossing heads on a coin and
   drawing a red tile from a bag that contains 2 red tiles and 3 green tiles.
   Use a tree diagram to check your probability.

Compare your results and rule with those of another pair of classmates.
Did you write the same rule? If not, do both rules work?
Does the order in which the events are performed matter? Why or why not?

Connect

Two events are **independent events** when one event does not affect the other event.

The pointer on this spinner is spun twice. Landing on red and landing on blue are examples of two independent events.

Use a table to find the probability of landing on red twice.

There are 9 possible outcomes:
RR, RB, RG, BR, BB, BG, GR, GB, GG
Only one outcome is RR.
So, the probability of landing on red twice is $\frac{1}{9}$.

The probability of landing on red on the first spin is $\frac{1}{3}$.

The probability of landing on red on the second spin is $\frac{1}{3}$.
Note that: $\frac{1}{9} = \frac{1}{3} \times \frac{1}{3}$

		First Spin		
		R	B	G
Second Spin	R	RR	RB	RG
	B	BR	BB	BG
	G	GR	GB	GG

probability of landing on red twice = probability of landing on red on the first spin × probability of landing on red on the second spin

This illustrates the rule below for two independent events.

Suppose the probability of event A is written as P(A).
The probability of event B is written as P(B).
Then, the probability that both A and B occur is written as P(A and B).
If A and B are independent events, then: P(A and B) = P(A) × P(B)

Example 1

A coin is tossed and a regular tetrahedron labelled 5, 6, 7, and 8 is rolled.
a) Find the probability of tossing heads and rolling an 8.
b) Find the probability of tossing heads or tails and rolling an even number.
Use a tree diagram to verify your answers.

▶ **A Solution**

Since the outcome of tossing the coin does not depend on the outcome of rolling the tetrahedron, the events are independent.

a) When the coin is tossed, there are 2 possible outcomes.
   One outcome is heads.
   So, P(heads) = $\frac{1}{2}$
   When the tetrahedron is rolled, there are 4 possible outcomes.
   One outcome is an 8.
   So, P(8) = $\frac{1}{4}$

   P(heads and 8) = P(heads) × P(8)
   $= \frac{1}{2} \times \frac{1}{4}$
   $= \frac{1}{8}$

b) When the coin is tossed, there are 2 possible outcomes.
   Two outcomes are heads or tails.
   So, P(heads or tails) = $\frac{2}{2} = 1$
   When the tetrahedron is rolled, there are 4 possible outcomes.
   Two outcomes are even numbers: 6 and 8
   So, P(even number) = $\frac{2}{4} = \frac{1}{2}$

   P(heads or tails and even number) = P(heads or tails) × P(even number)
   $= 1 \times \frac{1}{2}$
   $= \frac{1}{2}$

Use a tree diagram to check your answers.

Toss	Roll	Possible Outcomes
Heads	5	Heads/5
	6	Heads/6
	7	Heads/7
	8	Heads/8
Tails	5	Tails/5
	6	Tails/6
	7	Tails/7
	8	Tails/8

There are 8 possible outcomes.
One outcome is heads/8.
So the probability of tossing heads and rolling 8 is $\frac{1}{8}$.

Four outcomes have heads or tails and an even number: heads/6, heads/8, tails/6, tails/8
So, the probability of tossing heads or tails and rolling an even number is $\frac{4}{8}$, or $\frac{1}{2}$.

7.3 Probability of Independent Events

Example 2

The pocket of a golf bag contains 9 white tees, 7 red tees, and 4 blue tees. The golfer removes 1 tee from her bag without looking, notes the colour, then returns the tee to the pocket.
The process is repeated.
Find the probability of each event.
a) Both tees are red.
b) The first tee is not red and the second tee is blue.

▶ **A Solution**

Since the first tee is returned to the pocket, the events are independent.

a) There are 20 tees in the pocket.
$$P(\text{red}) = \frac{7}{20}$$
So, $P(\text{red and red}) = P(\text{red}) \times P(\text{red})$
$$= \frac{7}{20} \times \frac{7}{20}$$
$$= \frac{49}{400}$$

b) $P(\text{not red}) = P(\text{white or blue})$
$$= \frac{13}{20}$$
So, $P(\text{not red, blue}) = P(\text{not red}) \times P(\text{blue})$
$$= \frac{13}{20} \times \frac{4}{20}$$
$$= \frac{13}{20} \times \frac{1}{5}$$
$$= \frac{13}{100}$$

Discuss the ideas

1. In a word problem, what are some words that can be used to suggest the events are independent?
2. In *Example 1*, how can you find the probability of *not* rolling an 8?

Practice

Check

3. A spinner has 2 congruent sectors coloured blue and green. The pointer is spun once, and a coin is tossed.

Find the probability of each event:
a) blue and tails
b) blue or green and heads

4. Stanley has two sets of three cards face down on a table. Each set contains: the 2 of hearts, the 5 of diamonds, and the 8 of clubs. He randomly turns over one card from each set.

Find the probability of each event:
a) Both cards are red.
b) The first card is red and the second card is black.
c) Both cards are even numbers.
d) The sum of the numbers is greater than 8.
Which strategy did you use each time?

5. Raoul spins the pointer on each spinner. Find the probability of each event.

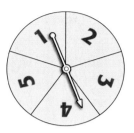 

a) green and a 2
b) red and an even number
c) green and a prime number

Use a tree diagram or a table to verify your answers.

Apply

6. Find the probability of each event:
a) i) The pointer lands on a blue spotted sector, then a solid red sector.
   ii) The pointer lands on a red sector, then a spotted sector.

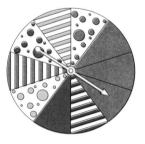

   iii) The pointer lands on a striped sector, then a solid blue sector.
   iv) The pointer lands on a blue or red sector, then a spotted sector.
b) Use a different strategy to verify your answers in part a.

7.3 Probability of Independent Events   **411**

7. Bart and Bethany play a game. They each roll a regular 6-sided die labelled 1 to 6. Find the probability of each event:
   a) Each player rolls a 6.
   b) Bart rolls a 6 and Bethany rolls a 2.
   c) Bart does not roll a 4 and Bethany rolls an even number.
   d) Bart rolls an even number and Bethany rolls an odd number.
   e) Bart rolls a number greater than 3 and Bethany rolls a number less than 4.

8. An experiment consists of rolling a die labelled 3 to 8 and picking a card at random from a standard deck of playing cards.
   a) What is the probability of each event?
      i) rolling a 6 and picking a spade
      ii) not rolling a 4 and picking an ace
   b) Use a tree diagram to verify your answer to part a, i.
   c) What is the probability of picking the ace of spades and rolling a 5? What is the advantage of using the rule instead of a tree diagram?

9. A game at a school carnival involves rolling a regular tetrahedron. Its four faces are coloured red, orange, blue, and green. A player rolls the tetrahedron twice. To win, a player must roll the same colour both times. Marcus has been watching the game. He says he has figured out the probability of a player winning. "The probability of rolling any colour is $\frac{1}{4}$. So, the probability of rolling the same colour again is $\frac{1}{4}$. Since the events are independent, the probability of rolling the same colour both times is $\frac{1}{4} \times \frac{1}{4} = \frac{1}{16}$."
   Do you agree with Marcus?
   Justify your answer.
   Use a tree diagram to show your thinking.

10. A dresser drawer contains five pairs of socks of these colours: blue, brown, green, white, and black. The socks in each pair are folded together.
    Pinto reaches into the drawer and takes a pair of socks without looking.
    He wants a black pair.
    a) What is the probability that Pinto takes the black pair of socks on his first try?
    b) What is the probability that Pinto takes the green pair of socks on his first and second tries?
    c) What assumptions do you make?

11. Suppose it is equally likely that a baby be born a boy or a girl.
    a) What is the probability that, in a family of 2 children, both children will be boys?
    b) Verify your answer to part a using a different method.

12. **Assessment Focus** A bag contains 6 red marbles, 4 blue marbles, and 2 yellow marbles. A student removes 1 marble without looking, records the colour, then returns the marble to the bag. The process is repeated.
    a) What is the probability of each outcome?
        i) a red marble, then a yellow marble
        ii) 2 blue marbles
        iii) not a blue marble, then a yellow marble
    b) Suppose the marbles are not returned to the bag. Could you use the rule for two independent events to find each probability in part a? Why or why not?

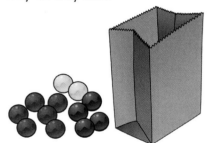

13. Luke and Salina play the card game "Slam." Each player has 10 cards numbered 1 to 10. Both players turn over one card at the same time. The player whose card has the greater value gets one point. If both cards are the same, a tie is declared and no point is given. After each round, the cards are returned to the pile and all the cards are shuffled.
    Find the probability of each event:
    a) Luke will get one point when he turns over a 3.
    b) Salina will get one point when she turns over a 10.
    c) Luke and Salina will tie.
    d) Salina will get one point when she turns over a 1.

14. **Take It Further** Neither Andrew nor David like to set the table for dinner. They each toss a coin to decide who will set the table. If both coins show heads, David sets the table. If both coins show tails, Andrew sets the table. If the coins show a head and a tail, both Andrew and David set the table.
    What is the probability David will set the table alone 2 days in a row?
    Show your work.

15. **Take It Further** A coin is tossed and a die labelled 1 to 6 is rolled.
    Write an event that has each probability below.
    a) $\frac{1}{2}$ b) $\frac{1}{6}$ c) $\frac{1}{3}$

Reflect

Which method of finding the probability of 2 independent events do you prefer? Why?
When might the rule not be the best method?
When might the tree diagram or table not be the best method?
Include an example in your explanation.

Doing Your Best on a Test

Have you ever written a test and have not been happy with the results? There are many things you can do to help you do well on a test.

Getting ready

- Find out what will be covered in the test.
- Give yourself plenty of time to review.
- Review your notes, the textbook, past quizzes, and tests. Try sample questions often.
- Think about what worked well for you on previous tests.
- Practise difficult questions with a friend or family member.
- Get a good night's sleep.
- Eat a nutritious breakfast.
- Bring all the supplies you need to the test.
- Think about what you are good at. Stay calm and confident.

When you first get the test

- Listen carefully to the instructions.
- Find out how much time you will have.
- Look over the whole test before you begin.
- Read the instructions. Ask questions about anything that is unclear.

Strategies for Success

Answering the questions

- Read each question carefully.
- Start by answering the quick and easy questions.
- Next, answer the questions that you know how to do but take more time.
- Leave for later any questions you get stuck on.
- Remember strategies for answering different kinds of questions.
- Look for key words, such as: compare, describe, determine, explain, and solve.
- Show your work.

When you have finished the questions

Use any time you have left to check your work.

For each question, ask yourself:

- Did I answer the question?
- Did I leave anything out?
- Did I give a complete solution?
- Did I show my work so someone else can follow my thinking?
- Is my answer reasonable? (Does it make sense?)

Strategies for Success: Doing Your Best on a Test

Empty the Rectangles

HOW TO PLAY

1. Each player draws 6 rectangles on a piece of paper.

   Label each rectangle from 0 to 5.

2. Each player places her 6 counters in any or all of the rectangles.
   You can place 1 counter in each rectangle, or 2 counters in each of 3 rectangles, or even 6 counters in 1 rectangle.

3. Take turns to roll the dice.
   Find the difference of the numbers.
   You remove counters from the rectangle labelled with that number.
   For example, if you roll a 6 and a 4, then $6 - 4 = 2$; so, remove all counters from rectangle 2.

4. The winner is the first person to have all rectangles empty.

> What strategies can you use to improve your chances of winning this game?

YOU WILL NEED
2 dice labelled 1 to 6; 12 counters

NUMBER OF PLAYERS
2

GOAL OF THE GAME
To remove all counters from all rectangles

7.4 Solving Problems Involving Independent Events

Focus: Solve a problem that involves finding the probability of independent events.

In Lesson 7.3, you learned that when A and B are independent events, the probability of both A and B happening is the product of the probability of event A and the probability of event B.
$P(A \text{ and } B) = P(A) \times P(B)$

Investigate

Work with a partner.

➤ A probability experiment involves tossing a coin, rolling a die labelled 1 to 6, and spinning the pointer on a spinner with 3 congruent sectors coloured pink, purple, and yellow.
Use a tree diagram or a table to find the probability of each event.
- tossing heads, rolling a 2, and landing on purple
- tossing tails, rolling an even number, and landing on yellow
- tossing heads, rolling a 1 or 2, and landing on pink

➤ Predict a rule to find the probability of three independent events.
Use your rule to verify the probabilities you found above.

Compare your results and your rule with those of another pair of classmates.
Use your rule to find the probability of getting heads on 3 consecutive tosses of a coin.

Connect

The rule for the probability of two independent events can be extended to three or more independent events.

Example 1

The students in a Grade 8 class were making buffalo horn beaded chokers.
Students could choose from dark green, yellow, and cobalt blue Crow beads.
Each student has the same number of beads of each colour.
Keydon, Patan, and Kada take their first beads without looking.
Find the probability that Keydon takes a yellow bead,
Patan takes a dark green bead, and Kada takes a yellow bead.

▶ **A Solution**

Since each student has her own set of beads, the events are independent.
Use a tree diagram.

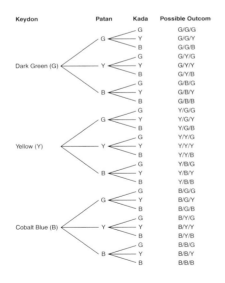

There are 27 possible outcomes. One outcome is Y/G/Y.
So the probability that Keydon takes a yellow bead,
Patan takes a dark green bead, and Kada takes a yellow bead is $\frac{1}{27}$.

In *Example 1*, the probability that Keydon takes a yellow bead is $P(Y) = \frac{1}{3}$.
The probability that Patan takes a dark green bead is $P(G) = \frac{1}{3}$.
The probability that Kada takes a yellow bead is $P(Y) = \frac{1}{3}$.
Note that: $\frac{1}{27} = \frac{1}{3} \times \frac{1}{3} \times \frac{1}{3}$
Suppose the probability of Event A is P(A), the probability of Event B is P(B), and the probability of Event C is P(C).
Then, the probability that all A, B, and C occur is P(A and B and C).
If A, B, and C are independent events,
then $P(A \text{ and } B \text{ and } C) = P(A) \times P(B) \times P(C)$.

Example 2

On a particular day in July, there is a 20% probability of rain in Vancouver, a 65% probability of rain in Calgary, and a 75% probability of rain in Saskatoon. What is the probability that it will rain in all 3 cities on that day?

▶ **A Solution**

The events are independent.
Write each percent as a decimal, then multiply the decimals.
P(Rain in Vancouver) = 20%, or 0.20
P(Rain in Calgary) = 65%, or 0.65
P(Rain in Saskatoon) = 75%, or 0.75
So, P(Rain V and C and S) = P(Rain V) × P(Rain C) × P(Rain S)
$\qquad = 0.20 \times 0.65 \times 0.75$
$\qquad = 0.0975$, or 9.75%

Use a calculator.

The probability that it will rain in all 3 cities on that day is 9.75%.

▶ **Example 2
Another Solution**

Assume the events are independent.
P(Rain in Vancouver) = 20%, or $\frac{20}{100} = \frac{1}{5}$
P(Rain in Calgary) = 65%, or $\frac{65}{100} = \frac{13}{20}$
P(Rain in Saskatoon) = 75%, or $\frac{75}{100} = \frac{3}{4}$
So, P(Rain V and C and S) = P(Rain V) × P(Rain C) × P(Rain S)
$\qquad = \frac{1}{5} \times \frac{13}{20} \times \frac{3}{4}$
$\qquad = \frac{39}{400}$

The probability that it will rain in all 3 cities on that day is $\frac{39}{400}$.   **Note that $\frac{39}{400} = 0.0975$.**

Discuss the ideas

1. How can the rule for the probability of 2 independent events be extended to 4 or 5 independent events?
2. In *Example 2*, how can you use the answer to find the probability of it not raining in all 3 cities on that day?
3. Why are the events in *Example 2* considered independent?

Practice

Check

4. One coin is tossed 3 times. Find the probability of each event:
   a) 3 heads
   b) 3 tails
   c) tails, then heads, then tails
   Use a tree diagram to verify your answers.

5. A red die, a blue die, and a green die are rolled. Each die is labelled 1 to 6. Find the probability of each event:
   a) a 2 on the red die, a 3 on the blue die, and a 4 on the green die
   b) a 4 on the red die, an even number on the blue die, and a number less than 3 on the green die

6. A spinner has 3 sectors coloured red, blue, and yellow. The pointer on the spinner is spun 3 times. Find the probability of each event:
   a) red, blue, and yellow
   b) blue, blue, and not red
   c) blue, blue, and blue

Apply

7. Stanley's bicycle lock has 4 dials, each with digits from 0 to 9. What is the probability that someone could guess his combination on the first try by randomly selecting a number from 0 to 9 four times?

8. A coffee shop has a contest. When you "lift the lid," you might win a prize. The probability of winning a prize is $\frac{1}{10}$. Suppose your teacher buys one coffee each day. Find the probability of each event:
   a) Your teacher will win a prize on each of the first 3 days of the contest.
   b) Your teacher will win a prize on the third day of the contest.
   c) Your teacher will not win a prize in the first 4 days of the contest.

9. **Assessment Focus** Nadine, Joshua, and Shirley each have a standard deck of playing cards. Each student randomly draws a card from the deck. Find the probability of each event:
   a) Each student draws a heart.
   b) Nadine draws a spade, Joshua draws a spade, and Shirley draws a red card.
   c) Nadine does not draw a heart, Joshua draws a black card, and Shirley draws an ace. Show your work.

10. Preet writes a multiple-choice test. The test has 5 questions. Each question has 4 possible answers. Preet guesses each answer. Find the probability of each event:
    a) She answers all 5 questions correctly.
    b) She answers only the first 3 questions correctly.
    c) She answers all the questions incorrectly.

11. Rocco chooses a 3-letter password for his e-mail account. He can use a letter more than once. What is the probability that someone else can access his e-mail by randomly choosing 3 letters?

12. Vanessa has 16 songs on a Classic Rock CD. Six of the songs are by the Beatles, 4 are by the Rolling Stones, 4 are by the Who, and 2 are by the Doors. Vanessa plays the CD. She selects a setting that randomly chooses songs to play. Find the probability of each event:
    a) The first 3 songs played are by the Beatles.
    b) The first 2 songs played are by the Rolling Stones and the next 2 songs are by the Beatles.
    c) The first 2 songs played are by the Doors, and the next song played is either by the Beatles or the Rolling Stones.

13. A bag contains 5 blue marbles and 1 white marble. Susan draws a marble from the bag without looking, then replaces it in the bag. This is done 5 times.
    a) What is the probability that the white marble is drawn 5 times in a row? Express your answer as a percent.
    b) Suppose the white marble is drawn 5 times in a row. What is the probability the white marble will be picked on the next draw? Explain.
    c) Is your answer to part b the same as the probability of drawing the white marble 6 times in a row? Why or why not?

14. Pancho wants to buy his teacher some flowers. The flower shop has 3 vases of cut flowers. One vase contains roses: 1 red, 4 yellow, and 3 white. A second vase contains carnations: 5 pink and 1 red. A third vase contains daisies: 1 yellow and 3 white. Pancho cannot decide so he closes his eyes and picks one flower from each vase.
Find the probability of each event.
  a) Pancho picks a red rose, a pink carnation, and a white daisy.
  b) Pancho picks a yellow or white rose, a red carnation, and a yellow daisy.
  c) Pancho picks a red rose, a red carnation, and a red daisy.

15. Take It Further In gym class, students take turns shooting a basketball at the net. Each successful shot is worth 1 point. A player is awarded a second shot only if the first shot is successful. The player can score 0, 1, or 2 points in this situation. Suppose a player shoots with 70% accuracy.
Find the probability of each event.
  a) She scores 0 points.
  b) She scores 1 point.
  c) She scores 2 points.

16. Take It Further A regular 6-sided die is rolled three times.
  a) What is the probability of rolling three 6s in a row?
  b) What is the probability of *not* rolling three 6s in a row?
  c) Find the sum of your answers in parts a and b. Explain the result.

Reflect

Give examples of random independent events outside the classroom. Why do you think the events are independent?

Math Link

Your World

The Coquihalla Highway is the only toll road in British Columbia. Those who run toll roads use probability to model the arrival of cars and trucks at the toll booths. This allows decisions to be made about how many toll booths should be open and how many toll booth operators are needed at any time of the day.

Using Technology to Investigate Probability

Focus Use virtual manipulatives to investigate the probability of independent events.

You have used manipulatives, such as spinners, dice, and coins. Many Web sites offer virtual manipulatives. These Web sites imitate, or *simulate*, spinning pointers, rolling dice, or tossing coins.

Your teacher will give you a Web address. Use the Web site to simulate tossing 3 coins 500 times. This is what you might see:

Number of trials required	500
Number of trials	500

heads	0	1	2	3
frequency	61	181	191	67

The theoretical probability of tossing 3 tails is:
$$P(T \text{ and } T \text{ and } T) = P(T) \times P(T) \times P(T)$$
$$= \tfrac{1}{2} \times \tfrac{1}{2} \times \tfrac{1}{2}$$
$$= \tfrac{1}{8}, \text{ or } 0.125$$

Recall:
The probability of tossing tails on the first coin is $\tfrac{1}{2}$, on the second coin is $\tfrac{1}{2}$, and on the third coin is $\tfrac{1}{2}$.

From the graph in this simulation, the experimental probability of tossing 3 tails (or 0 heads) is: $\tfrac{61}{500} = 0.122$
This experimental probability, 0.122, is very close to the theoretical probability, 0.125.

If you have forgotten the difference between experimental and theoretical probability, turn to the *Glossary*.

Combine your experimental results with those of 9 classmates. Find the experimental probability of tossing 3 tails. How do the experimental and theoretical probabilities compare?

Check

1. Use a Web site to simulate spinning 2 identical spinners 500 times. Each spinner has 4 congruent sectors coloured red, blue, green, and yellow. Combine your experimental results with those of 9 classmates.
   a) What is the experimental probability of each event?
      i) red and yellow
      ii) blue and green or yellow
   b) Use the rule for the probability of independent events to find the probability of each event in part a.
   c) How do your results in parts a and b compare? Explain.

Your teacher will give you the Web address.

Unit Review

What Do I Need to Know?

✓ Different graphs are used to visually represent data.

✓ Graphs can be used to misrepresent data.
For example, this can be done by changing: the scale on the vertical axis; the width of the bars; and the appearance of the graph.

✓ Suppose the probability of event A is P(A)
and the probability of event B is P(B).
Then, the probability that both A and B occur is P(A and B).
When A and B are independent events, then P(A and B) = P(A) × P(B).

✓ When A, B, and C are independent events, then
P(A and B and C) = P(A) × P(B) × P(C).

What Should I Be Able to Do?

LESSON

7.1

1. Each graph shows Canada's oil production from 1998 to 2005.

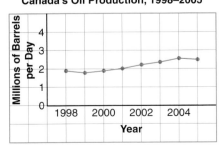

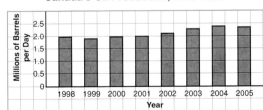

a) List 3 things you know from each graph.

b) Which graph is more appropriate to display these data? Justify your choice.

424 UNIT 7: Data Analysis and Probability

LESSON

2. Each graph below shows the snacks preferred by Chris' Grade 8 class.

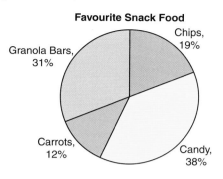

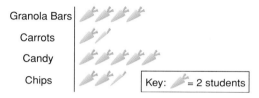

a) What are the advantages of each graph?
b) Describe a situation when the circle graph is the better graph to display these data.
c) Describe a situation when the pictograph is the better graph to display these data.

3. Each graph shows the best in show awards, by group, at the Westminster Dog Show, from 1924 to 2007.

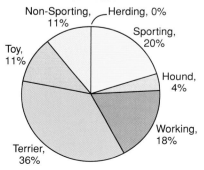

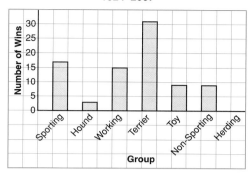

a) What can you tell from the bar graph that you cannot tell from the circle graph?
b) Which graph is more appropriate to display these data? Justify your choice.
c) Could you use a line graph to display these data? Why or why not?
d) Could you use a pictograph to display these data? Why or why not?

4. Paola owns a pizza parlour. This table shows the number of toppings on the pizzas she sold during the Grey Cup game.

Number of Toppings	Number of Pizzas Sold
2	26
3	48
4	60
5	50
6	16

a) Graph these data. Justify your choice of graph.
b) What are the advantages and disadvantages of the graph you drew?

LESSON 7.2

5. Stacy thought the pictograph in question 2 made it look like the students in her class were not making healthy choices. She redrew the graph.

Favourite Snack Food

Granola Bars
Carrots
Candy
Chips

Key: = 2 students

a) Did Stacy use the data from question 2? How do you know?
b) From the pictograph, which snack appears to be the favourite?
c) Why is this pictograph misleading?

6. How is this graph misleading? Explain.

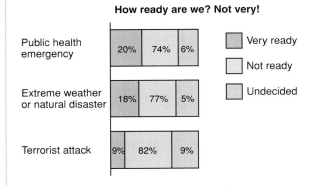

7. These graphs show the same data. Which graph is misleading? Explain why it is misleading.

Wrong Pizza Orders

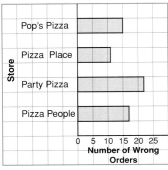

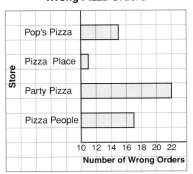

8. From this graph, Cleo concluded that the number of 15-year-old students with cell phones is about 5 times the number of 11-year-old students with cell phones. Is Cleo's conclusion correct? If yes, justify her conclusion. If not, explain how the graph may have led to the incorrect conclusion.

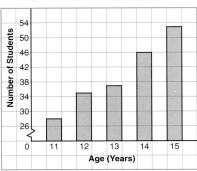

426 UNIT 7: Data Analysis and Probability

LESSON

9. Draw a graph to show how these data can be displayed in each way.

Bicycle Bonanza's Profit

Year	Profit ($)
2002	350 000
2003	400 000
2004	560 000
2005	610 000
2006	660 000

a) The employees want to show a large increase in profit each year with the hope of getting a raise.
b) The president wants to show a very small increase in profit each year to avoid having to pay his employees more money.

Explain how you created each impression.

7.3

10. A drawer contains one red shirt, two blue shirts, and one yellow shirt. Another drawer contains 2 pairs of brown pants and one pair of blue pants. You choose one item from each drawer without looking.
a) What is the probability of choosing a red shirt and a pair of brown pants?
b) Use a tree diagram to verify your answer.

11. A die has 2 faces labelled D, 2 faces labelled E, and 2 faces labelled F. The die is rolled twice. Find the probability of each event:
a) a D on the first roll
b) an E on the first roll
c) rolling a D followed by an E
d) rolling an E followed by a D
e) rolling 2 As

12. On her way to the market, Akita passes through 3 sets of traffic lights. For each set of lights, the probability that the light is green is $\frac{1}{3}$.
Find the probability of each event:
a) The first set of lights is green.
b) All the lights are green.
c) None of the lights is green.

7.4

13. Three cards are drawn from a standard deck of 52 playing cards. After each card is drawn, the card is replaced and the deck is shuffled. Find the probability of each event:
a) Three aces are drawn.
b) The ace of spades is drawn 3 times.
c) The first card is an ace, the second card is not a heart, and the third card is a face card.

14. One coin is tossed 10 times. Find the probability of tossing heads 10 times in a row.

15. The pointer on this spinner is spun 3 times. Find the probability of each event:
a) M, M, M
b) M, A, T
c) A, A, H
d) not A, not a consonant, A

Practice Test

1. This table shows the student population of H. J. Cambie Secondary School in 2007.

Grade	Number of Students
8	125
9	155
10	162
11	200
12	185

   a) What are the limitations of graphing these data with:
      i) a bar graph?
      ii) a circle graph?
   b) Would a pictograph be a good choice? Would a line graph be a good choice? Why or why not?
   c) Which graph is most appropriate to display these data? Justify your choice.

2. The new manager of a company created these two graphs to show shareholders how much better the company is doing since she took over in 2003.

   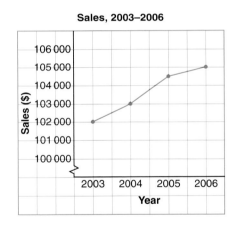

   a) What impressions do these graphs create?
   b) Describe how the graphs create a false impression. What features of the graphs cause this impression?
   c) How could the graphs be changed to present the data accurately?
   d) Is the company doing better since the new manager was hired? Explain.

3. A bag contains four 2007 pennies, two 2005 pennies, and six 2001 pennies. A student removes 1 penny without looking, records the year, then returns the penny to the bag. The process is repeated.
Find the probability of each event:
   a) two 2001 pennies
   b) a 2007 penny, then a 2005 penny
   c) a 2001 penny, then a 2005 penny
   d) not a 2001 penny, then a 2007 penny

4. This diagram shows part of a model railroad track.
At each of the junctions E, F, and G, the probability of the train going straight on is $\frac{3}{4}$ and the probability of it turning right is $\frac{1}{4}$. At any other junction, the train always goes straight on.
The train starts at A.
Find the probability of each event:
   a) The train reaches B.
   b) The train reaches C.
   c) The train reaches D.

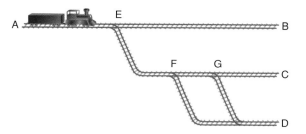

5. This table shows the sales, in thousands of dollars, of 2 employees at Electronics Warehouse over a 6-month period.
   a) Which graph do you think would be most appropriate to display Jamar's sales?
   b) Which graph would be most appropriate to compare Jamar's and Laura's sales?
   c) Both employees want to be named salesperson of the year.
      i) Graph these data to make Jamar's sales figures look much greater than Laura's.
      ii) Graph these data to make Laura's sales figures appear to be increasing much faster than Jamar's.
   Describe how you created each impression.

Electronics Warehouse Sales Figures

Month	Jamar ($1000s)	Laura ($1000s)
Jan.	117	124
Feb.	118	125
Mar.	119	124
Apr.	121	126
May	124	127
June	127	128

Unit Problem: Promoting Your Cereal

You have created your own cereal.
To promote your cereal, you will design two advertisements.
You will also develop a game for the back of your cereal box.

Part A
Creating Your Cereal

Create a name, logo, and slogan for your cereal.
Decide to whom you will market your cereal.
Create two data sets for your cereal that may include:
- nutritional information
- sales figures
- market research data

Part B
Promoting Your Cereal

To promote your cereal, create two advertisements for a local newspaper.
- Use one of the data sets in Part A.
- Create an advertisement that includes a graph that could be misinterpreted.
- Use the other data set from Part A.
- Create another advertisement that includes a graph that accurately represents these data.

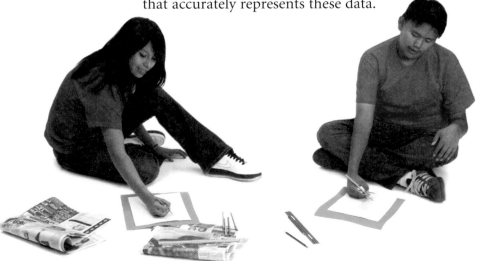

Part C
Creating a Game

To make your cereal packaging more appealing, create a game for the back of the box.
Your game must involve two or more independent events.
Your game can include spinners, cards, or anything else that can be cut out from the box.
Be sure to include the rules of the game.
Everything needed to play the game should be on the back of the box.

Determine the probability of winning the game.

Check List

Your work should show:
- ✓ data sets and two graphs
- ✓ a game for your box and its rules
- ✓ how you calculated the probability
- ✓ the correct use of math language

Reflect on Your Learning

Why do you think it is important to be able to represent a set of data in different ways? What are the advantages and disadvantages of representing a set of data in different ways?

UNIT 8 Geometry

First Nations artists use their artwork to preserve their heritage. Haida artist Don Yeomans is one of the foremost Northwest Coast artists. Look at this print called *The Benefit*, created by Don Yeomans. Describe any translations, reflections, or rotations you see.

What other art have you seen that demonstrates transformations?

What You'll Learn

- Draw and recognize different views of objects made from rectangular prisms.
- Identify shapes that will tessellate.
- Create tessellations using transformations.
- Identify tessellations in the environment.

Why It's Important

- We learn about the environment by looking at objects from different views. We can combine these views to get a betterunderstanding of these objects.
- Tessellations are found in the environment, in architecture, and in art.

The Benefit

Key Words

- isometric
- isometric drawing
- axis of rotation
- plane
- tessellate
- tessellation
- composite shape
- conservation of area

Gunarh and the Whale

8.1 Sketching Views of Objects

Focus Draw the front, top, and side views of objects from models and drawings.

Which rectangular prisms do you see in this picture?

Choose a rectangular prism.
What does it look like from the top?
From the side?
From the front?

Investigate

Work on your own.
Choose a classroom object that is a rectangular prism, or is made of more than one rectangular prism.

Sketch the object from at least 4 different views.
Label each view.
Use dot paper or grid paper if it helps.

Trade sketches with a classmate.
Try to identify the object your classmate chose.
What strategies did you use to identify it?

Connect

Use linking cubes to make the object at the right.
Rotate the object to match each photo below.

Left side

Front

Right side

Top

We can use square dot paper to draw
each view of the object.
We ignore the holes in each face.

To draw the views of an object:
- Place the top view above the front view,
  and the side views beside the front view.
  This way, matching edges are adjacent.
- Use broken lines to show how the views align.
- Show internal line segments only where the
  depth or thickness of the object changes.

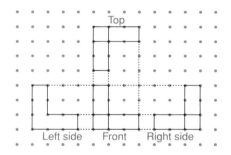

When a photo of an object is not available,
the object may be drawn on triangular dot paper.
This is called isometric paper.
Isometric means "equal measure."
The line segments joining 2 adjacent dots
in any direction are equal.

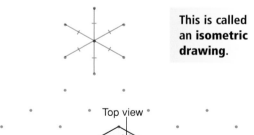

This is called
an **isometric
drawing**.

Here is an isometric drawing of the object
made from linking cubes at the top of this page.
Each vertical edge on the object is drawn
as a vertical line segment.
A horizontal edge is drawn as a line segment
going up to the right or down to the left.
The faces are shaded differently to give
a three-dimensional look.

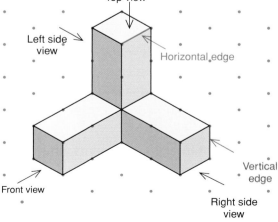

8.1 Sketching Views of Objects **435**

Example 1

Here is an object made from linking cubes.
Draw the front, top, and side views of the object.

▶ **A Solution**

Use linking cubes to make the object.
Use square dot paper.

Draw the front view first.
The front view is a rectangle 4 squares long.

Then draw the top view above the front view.
The top view is a rectangle 4 squares long.
Keep matching edges adjacent.

Then draw the side views beside the front view.
Each side view is a square.
Keep matching edges adjacent.

Use broken lines to show how the views align.

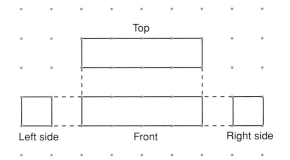

Example 2

Here is an isometric drawing of an object.
Draw the front, top, and side views of the object.

▶ **A Solution**

Visualize the object.
Draw the front view first.
Then draw the top view above the front view.
Then draw the side views beside the front view.
Keep matching edges adjacent.

Show internal line segments where the depth of the object changes.

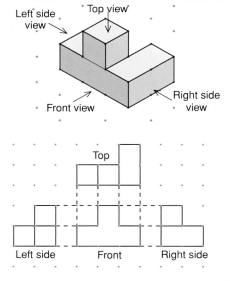

We can build a model of the object in *Example 2*, then rotate the model to compare the views of the object with the actual object.

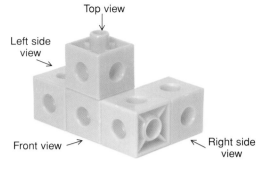

Left side

Front

Right side

Top

Discuss the ideas

1. In *Example 1*, why are there no internal line segments in any of the views drawn?
2. Do 4 views of an object made from linking cubes always provide enough information to make the object?
3. In what types of jobs or occupations might different views of objects be important?

Practice

Check

4. Use linking cubes. Make each object. Use square dot paper. Draw the front, top, and side views of each object.

   a)

   b)

   c)

8.1 Sketching Views of Objects   437

5. Use linking cubes. Make each object A to E. Figures J to Q are views of objects A to E. Match each view (J to Q) to each object (A to E) in as many ways as you can.

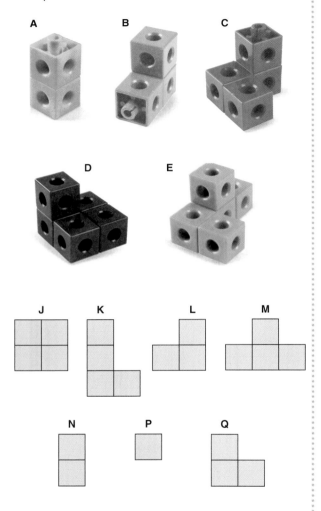

Apply

6. Sketch the top, front, and side views of this recycling bin. Label each view.

7. Sketch the front, top, and side views of each object in the classroom.
   a) filing cabinet
   b) whiteboard eraser
   c) teacher's desk

8. Sketch the front, top, and side views of this object drawn on isometric dot paper.

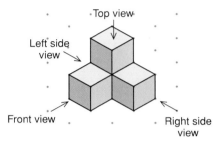

9. Sketch the front, top, and side views of this object.

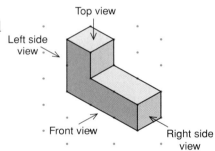

10. Sketch the front, top, and side views of this object.

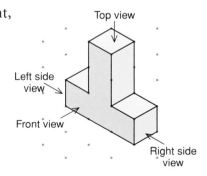

11. Use linking cubes. Make the objects in questions 8, 9, and 10. Rotate the objects to check that the views you drew are correct.

12. Sketch the front, top, and side views of this object.

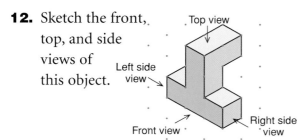

13. Sketch the front, top, and side views of this object.

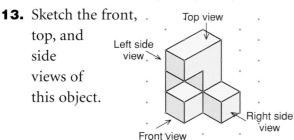

14. Sketch the front, top, and side views of this object.

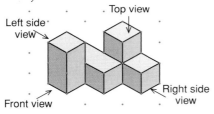

15. Assessment Focus Use 4 linking cubes. Make as many different objects as possible. Draw the front, top, and side views of each object. Label each view. Use your drawings to explain how you know all the objects you made are different.

16. Use linking cubes. Make the letter H. Sketch the front, top, and side views of your model.

17. Take It Further The front, back, top, and side views of a cube are the same. Use the objects in the classroom.
  a) Are there any other prisms with all views the same? Explain.
  b) Which object has 4 views the same?
  c) Which object has only 3 views the same?
  d) Which object has no views the same? If you cannot name an object for parts b to d, use linking cubes to make an object.

18. Take It Further Sketch two possible views of each object.
  a)
  b)
  c)

Reflect

Choose an object made from linking cubes.
What could be a problem with representing only one view?
How many views do you need to draw so someone else can identify the object?
Explain. Sketch the views you describe.

Using a Computer to Draw Views of Objects

Focus Use technology to sketch views of objects.

Geometry software can be used to draw different views of objects.
Use available geometry software.

Make this object with linking cubes.
Use the software to create views
of this object.

Front view

Open a new sketch.
Check that the distance units
are centimetres.

To make a "dot paper" screen, display a coordinate grid.
Change the grid lines to dotted lines.
Then hide the axes.

Front View

Use broken line segments to draw the front view.
Label the view.

Use the software to colour the view to match the object.

Draw and label the top and side views.
The views you drew should look similar to these.

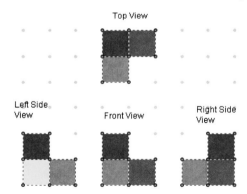

Top View

Left Side View   Front View   Right Side View

Check

1. Open a new sketch. Draw different views of the objects in Lesson 8.1, *Practice* question 4. Compare the hand-drawn views with the computer-drawn views.

8.2 Drawing Views of Rotated Objects

Focus Draw views of objects that result from a given rotation.

An object can be rotated:

horizontally　　　　　or　　　　　vertically

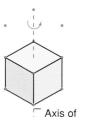
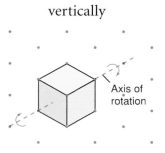

When an object is rotated horizontally, the **axis of rotation** is vertical. The object may be rotated clockwise or counterclockwise.

When an object is rotated vertically, the axis of rotation is horizontal. The object may be rotated toward you or away from you.

Investigate

Work with a partner.
You will need 6 linking cubes and square dot paper.

Use the linking cubes to build an object.
Draw the front, top, and side views of the object.

Predict each view when the object is rotated:
- horizontally 90° clockwise
- horizontally 90° counterclockwise
- horizontally 270° clockwise

Rotate the object to check your predictions.
Draw the new front, top, and side views after each rotation.

Were the views after a 90° clockwise rotation and a 90° counterclockwise rotation the same or different?
If they were the same, what do you notice about the object?
If they were different, talk to another pair of classmates who did get the same views, to see what they found out.

Connect

Here are the views of the object at the right.

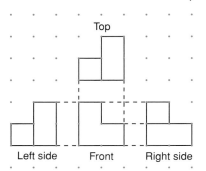

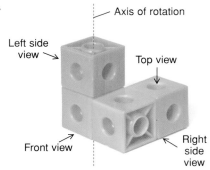

➤ Suppose the object is rotated horizontally 180°, about a vertical axis.
Here are the object and its views after the rotation.

Recall that a rotation of 180° clockwise is the same as a rotation of 180° counterclockwise.

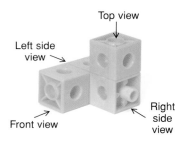

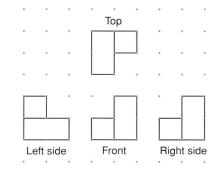

➤ Suppose the object is rotated vertically 180°, about a horizontal axis.

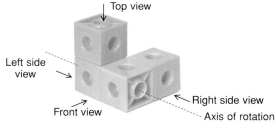

Here are the object and its views after the rotation.

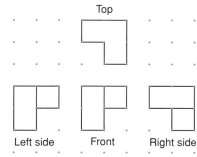

442   UNIT 8: Geometry

Example 1

Build this object.
It is rotated about the vertical axis shown.
Draw the front, top, and side views after a horizontal rotation of 270° clockwise.

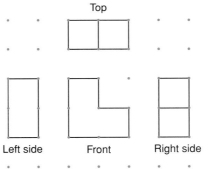

▶ A Solution

After a horizontal rotation of 270° clockwise, the object is in this position.

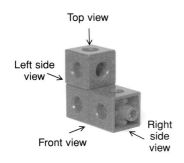

The views of the object are:

Example 2

Use the object in *Example 1*.
Draw the front, top, and side views after a horizontal rotation of 90° counterclockwise.

▶ A Solution

A rotation of 90° counterclockwise is the same as a rotation of 270° clockwise.
So, the views will match those in *Example 1*.

Example 3

Build this object.
It is rotated about the horizontal axis shown.
Draw the front, top, and side views after
a vertical rotation of 90° away from you.

▶ **A Solution**

After a vertical rotation of 90° away from you,
the object is in this position.

The views of the object are:

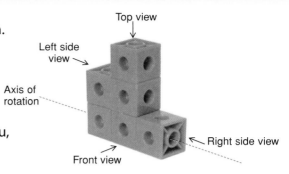

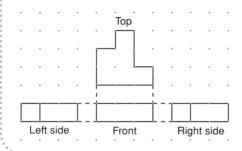

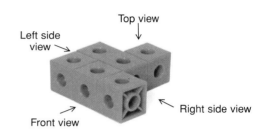

Discuss the ideas

1. An object is rotated vertically 90° toward you. The same object is rotated vertically 270° away from you. Will the views of the object after each rotation be the same? Justify your answer.
2. An object can be rotated vertically or horizontally 90°, 180°, or 270°. The views of the rotated object are always the same. What might the object be?

Practice

Check

Use linking cubes and square dot paper.

3. This object is rotated horizontally about the axis shown.

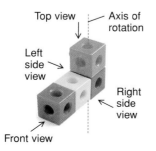

The front view after each rotation is shown below. Describe each rotation.

a)   b)

c)

444 UNIT 8: Geometry

4. Build each object. Rotate each object horizontally 90° clockwise through the axis shown. Match each view (A to G) to the front, top, and side views of each rotated object. A lettered view can be used more than once.

a)
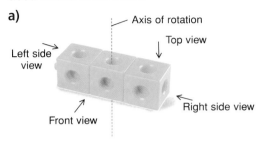

b)
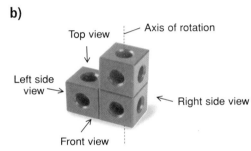

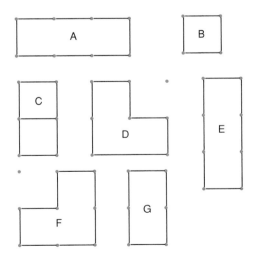

5. Use the objects in question 4. Suppose each object is rotated horizontally 180°. For which object will the views not change? How do you know?

Apply

6. Here is the front view of a ladder. Suppose the ladder is rotated horizontally 90° clockwise about the axis shown. Sketch the new front view.

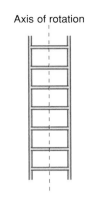

7. Here is an object made of linking cubes.

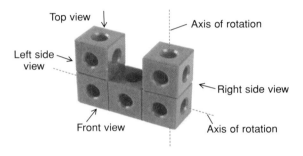

a) Predict the front, top, and side views of the object after each horizontal rotation about the vertical axis shown.
b) Build the object. Draw the views after each rotation to verify your predictions.
   i) 90° clockwise
   ii) 180°
   iii) 270° clockwise

8. Draw the front, top, and side views of the object in question 7 after each vertical rotation about the horizontal axis shown.
a) 90° toward you
b) 180°

8.2 Drawing Views of Rotated Objects **445**

9. Here is an object made of linking cubes.

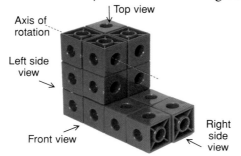

a) Predict the front, top, and side views of the object after each vertical rotation about the axis shown.
b) Build the object. Draw the views after each rotation to verify your predictions.
   i) 90° away from you
   ii) 180°
   iii) 270° away from you

10. **Assessment Focus** Use 5 linking cubes. Build an object.
a) Draw the front, top, and side views of the object.
b) Choose a horizontal rotation and a vertical axis. Rotate the object. Draw the new front, top, and side views.
c) Describe a different rotation that will have the same views as the ones you drew in part b.
d) Choose a vertical rotation and a horizontal axis. Rotate the object. Draw the new front, top, and side views.
e) Describe a different rotation that will have the same views as the ones you drew in part d.

11. **Take It Further** Here is an isometric drawing of an object. The object is rotated horizontally 180° about the vertical axis shown. Draw the new front, top, and side views of the object.

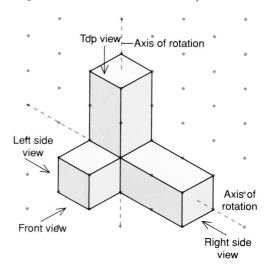

12. **Take It Further** Use the object in question 11. Suppose the object is rotated horizontally 90° clockwise. Then the rotated object is rotated vertically 180°. Draw the new front, top, and side views of the object.

Reflect

For which type of rotation–horizontal or vertical–did you find the views easier to draw? Justify your answer.

8.3 Building Objects from Their Views

Focus Build an object given different views of the object.

Here are the views of an object made from linking cubes.

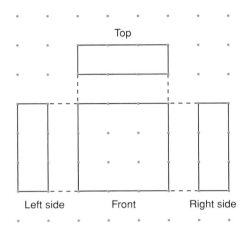

Which view or views can you use to find the height of the object?
The width of the object?
The length of the object?

Investigate

Work in groups of three.

Each student needs 16 linking cubes.
Each student chooses *one* of these views.

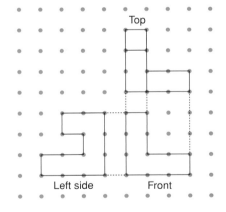

Use 8 cubes to build an object that matches the view you chose.
Use the other 8 cubes to build a different object that has the view you chose.

Compare your objects with those of the other members of your group.
Does any object match all 3 views?
If not, work together to build one that does.
What helped you decide the shape of the object?
Are any other views needed to help build the object? Explain.

Connect

Each view of an object provides information about the shape of the object. The front, top, and side views often provide enough information to build the object. Remember that, when you look at views of an object, internal line segments show changes in depth.

Example 1

Here are the views of an object made with linking cubes. Colours are labelled to show how the views relate. Build the object.

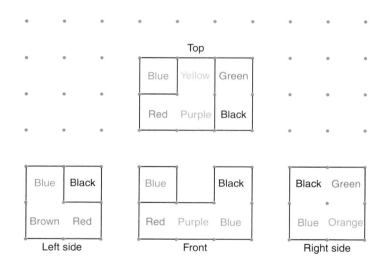

▶ **A Solution**

The left side view shows that the object is 2 cubes high and 2 cubes deep. The black cube is behind or in front of the other cubes.

The front view shows that the object is 3 cubes long and 2 cubes high. One blue cube is behind the other cubes. The black cube is in front.

So, in the left side view, the black cube is behind the other cubes. Use the left side and front views to build part of the object.

448 UNIT 8: Geometry

The top view shows that the red, purple, and yellow cubes are on the same level; and the blue, green, and black cubes are on a different level. Use the top view to continue building the object.

The right side view shows that the bottom right cube is orange.
Insert that cube to complete the object.

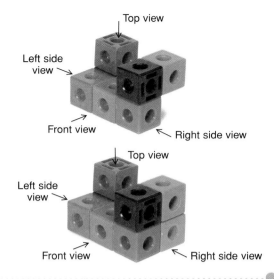

Example 2

Use linking cubes.
Build an object that has these views.

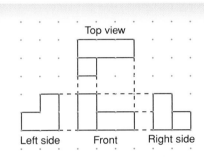

▶ **A Solution**

The left side view shows no change in depth. So, use linking cubes to build the left side first. The left side is shaped like a backward L.

Rotate the L shape horizontally 90° clockwise, about a vertical axis.
Compare the front views.
There is a change in depth of the object.
So, to match the given front view, add 2 cubes to the back of the object.

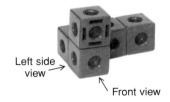

Check the top and right side views of the object.
The views match the given views.
So, the object is correct.

8.3 Building Objects from Their Views

Discuss the ideas

1. Build the object in *Example 1*.
   This time, start with the right side view.
   Which way was easier? Justify your choice.
2. When you build an object, how do you decide which view to start with? Explain.
3. Can you build more than one object for a given set of views? Justify your answer.

Practice

Check

You will need linking cubes.

4. Which object has these views? How do you know?

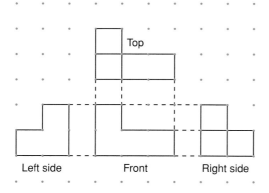

A

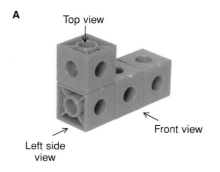

B

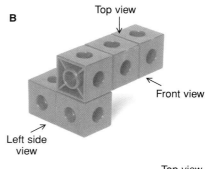

C

5. Use these views to build an object.

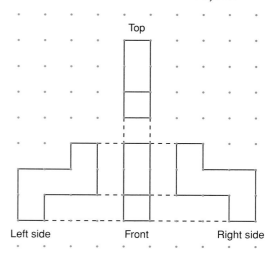

450   UNIT 8: Geometry

Apply

6. Use these views to build an object.

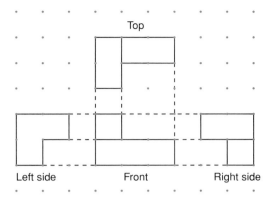

7. Use these views to build an object.

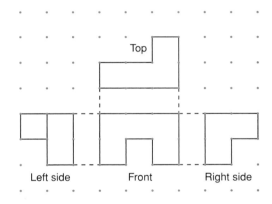

8. Use these views to build an object.

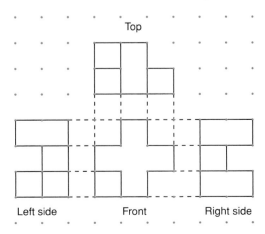

9. Assessment Focus For the set of views shown below:

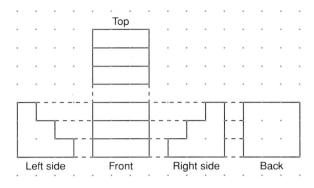

a) Use linking cubes to build the object.
b) Describe the object.
c) Suppose one side view had not been shown. Would you have been able to build the object? Explain.
d) Suppose the back view had not been shown. Could you have built a different object that matches the other views? If your answer is yes, build the object.

10. Use these views to build an object.

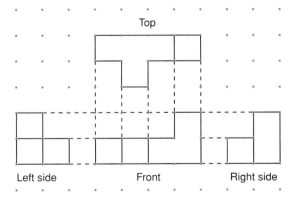

8.3 Building Objects from Their Views 451

11. The front, top, and side views of a cube are the same. Build another object that has the front, top, and side views the same.

12. Use 6 linking cubes to build an object.
   a) Draw the front, top, and side views of your object.
   b) Show a classmate 2 views in part a. Have her use the views to build an object. Does her object match your object? Explain.
   c) Show your classmate 3 views in part a. Have her use the views to build an object. Does her object match your object? Explain.
   d) Show your classmate all the views in part a. Have her use the views to build an object. Does her object match your object? Explain.
   e) Given all 4 views, is it possible that your classmate's object still does not match your object? Justify your answer.

13. Take It Further These four views show an object made with linking cubes.

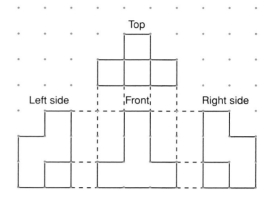

   a) Build the object.
   b) Assume each linking cube has edge length 2 cm. What are the surface area and volume of the object?

14. Take It Further
   a) Use these views to build an object.

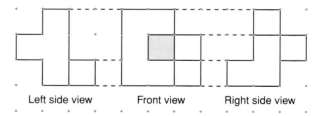

   A shaded region has no cubes.

   b) Draw the top view of the object.

15. Take It Further
a) Use these views to build an object. A shaded region has no cubes.
b) Explain the steps you took to build the object.

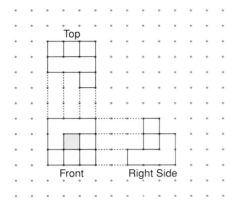

16. Take It Further Use these views to build an object.

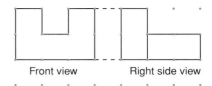

a) What is the greatest number of cubes you can use to make the object? How do you know?
b) What is the least number of cubes you can use to make the object? How do you know?
c) How many different possible objects can you make? Explain.

Reflect

Why do architects need to draw the front, top, back, and side views when planning the construction of a building?
Construct two different objects that have the same front, top, and side views.

Math Link

Art

An IMAX® 3-D film gives the illusion of three-dimensional depth. The scenes are filmed from two slightly different angles. One camera lens represents the right eye and the other lens represents the left eye. The specialized IMAX 3-D Projection system allows separate right- and left-eye images to be projected onto the screen alternately at a rate of 48 frames (pictures) per second, which your brain naturally fuses into one 3-D image.

© IMAX Corporation

Using a Computer to Construct Objects from Their Views

Focus Use a computer to build an object given its views.

A computer can be used to create an object from its views.
Use an interactive isometric drawing tool.

Here are the views of an object.

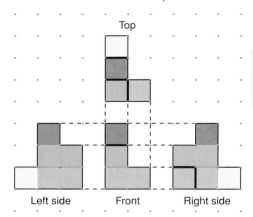

Should you need help at any time, use the Help or Instructions Menu.

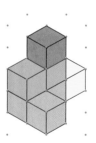

➤ Use the views to build the object on screen.
   Use the cube tool on the screen to build the object.
   Use the paintbrush tool to colour the cubes as indicated.
➤ When your object is complete, use the View icon to show the views of your object.

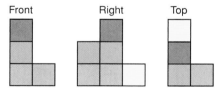

If the views match the views given, your object is correct.
If the views do not match, make changes to your object.
Continue to check the views until they match exactly those given.

Check

1. Use the set of views in *Example 1* of Lesson 8.3, and the interactive isometric drawing tool to build an object. Check that the views of your object match the views.

Mid-Unit Review

LESSON

8.1

1. Sketch the front, top, and side views of this object. Label each view.

2. Use linking cubes. Make each object. Use square dot paper. Draw the front, top, and side views of each object.

   a)

   b)

   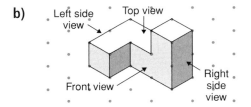

8.2

3. Here is an object made of linking cubes. Draw the front, top, and side views of the object after each rotation.
   a) horizontally 90° clockwise
   b) vertically 180°

   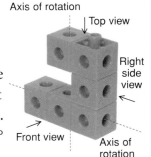

4. Suppose the object in question 3 is rotated horizontally 270° counterclockwise. Predict the new top, front, and side views of the object. Explain how you found your answer.

8.3

5. Suppose you had to build an object with linking cubes. Which would you prefer to build from?
   a) a drawing of the object on isometric dot paper
   b) the different views of the object on square dot paper

   Justify your choice.

6. Use linking cubes. Use each set of views to build an object.

   a)

   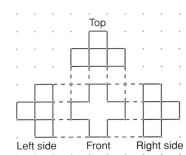

   b)

   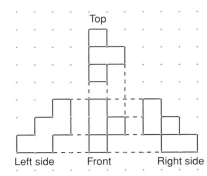

Mid-Unit Review   **455**

8.4 Identifying Transformations

Focus | Recognize transformation images.

Look around the classroom.
What transformations do you see?
How did you identify each transformation?

Investigate

Work with a partner.
Your teacher will give you a large copy of this design.

Tia used this design when she laid interlocking paving stones in her driveway.

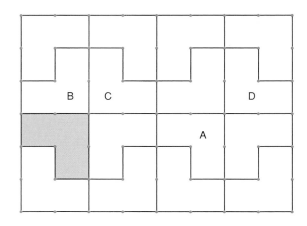

To create the design, Tia translated, rotated, and reflected the shaded shape.
Each labelled shape is the image after a transformation.
Identify a transformation that produced each image.
Explain how you know.

Discuss your strategies for identifying each transformation with another pair of classmates. How does the image relate to the original shape for each transformation?
- a reflection
- a translation
- a rotation

456 UNIT 8: Geometry

Connect

Here is a design that shows 3 different transformations.

Translation

The shaded shape is translated
6 units left. Its translation image is Shape A.
The translation arrow shows the
movement in a straight line.
The translation image and the shaded shape are
congruent and have the same orientation.

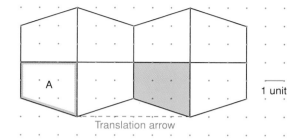

Reflection

The shaded shape is reflected in the
red line of reflection.
Its reflection image is Shape B.

The shaded shape is reflected in the
green line of reflection.
Its reflection image is Shape C.

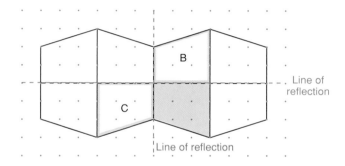

The shaded shape and each reflection image have opposite orientations.
Each reflection image and the shaded shape are congruent.

Rotation

The shaded shape is rotated 180° clockwise
about the point of rotation.
The rotation image is Shape D.

We get the same image if the shaded shape
is rotated 180° counterclockwise about
the point of rotation.
The rotation image and the shaded shape
are congruent and have the same orientation.

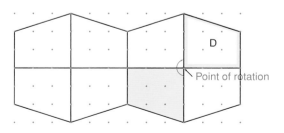

8.4 Identifying Transformations

Under any transformation, the original shape and its image are always congruent.

Example 1

Look at this design of squares.
Describe each transformation.
a) a translation for which Square R is an image of Square B
b) a reflection for which Square R is an image of Square B

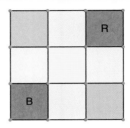

▶ **A Solution**

a) Square R is the image of Square B after a translation 2 units right and 2 units up. The translation arrow shows the movement.

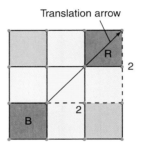

b) Square R is the image of Square B after a reflection in the slanted line. Use a Mira to verify the image.

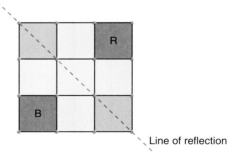

Example 1 shows that an image may be the result of more than one type of transformation.

Example 2

Look at this design of triangles.
Describe each rotation.
a) a rotation for which Triangle D is an image of Triangle C
b) a rotation for which Triangle F is an image of Triangle E

▶ *A Solution*

a) Triangle D is the image of Triangle C after a rotation of 90° clockwise about P. P is a vertex the two triangles share. The same image is also the result of a rotation of 270° counterclockwise about P.

b) Triangle F is the image of Triangle E after a rotation of 180° about R. The point of rotation, R, is *not* on the shape being rotated.

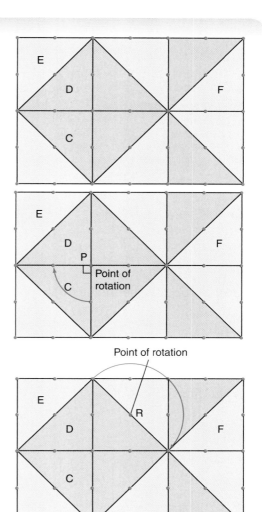

1. In *Connect*, what does it mean when we say that a shape and its reflection image have opposite orientations?
2. In *Example 1b*, how is a point on the image related to a point on the original shape?
   How is the line segment that joins these points related to the line of reflection?
3. In *Example 1*, identify another transformation for which Square R is an image of Square B.
4. Can you change the size of a shape by translating, reflecting, or rotating it? Justify your answer.

Practice

Check

5. In the design below, identify each transformation.

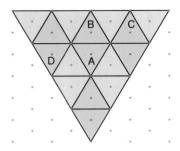

a) Shape B is the image of Shape A.
b) Shape C is the image of Shape A.
c) Shape D is the image of Shape A.
d) Shape C is the image of Shape B.

6. Use this design.

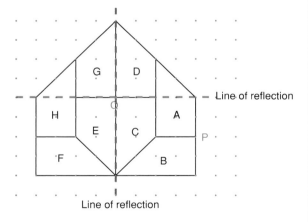

Match each transformation to a transformation image.
a) Rotate Shape A 90° counterclockwise about point P.
b) Reflect Shape C in the red line of reflection.
c) Translate Shape D 2 units right and 2 units down.
d) Rotate Shape G 180° about point Q.
e) Reflect Shape B in the blue line of reflection.

Apply

7. Identify each transformation.

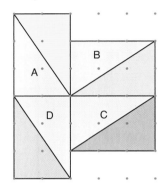

a) Shape A is the image of Shape B.
b) Shape B is the image of Shape C.
c) Shape C is the image of Shape D.
d) Shape D is the image of Shape A.

8. On grid paper, copy this square, the red and blue lines, and point P.

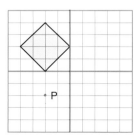

Draw the image of the original square after each transformation to create a design.
a) a translation 2 units right
b) a reflection in the red line
c) a rotation of 90° clockwise about P
d) a translation 2 units right and 4 units down
e) a reflection in the blue line

460   UNIT 8: Geometry

9. **Assessment Focus** How many different ways can each shape be described as a transformation of another shape? Explain.

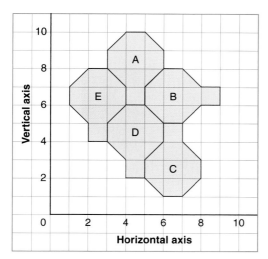

10. **Take It Further** Use grid paper. In each case, describe the shape you drew.
    a) Draw a shape for which a translation image is also a reflection image and a rotation image. Draw the translation image.
    b) Draw a shape for which a translation image is also a reflection image, but *not* a rotation image. Draw the translation image.
    c) Draw a shape for which a translation image is *not* a reflection image *nor* a rotation image. Draw the translation image.

11. **Take It Further** This is a photo of a silk-screen print, *Haida Frog*, by Northwest Coast artist Bill Reid. Describe as many transformations in the print as you can. Include at least one translation, one reflection, and one rotation.

12. **Take It Further** Describe Shape A as a transformation image of Shape B in as many different ways as possible.

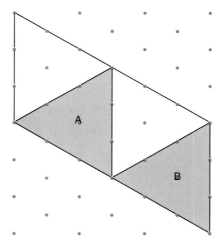

Reflect

When you see a shape and its transformation image in a design, how do you identify the transformation? Use diagrams in your explanation.

8.5 Constructing Tessellations

Focus Construct and analyse tessellations.

One of the basic ideas of Geometry is that of a *plane*. A plane is a flat surface. It has the property that a line joining any two points lies completely on its surface. What does this make you visualize?

Investigate

Work with a partner.
You will need tracing paper, plain paper, and a ruler.

One partner draws a triangle on his paper.
The other partner draws a quadrilateral on her paper.

Trace your shape.
Use the tracing to cover the paper with copies of your shape.
Try to do this with no overlaps or gaps.

> You can rotate or flip the shape to try to make it fit.

Compare your results with those of other classmates.
Can congruent copies of any triangle cover a plane with no overlaps or gaps? Justify your answer.
Can congruent copies of any quadrilateral cover a plane with no overlaps or gaps? Justify your answer.
What do you notice about the sum of the angles at a point where vertices meet?

Connect

When congruent copies of a shape cover a plane with no overlaps or gaps, we say the shape **tessellates**.
The design created is called a **tessellation**.

Not all polygons tessellate.
➤ This hexagon *does* tessellate.

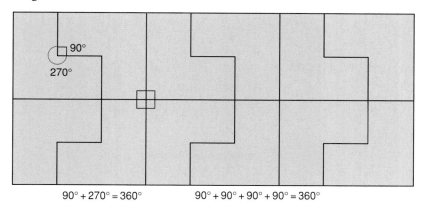

➤ This hexagon *does not* tessellate.
Here are two different pictures to illustrate this.

There are gaps among the hexagons.

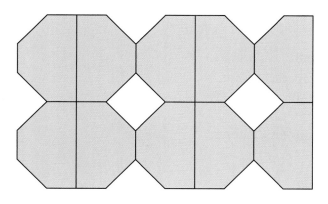
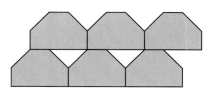

For copies of a polygon to tessellate, the sum of the angles at any point where vertices meet must be 360°. We say the *polygons surround a point*.

8.5 Constructing Tessellations   **463**

In *Investigate*, you found that triangles and quadrilaterals tessellate.

> At any point where vertices meet, the sum of the angle measures is 360°.

Acute triangle

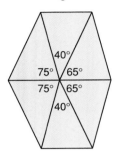

Obtuse triangle

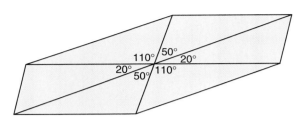

Six congruent triangles surround a point.

At each point:
75° + 40° + 65° + 65° + 40° + 75°
= 360°

At each point:
20° + 50° + 110° + 20° + 50° + 110°
= 360°

Convex quadrilateral

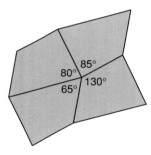

Concave quadrilateral

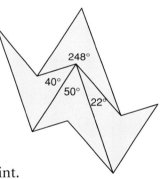

Four congruent quadrilaterals surround a point.

At each point:
80° + 85° + 130° + 65° = 360°

At each point:
50° + 40° + 22° + 248° = 360°

It is also possible for combinations of shapes to tessellate.

Example 1

Does each shape tessellate?
Justify your answer.

a) 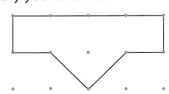   b)   c)

▶ **A Solution**

Trace each shape.
Try to cover a page with the shape so there are no gaps or overlaps.

a) The shape does not tessellate.
There are gaps that are triangles.

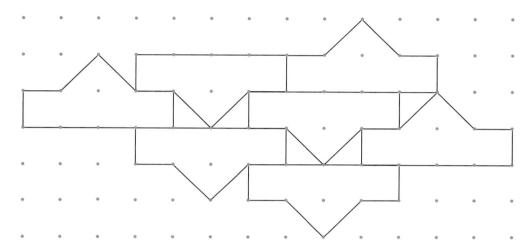

b) The shape does tessellate.
There are no gaps or overlaps.

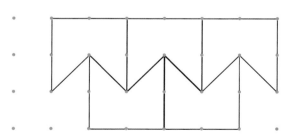

c) The shape does tessellate.
There are no gaps or overlaps.

Example 2

Look at each shape in *Example 1* that does not tessellate.
Try to combine it with other shapes in *Example 1* so that
the combined shape tessellates.
How many ways can you do this?

▶ *A Solution*

The shapes in parts a and b can be combined to make a T shape.
Tessellate with this new shape.

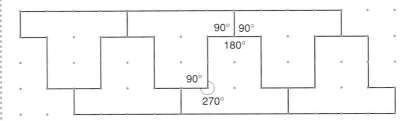 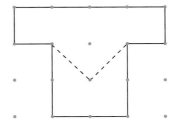

$90° + 90° + 180° = 360°$
$90° + 270° = 360°$
At each point where vertices meet, the sum of the angle measures is 360°.
So, the T shape tessellates.
The shape in part a and 2 of the shapes in part c can be combined
to make a rectangle.

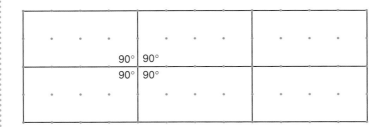 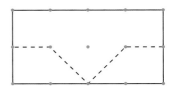

$90° + 90° + 90° + 90° = 360°$
At each point where vertices meet, the sum of the angle measures is 360°.
So, the rectangle tessellates.

The shape in part b and 2 of the shapes in part c can be combined
to make a rectangle.

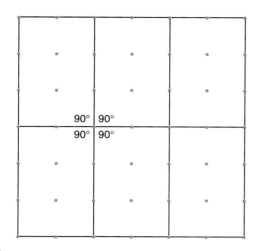
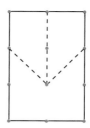

$90° + 90° + 90° + 90° = 360°$
At each point where vertices meet, the sum of the angle measures is 360°.
So, the rectangle tessellates.

Examples 1 and *2* illustrate that a shape that does *not* tessellate may be combined with one or more shapes to make a new shape that tessellates. This new shape is called a **composite shape**.
In the *Practice* questions, you will investigate to find which other shapes and combinations of shapes tessellate.

1. What is meant by "a shape tessellates"?
2. How can you tell that a shape does not tessellate?
3. Look around the classroom. Where do you see examples of tessellations? Describe each tessellation you see.
4. How do you know that all triangles tessellate?
5. How do you know that all quadrilaterals tessellate?

Practice

Check

Keep all the tessellations you create for Lesson 8.6.

6. a) Which of these designs are tessellations? Justify your answer.
   i)   ii)

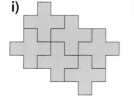
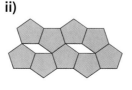

iii)   iv)

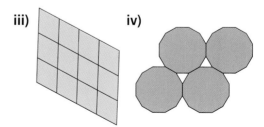

b) Which designs in part a are *not* tessellations? Justify your answer.

8.5 Constructing Tessellations   **467**

7. Use copies of each polygon. Does the polygon tessellate? If your answer is yes, create the tessellation. If your answer is no, explain how you know the shape does not tessellate.

a) b) c)

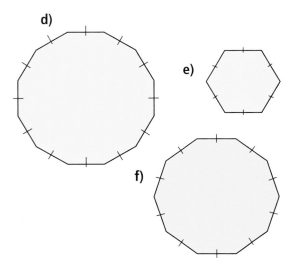

d)

e)

f)

8. a) Use angle measures to justify your answers in question 7.
   b) Which regular polygons tessellate? How do you know?

Apply

9. Use copies of the polygons in question 7. Find polygons that combine to make a composite shape that tessellates. How many different composite shapes can you find? Show your work.

10. Use Pattern Blocks. Does the composite shape of a square and a regular hexagon tessellate? If your answer is yes, show the tessellation. If your answer is no, explain why not.

11. Use copies of each polygon. Which of these irregular polygons tessellate? Use angle measures to justify your answer.

a)

b)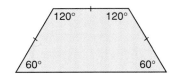

c)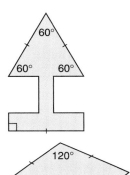

d)

e)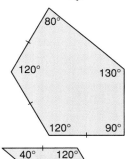

f)

468   UNIT 8: Geometry

12. **Assessment Focus**
    a) Which polygons in question 11 combine to make a composite shape that tessellates? Justify your answer in 2 ways.
    b) How many different composite shapes can you find that tessellate? Show your work.

13. Here is a regular octagon.
    Trace this octagon. Does this octagon tessellate? Justify your answer.

    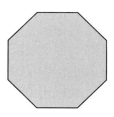

14. Identify the two shapes that combine to make the composite shape in this tessellation. Explain how you know that this composite shape tessellates.

    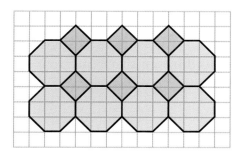

15. Where have you seen tessellations outside the classroom? What shapes were used in each tessellation?

16. **Take It Further** In question 13, you discovered that a regular octagon does not tessellate. Use dot paper. Draw an octagon that tessellates. Explain why it tessellates.

17. **Take It Further** A 7-sided shape is called a heptagon.
    a) Does a regular heptagon tessellate? Justify your answer.
    b) If your answer to part a is yes, draw the tessellation.
    c) If your answer to part a is no, draw a heptagon that tessellates.

18. **Take It Further** A 9-sided shape is called a nonagon.
    a) Does a regular nonagon tessellate? Justify your answer.
    b) If your answer to part a is yes, draw the tessellation.
    c) If your answer to part a is no, draw a nonagon that tessellates.

Reflect

How can you tell if a polygon or a composite shape tessellates? Use examples in your explanation.

Game

Target Tessellations

Players score points for each tessellation they create.
The first player to score 18 points wins.

HOW TO PLAY

1. Take turns to use the software to construct a shape that you think will tessellate.
   Use the software to check.

2. If the shape tessellates, the player scores 1 point for each side of the shape.
   For example, a 5-sided shape scores 5 points.
   If the shape does not tessellate, no points are scored.

3. Use the software to construct a different shape that you think will tessellate.
   The shape cannot have the same number of sides as the shape you used in *Step 1*.
   Use the software to check.

4. Play continues with a different shape each time.
   The first player to score 18 points wins.

TAKE IT FURTHER

Play the game again.
This time you choose the target number.
You cannot use shapes that have been used before.

YOU WILL NEED

Computer with geometry software, paper and pencil

NUMBER OF PLAYERS

2

GOAL OF THE GAME

To hit the target number exactly

What strategies did you use to hit the target number exactly?

470   UNIT 8: Geometry

8.6 Identifying Transformations in Tessellations

Focus Create and analyse tessellations using transformations.

In Lesson 8.5, you drew tessellations.
In this lesson, we will use transformations to describe some of these tessellations.

Investigate

Work on your own.
You will need isometric dot paper and Pattern Blocks.
Choose two Pattern Blocks that form a composite shape that tessellates.

Make a tessellation to cover a plane.
Copy your tessellation onto dot paper.
Label each composite shape in your tessellation.
Explain the tessellation in terms of transformation images.
That is, how do you rotate, translate, or reflect each composite shape to create the tessellation? Write your instructions carefully.

Trade instructions and the composite shape with a classmate.
Create your classmate's tessellation.
Check your tessellation with your classmate's tessellation.
How do they compare?

Connect

We can describe a tessellation in terms of transformations.

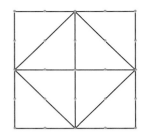

Label the shapes in the tessellation.

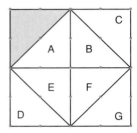

Start with the shaded shape.

Step 1 To get Shape A, reflect the shaded shape in the line of reflection shown.

Step 2 To get Shape D, reflect the shaded shape in the line of reflection shown.

Step 3 To get Shape E, reflect Shape D in the line of reflection shown.

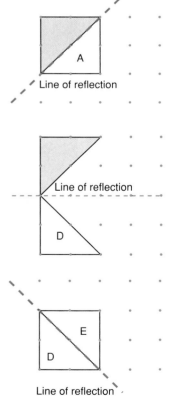

Repeat similar reflections to get shapes B, C, F, and G.

472 UNIT 8: Geometry

Under each transformation, the area of the shape does not change. This is known as **conservation of area**. This means that all the triangles in the tessellation have the same area.

Sometimes, a tessellation may be described by more than one type of transformation.

Example 1

In this tessellation, identify:
a) a translation
b) a reflection
c) a rotation

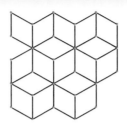

▶ A Solution

Label and shade Shape A as the original shape.
Label some more shapes.

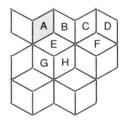

a) Shape C is a translation image of Shape A.
Shape A has been translated right to get Shape C.

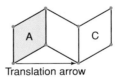

Translation arrow

b) Shape B is a reflection image of Shape A.
The line of reflection is the side the two shapes share.

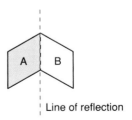

Line of reflection

c) Shape E is a rotation image of Shape A.
The point of rotation is a vertex they share.
Shape A is rotated 60° clockwise or 300° counterclockwise to its image, Shape E.

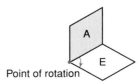

Point of rotation

8.6 Identifying Transformations in Tessellations   **473**

Example 2

a) Identify a combination of transformations in this tessellation.

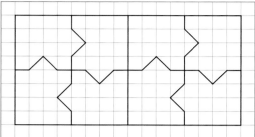

b) How do you know that the area of each shape is conserved?

▶ **A Solution**

a) Label the shapes in the tessellation.

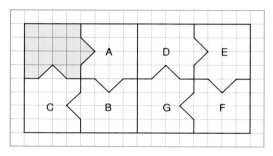

Start with the shaded shape.
To get Shape A, rotate the shaded shape
90° clockwise about P.

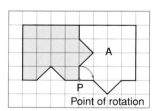

To get Shape B, rotate the
shaded shape 180° about P.

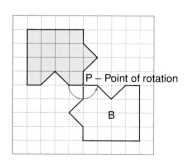

To get Shape C, rotate the shaded shape 90° counterclockwise about P.

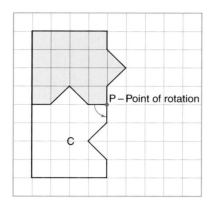

To get shapes D, E, F, and G, translate the square that contains the shaded shape, and shapes A, B, and C 8 units right.

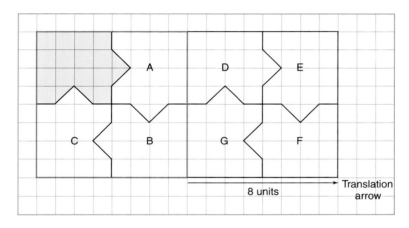

b) Each image is congruent to the original shape.
Congruent shapes are identical.
So, the area of each image is equal to the area of the original shape.
This means that the area of each shape is conserved.

1. Suppose the area of the shaded shape in *Example 1* is 8 cm². What is the area of Shape B? Shape E? How do you know?
2. In *Example 2*, identify a different combination of transformations in the tessellation.

8.6 Identifying Transformations in Tessellations   **475**

Practice

Check

3. In each tessellation, Shape A is the original shape. In each tessellation, identify:
   a) a translation
   b) a reflection
   c) a rotation

   i)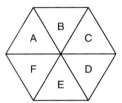

   ii)
A	B
C	D
E	F

   iii)
B	C	D
E	A	F
G	H	J

4. In each tessellation, Shape A is the original composite shape. In each tessellation, identify:
   a) a translation
   b) a reflection
   c) a rotation

   i)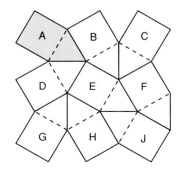

   ii)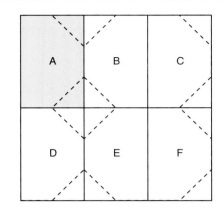

5. Here are three patterns. Describe the transformations that can be used to create each pattern. Start with the shaded shape.

   a)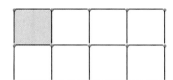

   b)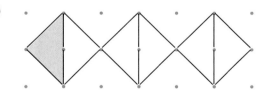

   c)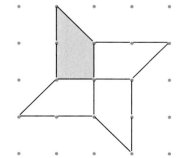

Apply

6. Here is a tessellation one student drew for *Practice* question 9 in Lesson 8.5.

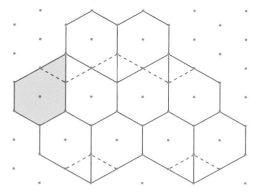

   a) Describe the tessellation in terms of translations.
   b) Describe the tessellation in terms of reflections.
   c) Is area conserved when each shape is transformed? How do you know?

7. Look at the picture called *Knights on Horseback*, by M.C. Escher. Identify different transformations that may have been used to create the tessellation. Start with the red shape.

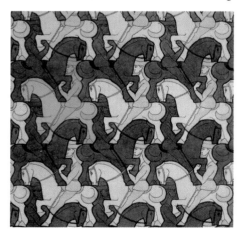

8. Use this shape and transformations to create a tessellation on grid paper.

Describe the tessellation in terms of transformations and conservation of area.

9. Here is a quilt design. Use a copy of the design.

Find as many transformations in the design as you can. Ignore the different patterns on the material. Consider only the shapes.

10. Use transformations to create your own quilt design. Describe the transformations you used.

11. Choose 2 more tessellations you created in Lesson 8.5. Describe each tessellation in terms of transformations and conservation of area.

8.6 Identifying Transformations in Tessellations **477**

12. **Assessment Focus** Look at the tessellations you created in Lesson 8.5. Choose a tessellation that can be described in more than one way. Copy the tessellation onto dot paper. Label the shapes. Colour the tessellation. Describe the tessellation in terms of transformations and conservation of area. Describe the tessellation in as many ways as you can.

13. **Take It Further** Here is a flooring pattern. Use a copy of this pattern. Use transformations to describe the patterns in one square.

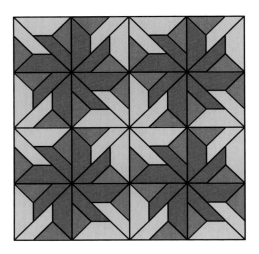

14. **Take It Further** The Alhambra is a walled city and fortress in Granada, Spain. It was built in the 14th century.

Here is part of one of its many tiling patterns. The pattern is a tessellation. Copy this tessellation onto dot paper.

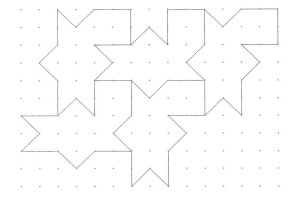

a) Identify the composite shape that tessellates.
b) Continue the tessellation to cover the page.
c) Use transformations to describe the tessellation.

Reflect

When you use transformations to describe a tessellation, how do you decide which transformations to use?
Include a tessellation in your explanation.

Using a Computer to Create Tessellations

Focus | Use technology to create and analyse tessellations.

Geometry software can be used to create tessellations.
Use available geometry software.

Open a new sketch. Check that the distance units are centimetres. To display a "grid paper" screen, display a coordinate grid. Then hide the axes.

> Should you need help at any time, use the software's Help Menu.

To create a tessellation:

Construct a triangle ABC.

Select the triangle.
Use the software to translate, reflect, or rotate the triangle.

Continue to transform the triangle or an image triangle to create a tessellation.
Colour your design to make it attractive.
Print your design.

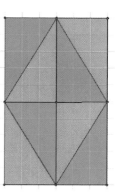

Check

Construct two different shapes.
Use these shapes to make a composite shape.
Use any or all of the transformations you know to create a tessellation that covers the screen.
Colour your tessellation.
Describe your tessellation in terms of transformations.
What can you say about the area of each composite shape in your tessellation?

Explaining Your Answer

How many times have you been asked to "explain your answer" when you answer a math question?

The explanation for your answer
shows your thinking.
It also helps you to review the solution
when you study for a test.

Part of explaining your answer to a question
is showing how you know your answer is correct.

Here is a problem.

What is the ones digit when 9 is multiplied by itself 500 times?

Compare these two solutions.

Solution 1

When 9 is multiplied by itself 500 times means...
9 × 9 × 9 × 9... × 9 (There are 500 9's.)
I know 9 × 9 = 8<u>1</u>
      9 × 9 × 9 = 729
      9 × 9 × 9 × 9 = 656<u>1</u> (Use a calculator.)
    9 × 9 × 9 × 9 × 9 = 59049
  9 × 9 × 9 × 9 × 9 × 9 = 53144<u>1</u>
Every time 9 is multiplied an even number of times, the ones digit is 1. Since 500 times is an even number of times, the ones digit will be 1.

The ones digit when 9 is multiplied by itself 500 times is 1.

Solution 2

Answer: 1

How does the solution on the right explain the answer?

Here are some things you can do to show how you know your answer is correct.

Strategies for Success

- Show all the steps in a logical order so that someone else can follow your thinking.
- Show all calculations.
- When a question involves one of the four operations, use estimation to check.
- Use a different strategy. For example: if the question involves subtraction, use addition to check.
- Verify the solution. For example: when solving an equation, substitute the solution into the original equation to check.
- Use thinking words and phrases such as:
  – because
  – so that means . . .
  – as a result
  – if you . . . then . . .
- Include labelled sketches, diagrams, or tables to help explain your answer.

Practice

Answer these questions.
Write a complete solution that explains your answer.

1. Hori is buying carpet for his living room. The living room is rectangular, with dimensions 4 m by 5 m. The regular price of the carpet is $9.99 per square metre. How much will Hori save when he buys the carpet on sale for 20% off?

2. Suppose you are in charge of setting up the cafeteria for a graduation dinner. One hundred twenty-two people will attend.
   The tables seat either 8 or 10 people. You do not want empty seats.
   How many of each size table will you need to make sure everyone has a seat? List all possible combinations.

Unit Review

What Do I Need to Know?

✓ Each view of an object provides information about the shape of the object.

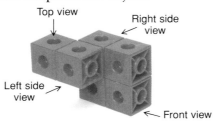
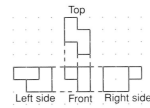

✓ An isometric drawing shows the three dimensions of an object.

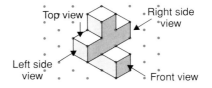

✓ An object may be rotated: horizontally about a vertical axis   or   vertically about a horizontal axis

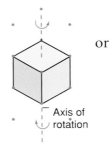
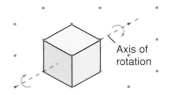

✓ Under a transformation, the area of a shape does not change. This is called *conservation of area*.

✓ When congruent copies of a shape cover a plane with no overlaps or gaps, we say the shape *tessellates*. The design created is called a *tessellation*. The tessellation can be described using transformations.

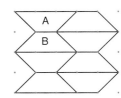

For a polygon or a composite polygon to tessellate, the sum of the angle measures where vertices meet must be 360°. The only regular polygons that tessellate are triangles, squares, and hexagons.

482   UNIT 8: Geometry

What Should I Be Able to Do?

LESSON

8.1

1. Sketch the front, top, and side views of each object.
   a)
   b)

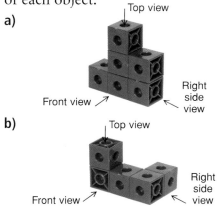

2. Use 3 linking cubes. Use square dot paper.
   a) Build an object. Draw the front, top, and side views of the object.
   b) How many different objects can you make with 3 linking cubes? Make each object. Draw its front, top, and side views.

8.2

3. Here is an object made from linking cubes.

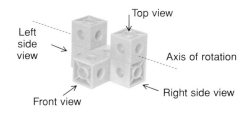

   The object is rotated 180° vertically toward you.
   a) Predict the front, top, and side views after the rotation. Sketch your predictions.
   b) Build the object. Rotate the object to check your predictions. Draw each view after the rotation.

4. This object is rotated horizontally.

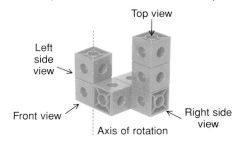

   A new view is shown. Describe the rotation that produced each view.
   a) front view

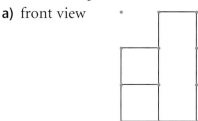

   b) top view

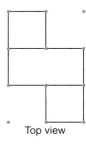

   c) right side view

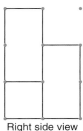

Unit Review   483

LESSON

8.3

5. Use linking cubes. Use this set of views to build an object. How can you check that your object is correct?

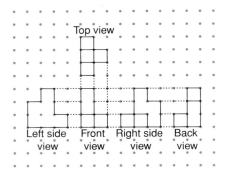

6. Which object below has these views? How do you know?

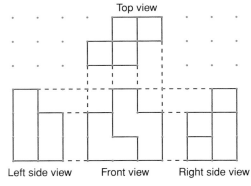

a)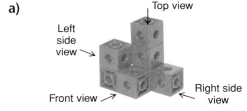

b)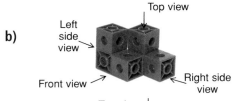

c)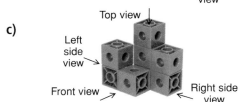

7. Two objects in question 6 did not have views that matched those given. Draw the top, front, and side views of these objects.

8.4

8. Use a copy of this tessellation.

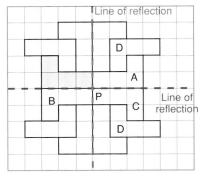

Match each transformation of the shaded shape to a transformation image.
   a) a translation 3 units down and 4 units right
   b) a rotation of 180° about point P
   c) a reflection in the red line
   d) a reflection in the blue line

9. On grid paper, copy this shape, the red line, and point P.

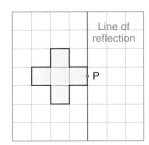

Draw the image of the shape after each transformation. What do you notice?
   a) a rotation of 180° about P
   b) a reflection in the red line
   c) a translation 3 units right

LESSON

8.5

10. Copy this shape on grid paper.
   a) How many ways can you use the shape to tessellate? Show each way.
   b) Use transformations to describe each tessellation in part a.

11. Is it possible to create a tessellation using each polygon below? If it is possible, use dot paper to draw the polygon and show the tessellation.
   If it is not possible, explain why not.
   a) a regular octagon
   b) a hexagon
   c) a parallelogram

12. Use grid paper. Find a concave polygon that tessellates. Explain how it tessellates.

13. Use these shapes.

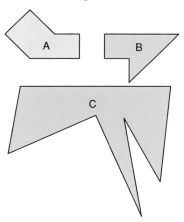

   a) Try to tessellate with each shape. Identify the shapes that tessellate.
   b) For each shape that does not tessellate:
      • Find another shape that forms a composite shape that tessellates.
      • Tessellate with the composite shape.

8.6

14. Here is a quilt design.

Use as many different transformations as you can to describe the design. Ignore the different patterns, colours, and textures on the material. Consider only the shapes.

15. Look at the picture called *Reptiles*, by M.C. Escher. What transformations of the red reptile are needed to create this tessellation?

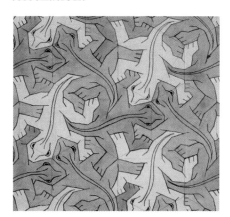

16. Draw two shapes that combine to make a composite shape that tessellates. Use the composite shape to make a tessellation. Describe the tessellation in terms of transformations and conservation of area.

Practice Test

1. Draw the front, top, and side views of this object.

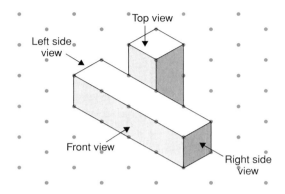

2. Use linking cubes to build the object at the right.
   a) Draw the front, top, and side views of the object.
   b) Rotate the object horizontally 90° clockwise. Draw the new front, top, and side views of the object.
   c) Rotate the object vertically 180° away from you. Draw the new front, top, and side views of the object.

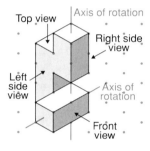

3. Use linking cubes.
   Use these views to build an object.
   Explain how you did this.

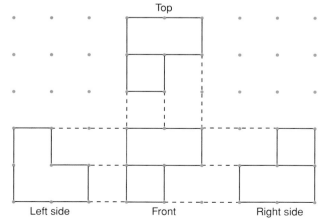

4. Use point P as a point of rotation.
   Use the blue line and red line as lines of reflection.
   Each of A, B, C, and D is a transformation image of the shaded shape. Identify each transformation.

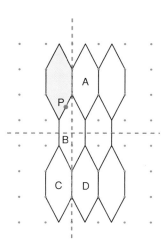

5. Use copies of each polygon.
   Which of these polygons tessellate?
   Use angle measures to justify your answer.

   a) 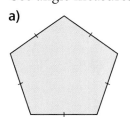   b) 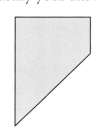   c)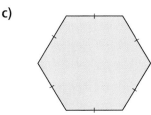

6. Choose a polygon in question 5 that tessellates.
   Create the tessellation.
   Colour the tessellation.
   Describe the tessellation in terms of transformations and conservation of area.
   Try to describe the tessellation in two different ways.

7. Three students looked at a tessellation.
   Igal used translations to describe the tessellation.
   Shaian used rotations to describe the tessellation.
   Cherie used reflections to describe the tessellation.
   All three students were correct. What might the tessellation look like?
   Draw a diagram to show your thinking.

8. Julie will use both of these tiles to cover her floor.
   Use isometric dot paper.
   Draw 2 different tessellations Julie could use.
   For each tessellation, use transformations to explain how to create the tessellation using a composite shape.

 10 cm   10 cm

Practice Test   487

Unit Problem: Creating Tessellating Designs

M.C. Escher was a famous Dutch graphic artist.
He designed many different tessellations.

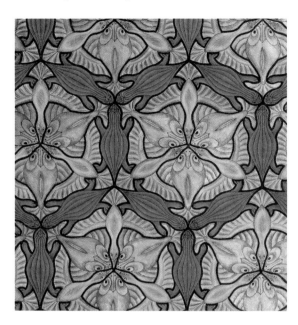

You will create two designs in the Escher style.
The first design is in the style of *Knights on Horseback*, on page 477.

Part 1

Use square dot paper or grid paper.
Tessellate with a shape of your choice.
Sketch a design on one shape.
Repeat the sketch until every shape in the plane has the design.
Use transformations to describe how to create the tessellation beginning with one shape.

Part 2

Instead, you could start with a rectangle, parallelogram, or regular hexagon.

Start with a square. Draw congruent curves on 2 sides.
A curve that goes "in" on one side must go "out" on the other side.
Draw different congruent curves on the other two sides.

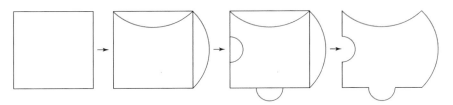

Check List

Your work should show:
- ✓ the initial shape you created for each tessellation
- ✓ the tessellations you created
- ✓ how you used transformations to create the tessellations
- ✓ the correct use of mathematical language

Trace the new shape on cardboard.
Cut out the shape.
Use it to tessellate.

Add details to your shape so it represents an animal or an object of your choice.

Reflect on Your Learning

You have worked in three dimensions with objects and in two dimensions with shapes. Which do you prefer? Give reasons for your choice.
Include a paragraph about what you have learned.

Investigation   A Population Simulation

Work with a partner.

An animal population changes from year to year depending on the rates of birth and death, and on the movement of the animals. A **simulation** is a model of a real situation. You will use a simulation to investigate how an animal population might decline.
As you complete this *Investigation*, include your completed table, graph, and written answers to the questions. You will hand these in.

Materials:
- a paper cup
- 30 to 40 two-sided counters
- 0.5-cm grid paper

➤ The counters are the animals in your population.
Count the number of animals.
Record that number for Year 0 in the *Population* table.

➤ Put the counters in the cup.
Choose which colour will be "face up."
Pour the counters from the cup.
Counters that land face up represent animals that died or moved away during the first year. Set them aside.
Count the number of animals left.

Record that number as the population for Year 1.
Calculate and record the decrease in population.
Write the decrease as a fraction of the previous year's population.
Write the fraction as a percent.

➤ Place the counters representing live animals back in the cup.
Repeat the experiment for Year 2.
Record the data for Year 2 in the table.

➤ Continue the simulation. Record your data for up to 8 years, or until you run out of animals.

➤ Graph the population data.
Plot *Year* horizontally and *Number of Animals* vertically.
Explain your choice of graph.
What trends do you see in the graph?

Population									
Year	0	1	2	3	4	5	6	7	8
Number of animals									
Decrease in population									
Change as a fraction									
Change as a percent									

➤ Describe the patterns you see in the table. Approximately what fraction and percent of the population remain from year to year?
➤ Use the patterns to predict the population for Year 9.
➤ How long does it take the population of animals to decrease to one-half its original size?
➤ What would happen to a population if there were no births to add to the population each year?

Take It Further

➤ Suppose you repeated this experiment, beginning with more animals. Would you see the same pattern? Explain.
Combine your data with data of other students to find out.
➤ Which environmental factors may cause an animal population to change?

A Population Simulation

Units 1–8 Cumulative Review

UNIT 1

1. Jasmine and Logan are swimming in a rectangular pool. The pool has dimensions 23 m by 9 m. They decide to race from one corner of the pool to the opposite corner. Jasmine swims diagonally across the pool. Logan is not a good swimmer so he swims along two sides of the pool.
   a) Who swims farther?
   b) How much farther does that person swim?

2. Is each statement true or false? Justify your answers.
   a) $\sqrt{5} + \sqrt{2} = \sqrt{7}$
   b) $\sqrt{46}$ is between 6 and 7, and closer to 7.
   c) $\sqrt{36} + \sqrt{64} = 14$

UNIT 2

3. Vernon used his allowance to buy lunch at school each day. Suppose Vernon spent $4 each day for one week. How much more allowance money did he have one week ago?

4. Evaluate. Show all steps.
   a) $(-4) \times (-3) + 3(-6)$
   b) $\dfrac{-18}{(-4) + (-25) \div 5}$
   c) $\dfrac{[11-(-5)] \div [2 \times (-2)]}{4 \div (-2)}$
   d) $-12 + (-21) \div 3 + 3 \times 6$

UNIT 3

5. Shantel worked a $4\frac{1}{2}$-h shift at the coffee shop. She spent $\frac{5}{6}$ of this time cleaning tables. How many hours did Shantel spend cleaning tables during her shift?

6. Find each quotient. Use number lines to illustrate the answers.
   a) Austin worked for $2\frac{1}{3}$ h at the pet store and cleaned 4 fish tanks. How long did Austin spend cleaning each tank?
   b) Riley has $5\frac{1}{4}$ cups of chocolate chips. She needs $\frac{3}{4}$ cup of chocolate chips to make one batch of cookies. How many batches of chocolate-chip cookies can Riley make?

UNIT 4

7. A rectangular prism has three equal dimensions. Sketch the prism. Show the equal dimensions. What is another name for this prism?

8. Identify each object from the description of its net.
   a) 2 congruent triangles and 3 rectangles
   b) 2 congruent circles and 1 rectangle
   c) 2 congruent hexagons and 6 congruent rectangles

9. Use 60 linking cubes. Assume the face of each cube has an area of 1 square unit.
   a) Use all the cubes. Create a rectangular prism with the greatest possible surface area.
   b) Use all the cubes. Create a rectangular prism with the least possible surface area.
   Explain the strategy you used each time.

492 UNITS 1–8

UNIT 5

10. Write each percent as a fraction and as a decimal.
   a) 63.25%
   b) $1\frac{1}{8}\%$
   c) 0.28%
   d) 0.7%

11. The cost of an adult bus ticket in 2006 was $2.00. Because of the rising cost of fuel, the price was increased by $7\frac{1}{2}\%$ in 2007. The cost is expected to increase by about 12% in 2008.
   a) Calculate the cost of a bus ticket in 2008. Describe the strategy you used.
   b) Could you have found the cost by finding $119\frac{1}{2}\%$ of the cost of a ticket in 2006? Justify your answer.

12. There are 3 goldfish and 5 guppies in Tank A. There are 5 goldfish and 7 guppies in Tank B.
   a) What is the ratio of goldfish to all the fish in each tank?
      i) Tank A   ii) Tank B
   b) Write each ratio in part a as a fraction.
   c) Suppose all the fish were moved into one large tank. What is the ratio of goldfish to all the fish in the large tank?

13. Scott delivers 14 newspapers in 10 min. This is 28% of his round.
   a) How many more papers does Scott have to deliver?
   b) Scott continues to deliver papers at the same rate. How long will it take him to deliver all his papers?

14. Express as a unit rate.
   a) Paula picked 120 apples in 15 min.
   b) Vince painted 6 fence posts in 30 min.
   c) Jay ran 42 km in 3.5 h.

UNIT 6

15. Expand.
   a) $4(13 + 3d)$
   b) $-7(5 - 6c)$
   c) $8(-9d + 7)$
   d) $6(8e - 1)$

16. Felix used algebra to solve this equation:
$3x + 5 = -7$
Look at Felix's work.
$$3x + 5 = -7$$
$$3x + 5 + 7 = -7 + 7$$
$$3x + 12 = 0$$
$$3x + 12 - 12 = 0 - 12$$
$$3x = -12$$
$$\frac{3x}{3} = \frac{-12}{3}$$
$$x = -4$$

   a) Felix made an unnecessary step at the beginning of the solution. What was this step?
   b) Did Felix get the correct solution? How can you find out?
   c) Explain why the unnecessary step did or did not affect the solution.

17. Copy and complete each table of values.
   a) $y = -3x$

x	-2	-1	0	1	2
y					

   b) $y = -x + 3$

x	-2	-1	0	1	2
y					

UNIT

18. Troy is making baseball-cap organizers to sell at a craft sale. He needs 8 clothespins for each organizer. An equation for this relation is $c = 8n$, where n represents the number of organizers, and c represents the number of clothespins needed.
   a) Use the equation to create a table of values.
   b) Suppose Troy makes 12 organizers. How many clothespins does he need?
   c) Suppose Troy has 144 clothespins. How many organizers can he make?
   d) Construct a graph for the data in part a.
   e) Describe the relationship between the variables in the graph.
   f) Find the ordered pair on the graph that shows how many organizers can be made with 48 clothespins.

7

19. Each graph below shows how water is used in Canadian homes.

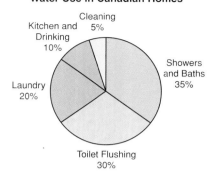

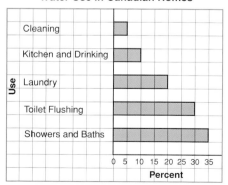

   a) What are the advantages of each graph?
   b) What are the disadvantages of each graph?
   c) Which type of graph is more appropriate to display these data? Justify your choice.
   d) Could you use a line graph to display these data? Why or why not?

20. Two groups debated how much money the government was spending on social programs. Each group used the same data to draw a graph.

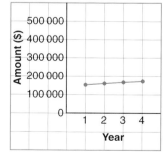
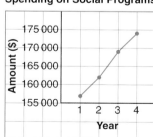

   a) What impression does each graph give?
   b) Describe how the graphs create a false impression.
   c) Who might use each graph? Justify your answer.

UNIT

21. Gavin bought a bag of gift wrapping bows. The bag contains 4 red bows, 7 gold bows, 3 white bows, and 6 silver bows. Gavin takes one bow without looking, records the colour, then returns the bow to the bag.
The process is repeated.
Find the probability of each event:
a) 2 gold bows
b) a red bow, then a white bow
c) a gold bow, then a silver bow
d) not a gold bow, then a white bow

22. On a particular day in January, there is a 40% probability of snow in Whitehorse, a 55% probability of snow in Iqaluit, and a 35% probability of snow in Fort McMurray. What is the probability that it will snow in all 3 cities on that day?

23. A true/false test has 6 questions. Jayla answered all the questions by guessing.
a) What is the probability that Jayla answered all the questions correctly?
b) Use a tree diagram to verify your answer in part a.

8

24. Sketch the top, front, and side views of this object.

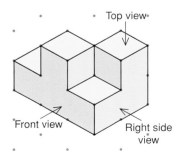

25. Use linking cubes. Use this set of views to build an object.
How can you check that your object is correct?

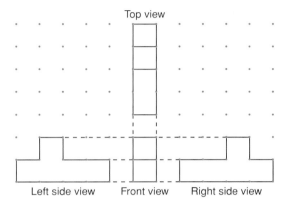

26. Copy this shape. Use transformations to create a tessellation on grid paper. Describe the

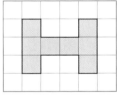

tessellation in terms of transformations and conservation of area.

27. Here is a regular hexagon. Does this hexagon tessellate?
If your answer is yes, create the tessellation.

If your answer is no, explain how you know the shape does not tessellate. Then find a polygon that combines with the hexagon to make a composite shape that tessellates. Create the tessellation.

Cumulative Review   495

Answers

Unit 1 Square Roots and the Pythagorean Theorem, page 4

1.1 Square Numbers and Area Models, page 8

4.a) ii)   **b)** i)   **c)** iii)
5.a) 64 square units   **b)** 100 square units
   c) 9 square units
6. Yes; $36 = 6^2$

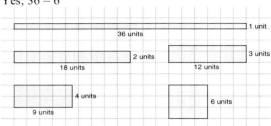

7. No; 28 cannot be modelled using a square.

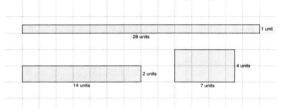

8. $25 = 5^2$
9. 12 cannot be modelled using a square.
10.a)   **b)**
   c)   **d)**
11.a) 10 m   **b)** 8 cm   **c)** 9 m   **d)** 20 cm
12.c) $81 = 9^2$
13. Answers may vary. For example:

Base (cm)	Height (cm)	Perimeter (cm)
64	1	130
32	2	68
16	4	40
8	8	32

A square with side length 8 cm has the smallest perimeter.

14. Answers may vary.
For example: 9, 36, 81, 144, 225, 324
15.a) 9 and 16   **b)** 36 and 49
   c) 64 and 81   **d)** 196 and 225
16.a) 12 m   **b)** 48 m
   c) 20 pieces; Assumptions may vary. For example: You must buy whole pieces.
17. 5 m
18. $64 = 8^2$, $81 = 9^2$, $100 = 10^2$, $121 = 11^2$, $144 = 12^2$, $169 = 13^2$, $196 = 14^2$
19.b) 49 m^2 **c)** 7 m   **d)** 28 m
   e) $280; Assumptions may vary. For example: There is no tax on fencing.
20.a) 30 cm   **b)** 2 cuts
21.a) 1, 4, 9, 7, 7, 9, 4, 1, 9, 1, 4, 9, 7, 7, 9
   b) The digital root of a square number is either 1, 4, 7, or 9.
   c) 2809, 4225, 625

1.2 Squares and Square Roots, page 15

5.a) 16   **b)** 36   **c)** 4   **d)** 81
6.a) 64   **b)** 9   **c)** 1   **d)** 49
7.a) 5   **b)** 9   **c)** 8   **d)** 13
8.a) i) 1   **ii)** 100   **iii)** 10 000   **iv)** 1 000 000
   b) i) 100 000 000   **ii)** 1 000 000 000 000

9.a)

i) Number = 50	ii) Number = 100	iii) Number = 144	iv) Number = 85
1, 50	1, 100	1, 100	1, 85
2, 25	2, 50	2, 72	5, 17
5, 10	4, 25	3, 48	
	5, 20	4, 36	
	10, 10	6, 24	
		8, 18	
		9, 16	
		12, 12	

$100 = 10^2$, $144 = 12^2$
   b) ii) 10   **iii)** 12
10.a) 1, 2, 4, 8, 16, 32, 64, 128, 256; 16
   b) 1, 5, 25, 125, 625; 25
   c) 1, 11, 121; 11
11. 225 and 324; Each has an odd number of factors.
12.a) i) 1, 2, 3, 4, 6, 8, 12, 16, 24, 32, 48, 96
   ii) 1, 2, 4, 11, 22, 44, 121, 242, 484
   iii) 1, 2, 3, 4, 5, 6, 8, 10, 12, 15, 16, 20, 24, 30, 40, 48, 60, 80, 120, 240
   iv) 1, 2, 4, 8, 19, 38, 76, 152
   v) 1, 3, 7, 9, 21, 49, 63, 147, 441
   vi) 1, 2, 3, 6, 9, 18, 27, 54
   b) 484 and 441; Each has an odd number of factors.
13.a) 1   **b)** 7   **c)** 12   **d)** 3

e) 4   f) 10   g) 25   h) 15
14. a) 3   b) 6   c) 10   d) 117
15. a) 4   b) 121   c) 225   d) 676
16. a) 13   b) 6   c) 14
17. 529
18. If I square a number then take the square root, I end up with the original number.
$3^2 = 9$ and $\sqrt{9} = 3$
19. a) $\sqrt{9}, 4, \sqrt{36}, 36$ b) $\sqrt{100}, 15, 19, \sqrt{400}$
c) $\sqrt{81}, \sqrt{100}, 11, 81$ d) $\sqrt{36}, \sqrt{49}, \sqrt{64}, 9$
20. $1^2 = 1, 2^2 = 4, 3^2 = 9, 4^2 = 16, 5^2 = 25, 6^2 = 36,$
$7^2 = 49, 8^2 = 64, 9^2 = 81, 10^2 = 100, 11^2 = 121,$
$12^2 = 144, 13^2 = 169, 14^2 = 196, 15^2 = 225,$
$16^2 = 256, 17^2 = 289, 18^2 = 324, 19^2 = 361,$
$20^2 = 400$
21. a) i) 11   ii) 111   iii) 1111   iv) 11 111
b) $\sqrt{12\,345\,654\,321} = 111\,111$
$\sqrt{1\,234\,567\,654\,321} = 1\,111\,111$
$\sqrt{123\,456\,787\,654\,321} = 11\,111\,111$
$\sqrt{12\,345\,678\,987\,654\,321} = 111\,111\,111$
22. a) i) 4   ii) 9   iii) 16   iv) 25
b) i) 13   ii) 25   iii) 20   iv) 34
c) $3^2 + 4^2$ is a square number. The sum of two square numbers may or may not be a square number.

1.3 Measuring Line Segments, page 20
3. a) 9   b) 16   c) 49   d) 100   e) 36   f) 144
4. a) 1   b) 8   c) 12   d) 13   e) 11   f) 25
5. a) 6 cm   b) 7 m   c) $\sqrt{95}$ cm   d) $\sqrt{108}$ m
6. a) 64 cm²   b) 44 cm²   c) 7 m²   d) 169 m²
7. a) $A = 18$ square units, $s = \sqrt{18}$ units
b) $A = 53$ square units, $s = \sqrt{53}$ units
c) $A = 34$ square units, $s = \sqrt{34}$ units
8. a) Square A
b) Squares C and D have the same area.
9. d < b < a < c
a) $A = 37$ square units, $s = \sqrt{37}$ units
b) $A = 32$ square units, $s = \sqrt{32}$ units
c) $A = 41$ square units, $s = \sqrt{41}$ units
d) $A = 20$ square units, $s = \sqrt{20}$ units
10. a) $A = 25$ square units, $s = 5$ units
b) $A = 13$ square units, $s = \sqrt{13}$ units
c) $A = 26$ square units, $s = \sqrt{26}$ units
d) $A = 29$ square units, $s = \sqrt{29}$ units

11. Line segment b is shorter.
If you draw a square on each segment, the area of the square in part b is smaller, so its side length is also smaller.
12. a) $\sqrt{20}$ units   b) $\sqrt{41}$ units
c) $\sqrt{10}$ units   d) $\sqrt{8}$ units
13. Divide the square into 4 congruent triangles. The area of each triangle is $\frac{1}{2}$ square units. So, the area of the square is 2 square units.
14. The area must be 25 cm². Draw 4 triangles each with area 6 cm² and place a small square with area 1 cm² in middle.

1.4 Estimating Square Roots, page 25
4. a) 15   b) 3   c) 22   d) 1
5. a) 2 and 3   b) 3 and 4   c) 7 and 8
d) 6 and 7   e) 13 and 14   f) 10 and 11
6. 2.6
7. a) $\sqrt{30}$; 30 is about halfway between 25 and 36.
$\sqrt{64}$; 64 is exactly 8.
$\sqrt{72}$; 72 is about halfway between 64 and 81.
b) $\sqrt{23}$ is about 4.8. $\sqrt{50}$ is about 7.1.
8.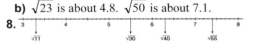
9. a) Greater than   b) Greater than
c) Equal to   d) Less than
10. a) 8   b) 8   c) 9   d) 12
11. a) False   b) False   c) True
12. a) 4.80   b) 3.61   c) 8.83
d) 11.62   e) 7.87   f) 6.71
13. a) 9.6 cm   b) 20.7 m   c) 12.2 cm   d) 5.4 m
14. a) Bad estimate   b) Good estimate
c) Good estimate   d) Good estimate
15. a) 2.24 m by 2.24 m
b) 10 m since perimeter is about 8.96 m
16. a) 12.33 m by 12.33 m   b) 49.32 m
17. 7.35
18. Answers will vary. For example:
My classroom is 10 m by 7 m. The area is 70 m².
If my classroom were a square: $s = 8.37$ m
19. a) 6.93 m by 6.93 m   b) 16 m²
20. Always a perfect square
21. Since 7.67 is closer to 8 than 7, the whole number is closer to 64.
22. $9^2 = 81$ and $10^2 = 100$
Any number between 81 and 100 has a square root between 9 and 10.
23. a) 13   b) 9.85   c) 5   d) 9.22   e) 3.61

Unit 1 Technology: Investigating Square Roots with a Calculator, page 29
1. a) 21   **b)** 4.36; Approximation
   c) 7.94; Approximation   **d)** 23

Unit 1 Mid-Unit Review, page 30
1. 100
2. a) 4   **b)** 7   **c)** 14   **d)** 20
3. a) 121   **b)** 8   **c)** 13   **d)** 15
4. a) i) $A = 16$ cm²   **ii)** $s = \sqrt{16}$ cm
   b) i) $A = 36$ cm²   **ii)** $s = \sqrt{36}$ cm
5. a) 1, 2, 3, 4, 6, 8, 9, 12, 18, 24, 27, 36, 54, 72, 108, 216; Not a square number since even number of factors.
   b) 1, 2, 4, 7, 13, 14, 26, 28, 52, 91, 182, 364; Not a square number since even number of factors
   c) 1, 3, 9, 27, 81, 243, 729; Square number since odd number of factors; $\sqrt{729} = 27$
6. Let the square number represent the area of a square, then its square root is the side length.
7. a) $\sqrt{24}$; 24 is not a square number.
   b) 81 cm²
8. a) 72 cm²   **b)** $\sqrt{72}$ cm   **c)** 8.5 cm
9. a) 12   **b)** 34
10. a) 1 and 2   **b)** 8 and 9
   c) 8 and 9   **d)** 7 and 8
11. a) 4.12   **b)** 10.39   **c)** 5.74   **d)** 8.89

1.5 The Pythagorean Theorem, page 34
3. a) 50 cm²   **b)** 52 cm²
4. a) 64 cm²   **b)** 28 cm²
5. a) 10 cm   **b)** 13 cm   **c)** 4.5 cm   **d)** 5.8 cm
6. a) 9 cm   **b)** 24 cm   **c)** 9.8 cm   **d)** 6.7 cm
7. a) 7.6 cm   **b)** 20 cm   **c)** 20 cm
8. a) 8.06 cm   **b)** 11.66 cm   **c)** 25 cm
9. a) 5 cm   **b)** 10 cm   **c)** 15 cm
Compared to rectangle a, the lengths are two times greater in rectangle b and three times greater in rectangle c. The next rectangle has dimensions 12 cm by 16 cm and diagonal 20 cm.
10. The longest side is the hypotenuse.
11. The two right triangles formed by the diagonals both have legs 12 cm and 16 cm. So the diagonals must be the same length.
12. 1 and $\sqrt{17}$, $\sqrt{2}$ and 4, $\sqrt{3}$ and $\sqrt{15}$, 2 and $\sqrt{14}$, $\sqrt{5}$ and $\sqrt{13}$, $\sqrt{6}$ and $\sqrt{12}$, $\sqrt{7}$ and $\sqrt{11}$, $\sqrt{8}$ and $\sqrt{10}$, 3 and 3
For each answer, the sum of the squares is 18.
13. a) 6 units   **b)** 8 units   **c)** 4 units
14. a)   **b)**

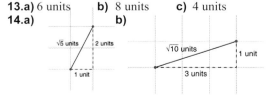

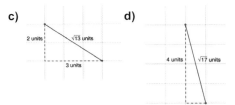

15. $15^2 = 12^2 + 9^2$; Length of legs: 12 cm and 9 cm
16. 3.535 cm², 6.283 cm², 9.817 cm²
The sum of the areas of the semicircles on the legs is equal to the area of the semicircle on the hypotenuse.
17. a)   **b)**
   c)

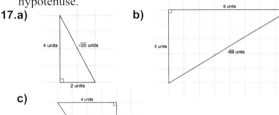

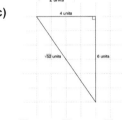

18. a) $\sqrt{2}$ cm, $\sqrt{3}$ cm, $\sqrt{4}$ cm, $\sqrt{5}$ cm, $\sqrt{6}$ cm, $\sqrt{7}$ cm
   b) 1.4, 1.7, 2.0, 2.2, 2.4, 2.6
   c) 1.4 cm, 1.7 cm, 2.0 cm, 2.2 cm, 2.4 cm, 2.6 cm
   d) The lengths of the hypotenuses are the square roots of consecutive whole numbers.

1.6 Exploring the Pythagorean Theorem, page 43
3. a) Yes; $38 + 25 = 63$ **b)** No; $38 + 25 \neq 60$
4. a) No; $10^2 + 1^2 \neq 13^2$ **b)** No; $7^2 + 5^2 \neq 8^2$
   c) Yes; $15^2 + 8^2 = 17^2$
5. No, since it is not a right triangle
6. a) Yes   **b)** No   **c)** Yes   **d)** Yes
   e) No   **f)** No   **g)** No   **h)** Yes
7. a, c, d, f
8. Yes, it is a right angle since $9^2 + 12^2 = 15^2$.
9. Yes, the triangle is a right triangle;
$$7^2 = 6^2 + \left(\sqrt{13}\right)^2$$
No, the side lengths do not form a Pythagorean triple since $\sqrt{13}$ is not a whole number.
10. $3^2 + 5^2 \neq 7^2$; Not a right triangle
11. a) Legs: 3, 4; 6, 8; 9, 12; 12, 16; 15, 20
Hypotenuse: 5; 10; 15; 20; 25
   b) All triples are multiples of first triple 3, 4, 5.
   c) 10, 24, 26; 15, 36, 39; 20, 48, 52; 25, 60, 65
12. a) 50, since $14^2 + 48^2 = 50^2$

b) 40, since $32^2 + 24^2 = 40^2$
c) 35, since $12^2 + 35^2 = 37^2$
d) 99, since $20^2 + 99^2 = 101^2$

13. Hold the 1st, 4th, and 8th knots to form a right triangle with side lengths 3 units, 4 units, and 5 units.

14. Yes; Since $48^2 + 55^2 = 73^2$; all angles are right angles.

15. 40 m and 9 m, since $9 + 40 + 41 = 90$ and $9^2 + 40^2 = 41^2$

16. a) For obtuse triangles, the area of the square on the longest side is greater than the sum of the areas of the squares on the two smaller sides.
  b) For acute triangles, the area of the square on the largest side is less than the sum of the areas of the squares on the two smaller sides.
  c) In question 6,
  • the acute triangle is: b
  • the right triangles are: a, c, d, h
  • the obtuse triangles are: e, f, g

17. Answers will vary. For example:
Lesser number: 8; Greater number: 14
Triple: 224, 132, 260

1.7 Applying the Pythagorean Theorem, page 49

4. a) 29 cm **b)** 12.2 cm **c)** 15.8 cm
5. a) 24 cm **b)** 15 cm **c)** 5.7 cm
6. 4 m
7. a) 26 cm or 21.8 cm
  b) The unknown side could be a leg or the hypotenuse of the right triangle.
8. a) 6.7 units **b)** 7.8 units
9. 65 cm
10. 91 m
11. 38.18 m
12. a) The area of the square on the hypotenuse is equal to the sum of the areas of the squares on the legs.
  b) The square of the length of the hypotenuse is equal to the sum of the squares of the lengths of the legs.
13. 57.4 cm
14. F; I drew two right triangles with hypotenuses AB and AF. The legs of both triangles were 4 units and 3 units.
15. 5.8 units **16.** 216.9 m
17. Yes; $650^2 + 720^2 = 970^2$
18. 403.1 km **19.** 7.6 cm **20.** 17 cm
21. 37.3 m **22.** 291.2 km

Unit 1 Unit Review, page 54

1. Rectangles: 1 unit by 24 units, 2 units by 12 units, 3 units by 8 units, 4 units by 6 units
Not a perfect square since 24 cannot be modelled by a square

2. 25
3. Answers may vary. For example: 16, 25, 1024, 1600, 2401, 2500
4. a) 25 **b)** 49 **c)** 81 **d)** 169
5. a) 7 **b)** 17 **c)** 20
6. a) i) 1, 2, 3, 4, 6, 9, 12, 18, 27, 36, 54, 108
  ii) 1, 19, 361
  iii) 1, 2, 3, 5, 6, 10, 15, 25, 30, 50, 75, 150
  iv) 1, 2, 11, 13, 22, 26, 143, 286
  v) 1, 2, 3, 4, 6, 9, 12, 18, 27, 36, 54, 81, 108, 162, 324
  vi) 1, 2, 4, 7, 8, 14, 28, 56
  b) 361 and 324; Both have an odd number of factors.
7. 44 cm
8. $A = 17$ square units; $s = \sqrt{17}$ units
9. a) $\sqrt{75}$ cm **b)** $\sqrt{96}$ cm **c)** 9 cm
10. b; I drew a square on each line segment and found the area. Square b has a greater area.
11. a) 26 **b)** 5 **c)** 50 **d)** 13
12. a) 6 and 7 **b)** 9 and 10
  c) 10 and 11 **d)** 34 and 35
13. a) 2 **b)** 3 **c)** 5 **d)** 6 **e)** 8 **f)** 9
14. a) 7.4 **b)** 8.7 **c)** 9.7 **d)** 10.2 **e)** 6.8 **f)** 10.7
15. 8.49, since $8.48^2 = 71.9104$ and $8.49^2 = 72.0801$
16. 130 cm
17. a) False **b)** True **c)** True
18. a) 34 cm **b)** 28 cm **c)** 16.2 cm
19. a) 8.5 cm **b)** 7.8 cm
20. Yes, since $24 + 57 = 81$
21. No; $7^2 + 12^2 \neq 15^2$
22. a and c
23. 21; One solution, because in a Pythagorean triple all three numbers must be whole numbers
24. 40 km
25. 42 cm
26. The distance from each possible position to x is the hypotenuse of a right triangle with legs lengths 2 units and 3 units.
27. 31.2 km

Unit 1 Practice Test, page 58

1. a) 11 **b)** 196 **c)** 6.32 **d)** 81
2. $\sqrt{1} = \sqrt{1 \times 1} = 1$
3. $s = 8$ cm, $A = 64$ cm^2
4. a) 25 square units **b)** 5 units
5. a) Yes; $15 + 9 = 24$ **b)** No; $11 + 7 \neq 20$
6. a) 14.2 cm **b)** 16 cm
7. a) No; $20^2 + 48^2 \neq 54^2$ **b)** Yes; $18^2 + 24^2 = 30^2$
8. a) 16.2 m **b)** 81 m
9. a) 3.6 cm, 2.2 cm, 2.0 cm

ANSWERS **499**

b) The line segments could form a triangle because 2.0 + 2.2 > 3.6.
They could not form a right triangle because $2.0^2 + 2.2^2 \neq 3.6^2$.

10. 19 times

Unit 1 Unit Problem: The Locker Problem, page 61

2. 1, 4, 9, 16, and 25; All numbers are perfect squares.
3. 1, 4, 9, 16, 25, 36, 49, 64, 81, 100
4. 1, 4, 9, 16, 25, 36, 49, 64, 81, 100, 121, 144, 169, 196, 225, 256, 289, 324, 361, 400
5. Open lockers have perfect squares as numbers. The number of students that change the locker corresponds to the number of factors for that locker number. All perfect squares have an odd number of factors, so those lockers are open.
6. The numbers in the third column are the consecutive odd numbers beginning at 3.
7.a) Every odd square number will appear in the third column if you continue the table far enough.
   b) No, the difference between 2 consecutive odd numbers is always an odd number.
   c) 25   **d)** 12, 13, 5 and 24, 25, 7
   e) 9, 40, 41

Unit 2 Square Roots and the Pythagorean Theorem, page 62

2.1 Using Models to Multiply Integers, page 68

5.a) $(+3) \times (-1) = -3$ **b)** $(+5) \times (-2) = -10$
   c) $(+4) \times (+11) = +44$
6.a) $(-4) + (-4) + (-4) + (-4) + (-4) + (-4) + (-4) = -28$
   b) $(+3) + (+3) + (+3) + (+3) + (+3) + (+3) = +18$
   c) $(+6) + (+6) + (+6) + (+6) = +24$
   d) $(-6) + (-6) + (-6) + (-6) + (-6) = -30$
7.a) $(+3) \times (+3) = +9$ **b)** $(+4) \times (-2) = -8$
8.a) -6 **b)** $+27$ **c)** $+12$ **d)** -20
9.a) $(+5) \times (-2) = -10$ **b)** $(+5) \times (+2) = +10$
   c) $(-7) \times (-3) = +21$ **d)** $(-9) \times (+4) = -36$
   e) $(+11) \times (+3) = +33$ **f)** $(-10) \times (-5) = +50$
10.a) $+5$ **b)** $+24$ **c)** -10 **d)** -24 **e)** -30 **f)** -32
11.a) $+8$ **b)** $+8$ **c)** $+16$ **d)** -30 **e)** -24 **f)** $+21$
12. $(+2) \times (+9) = +18$; The temperature rose 18°C.
13. $(-3) \times (+11) = -33$; The water level dropped 33 cm.
14.a) Answers will vary. For example:
   Olinga withdraws $6 from his bank account every day for 8 days. $(+8) \times (-6) = -48$

15. Use tiles: withdraw 7 sets of 8 red tiles.
Use a number line: Face the negative end and take 7 steps backward each of size 8.
$(-7) \times (-8) = +56$
16. $(-4) + (+4) = -16$
17.a) $(+8) \times (-5) = -40$; He will have $40 less.
   b) $(-2) \times (-5) = +10$; He had $10 more.
18.a) -40 or 40 cm to the left
   b) $+12$ or 12 cm to the right
   c) $(-4) \times (+10) = -40$
      $(-3) \times (-4) = +12$
19. Answers may vary. For example:
Hugh threw out 7 cartons each with half a dozen eggs. How many eggs did he throw out?
$(-7) \times (+6) = -42$
20.a) -24 **b)** $+15$ **c)** -30 **d)** $+36$

2.2 Developing Rules to Multiply Integers, page 73

3.a) Negative   **b)** Positive
   c) Negative   **d)** Positive
4.a) -24 **b)** $+20$ **c)** -27 **d)** -42 **e)** -30
   f) $+42$ **g)** 0   **h)** -10 **i)** $+56$ **j)** -81
5.a) i) $-21, -21$ **ii)** $+32, +32$
   iii) $+45, +45$ **iv)** $-60, -60$
   b) No
6.a) $+300$ **b)** $+780$ **c)** -1600 **d)** -840
   e) -780 **f)** -2640 **g)** $+3290$ **h)** $+4680$
7.a) -300 **b)** -945 **c)** $+544$ **d)** -606
   e) $+221$ **f)** -3024 **g)** $+1275$ **h)** $+667$
8.a) $+4$ **b)** -3 **c)** $+6$ **d)** -6
   e) -4 **f)** -12 **g)** -30 **h)** -6
9.a) $+16, +32, +64$; Multiply by $+2$ each time.
   b) $+1296, -7776, +46\ 656$; Multiply by -6 each time.
   c) $-81, +243, -729$; Multiply by -3 each time.
   b) $-4, +4, -4$; Multiply by -1 each time.
10. $(+17) \times (-26) = -442$; Gaston withdrew $442.
11.a) -8 and -5 **b)** $+9$ and -8
12.a) i) $+6$ **ii)** -24 **iii)** $+120$
   iv) -720 **v)** $+5040$ **vi)** $-40\ 320$
   vii) $+362\ 880$ **viii)** $-3\ 628\ 800$
   b) i) Positive   **ii)** Negative
   c) Yes
13.a) $(+60) \times (-20) = -1200$; Amelie wrote $+1200$ instead of -1200.
   b) -1080
14. Answers will vary. For example:
Gavin ate 15 handfuls of 8 jelly beans. How many jelly beans did he eat?
$(+15) \times (-8) = -120$; Gavin ate 120 jelly beans.
15. When you multiply an integer by itself, you multiply two integers with the same sign. This always gives a positive product.

16. Answers may vary. For example:
Bridget's van uses 12 L of gas every day. How much gas does Bridget use in 7 days?
(–12) × (+7) = –84; Bridget uses 84 L of gas.

17. For example:
Two factors: (+2) × (–18); (–2) × (+18)
Three factors: (–2) × (–2) × (–9);
(–2) × (+2) × (+9)
Four factors: (–2) × (–2) × (–3) × (+3)

18. +9 and –16

19. No; The product of the positive integer and a negative integer is always less than or equal to each of the integers.

20. a) –5   **b)** –9

21. Multiply the integers from left to right.
The product is positive when there is an even number of negative factors.
The product is negative when there is an odd number of negative factors.
For example: (–1) × (+2) × (–4) has 2 negative factors, so the answer is +8.

2.3 Using Models to Divide Integers, page 80

3. a) (+5) × (+5) = +25   **b)** (–12) × (–2) = +24
  c) (+2) × (–7) = –14   **d)** (–3) × (+6) = –18
4. a) (–20) ÷ (–4) = +5   **b)** (+21) ÷ (+3) = +7
  c) (–26) ÷ (+2) = –13
5. (–24) ÷ (+4) = –6
6. a) +8 **b)** +3 **c)** –2 **d)** +3 **e)** –5 **f)** –10
7. a) i) 2   **ii)** 4
   b) i) 5   **ii)** 3
8. a) +3 **b)** –2 **c)** +4 **d)** –3 **e)** –3 **f)** +2
9. a) +2 **b)** +2 **c)** –2 **d)** –2
10. a) +3 **b)** +4 **c)** –4 **d)** –5 **e)** –7 **f)** –2
11. (+12) ÷ (+3) = +4; 4 hours
12. (–20) ÷ (–4) = +5; 5 hours
13. (–148) ÷ (+4) = –37; 37 m
14. Answers will vary. For example:
Heather returned 5 towels to a store and received $45. How much had each towel cost?
(+45) ÷ (–5) = –9; Each towel cost $9.
15. Answers will vary. For example:
A scuba diver descends a total of 12 m over the course of 6 descents. How far did he descend each time?
(–12) ÷ (+6) = –2; 2 m
16. a) (–36) ÷ (–6) = +6; After 6 minutes
   b) (+18) ÷ (–6) = –3; 3 minutes ago
17. a) –4°C   **b)** +3°C
18. a) 6 weeks   **b)** $8

Unit 2 Mid-Unit Review, page 83

1. a) –36   **b)** +35   **c)** +32   **d)** –15
2. (–2) × (+7) = –14; 14 m
3. (+4) × (+5) = +20; 20°C
4. a) Negative   **b)** Positive
   c) Negative   **d)** Positive
5. a) –40   **b)** +15   **c)** –48   **d)** +72
6. a) –280   **b)** +456   **c)** +1080 **d)** –403
7. (–35) × (+30) = –1050; 1050 L
8. a) –8   **b)** –9   **c)** +7   **d)** –12
9. a) (+9) × (+3) = +27; (+3) × (+9) = +27
   b) (–2) × (–7) = +14; (–7) × (–2) = +14
   c) (+7) × (–3) = –21; (–3) × (+7) = –21
   d) (–13) × (+2) = –26; (+2) × (–13) = –26
10. a) +5   **b)** +4   **c)** –4   **d)** –9
11. (–30) ÷ (–5) = +6; 6 hours
12. Answers will vary. For example:
A stock dropped 18 points steadily over 3 days. How much did it drop each day?
(–18) ÷ (+3) = –6; 6 points
13. To go from 0 to +64 take steps backwards of size 8. It will take 8 steps and you will face the negative direction. (+64) ÷ (–8) = –8

2.4 Developing Rule to Divide Integers, page 87

4. a) Negative   **b)** Positive
   c) Negative   **d)** Positive
5. a) +3 **b)** +5 **c)** –2 **d)** –9 **e)** –9
   f) +8 **g)** –14 **h)** –9 **i)** +9 **j)** +8
6. a) (0) ÷ (+3) = 0; (+3) ÷ (+3) = +1;
       (+6) ÷ (+3) = +2; (+9) ÷ (+3) = +3
   The quotient of two integers with opposite signs is negative.
   The quotient of two integers with the same signs is positive.
   b) (–15) ÷ (+3) = –5; (–25) ÷ (+5) = –5;
       (–35) ÷ (+7) = –5; (–45) ÷ (+9) = –5
   The quotient of two integers with opposite signs is negative.
   c) (0) ÷ (+2) = 0; (–2) ÷ (+2) = –1;
       (–4) ÷ (+2) = –2; (–6) ÷ (+2) = –3
   The quotient of two integers with opposite signs is negative.
   The quotient of two integers with the same signs is positive.
   d) (–2) ÷ (–1) = +2; (–6) ÷ (–3) = +2;
       (–10) ÷ (–5) = +2; (–14) ÷ (–7) = +2
   The quotient of two integers with the same signs is positive.
   e) (+2) ÷ (–1) = –2; (+6) ÷ (–3) = –2;
       (+10) ÷ (–5) = –2; (+14) ÷ (–7) = –2
   The quotient of two integers with opposite signs is negative.
   f) (+10) ÷ (–5) = –2; (+15) ÷ (–5) = –3;
       (+20) ÷ (–5) = –4; (+25) ÷ (–5) = –5
   The quotient of two integers with opposite signs is negative.
   The quotient of two integers with the same signs is positive.

7. a) i) +8   ii) –5   iii) –7
   b) i) (+24) ÷ (+8) = +3 ii) (+45) ÷ (–5) = –9
   iii) (–28) ÷ (–7) = +4
8. a) (–30) ÷ (–6) = +5 and (–30) ÷ (+5) = –6
   b) (+42) ÷ (+7) = +6 and (+42) ÷ (+6) = +7
   c) (–36) ÷ (+9) = –4 and (–36) ÷ (–4) = +9
   d) (+32) ÷ (–4) = –8 and (+32) ÷ (–8) = –4
9. a) +4   **b)** –3   **c)** –9   **d)** 0
10. a) +5   **b)** –90   **c)** +9   **d)** –21
   e) –60   **f)** +49   **g)** –48   **h)** +44
11. a) (–56) ÷ (–7)   **b)** +8; 8 days
12. a) (–15) ÷ (+5) **b)** –3; Dropped 3°C per day
13. $11
14. a) (–24) ÷ (–6)   **b)** +4; 4 performances
15. a) +81, –243, +729; Multiply by –3.
   b) +30, –36, +42; Add and multiply by –1.
   c) –40, –160, –80; Alternately multiply by +4, divide by –2.
   d) +8, –9, +2; Divide by –2.
   e) –100, +10, –1; Divide by –10.
16. a) When the dividend is positive and the divisor is greater than the quotient; (+8) ÷ (+4) = +2
   b) When the dividend and divisor are negative; (–8) ÷ (–4) = +2
   c) When the dividend is positive and the divisor is less than the quotient; (+8) ÷ (+2) = +4
   d) When the dividend and divisor are equal and non-zero; (+8) ÷ (+8) = +1
   e) When the dividend and divisor are opposite integers; (+8) ÷ (–8) = –1
   f) When the dividend is 0 and the divisor is non-zero; 0 ÷ (–4) = 0
17. (–32) ÷ (–1) = +32; (–32) ÷ (+1) = –32
   (–32) ÷ (–2) = +16; (–32) ÷ (+2) = –16
   (–32) ÷ (–4) = +8; (–32) ÷ (+4) = –8
   (–32) ÷ (–8) = +4; (–32) ÷ (+8) = –4
   (–32) ÷ (–16) = +2; (–32) ÷ (+16) = –2
   (–32) ÷ (–32) = +1; (–32) ÷ (+32) = –1
18. a) –5 **b)** +6 **c)** –7 **d)** +4 **e)** +4 **f)** –5
19. +$9
20. Answers may vary. For example:
   A squirrel had 78 acorns and ate 13 each week. How many weeks ago did the squirrel have 78 acorns? (+78) ÷ (–13) = –6; 6 weeks ago
21. a) 2°C/hour
   b) The temperature must have been less than –20°C.
22. –2, +2, +4; –6, +6, +4; –8, +8, +4; –4, +2, +6; –6, +2, +8
23. (–36) ÷ (+1) = –36; (–36) ÷ (–1) = +36
   (–36) ÷ (+2) = –18; (–36) ÷ (–2) = +18
   (–36) ÷ (+3) = –12; (–36) ÷ (–3) = +12
   (–36) ÷ (+4) = –9; (–36) ÷ (–4) = +9
   (–36) ÷ (+6) = –6; (–36) ÷ (–6) = +6
   (–36) ÷ (+9) = –4; (–36) ÷ (–9) = +4
   (–36) ÷ (+12) = –3; (–36) ÷ (–12) = +3
   (–36) ÷ (+18) = –2; (–36) ÷ (–18) = +2
   (–36) ÷ (+36) = –1; (–36) ÷ (–36) = +1
   –36 is not a square number; Each pair of factors has opposite signs.
24. Represent the mean with the division equation: (–140) ÷ (+7) = –20
   Find 7 numbers whose sum is –140. Use guess and check.

2.5 Order of Operations with Integers, page 92

3. a) Multiply   **b)** Divide   **c)** Add
   d) Add   **e)** Multiply   **f)** Divide
4. a) 10 **b)** 7 **c)** 4 **d)** –4 **e)** –12 **f)** 9
5. No; Elijah added before subtracting
   Correct answer: 24
6. a) i) 0   ii) 6
   b) The brackets are in different positions.
7. a) Multiply; 23   **b)** Add; –18
   c) Multiply; 25   **d)** Multiply; –14
   e) Divide numbers in brackets; –3
   f) Divide; –54
8. a) –15 **b)** 10 **c)** 2 **d)** 14 **e)** 10 **f)** 14
9. a) –5 **b)** 1 **c)** –2 **d)** –1
10. a) 8 **b)** 2 **c)** –4 **d)** –2
11. a) Robert
   b) Christian calculated (–8) ÷ 2 to be +4 instead of –4. Brenna subtracted (–40) – 2 first.
12. a) –8; –20 ÷ [2 – (–2)]
   b) –19; [(–21) + 6] ÷ 3
   c) 9; 10 + 3 × [2 – 7]
13. 405 + 4(–45); $225
14. Answers may vary. For example:
   a) (–4) + (–4) + (–4)   **b)** (–4) – (–4) + (–4)
   c) (–4)[(–4) – (–4)]   **d)** (–4) ÷ (–4) + (–4)
   e) (–4) ÷ (–4) – (–4)   **f)** $\frac{(-4) + (-4)}{-4}$
15. –5°C
16. a) [(–24) + 4] ÷ (–5) = 4
   b) [(–4) + 10] × (–2) = –12
   c) [(–10) – 4] ÷ (–2) = 7
17. a) (–10) × (–2) + 1 = 21
   b) (–5) – (–2) + 4 = 1 **c)** 6 × (–7) – 2 = –44
   d) (–2)(–2) – 8 = –4

Unit 2 Strategies for Success: Understanding the Problem, page 95

1. Answers may vary. For example:
   1 – 5 = –4; 2 – 6 = –4; 3 – 7 = –4; 4 – 8 = –4
2. (–5) + [(+4) – (–2) ÷ (–1)] × (–3) = –11
3. 10°C

Unit 2 Unit Review, page 97

1. a) (–1) + (–1) = –2   **b)** (+9) + (+9) = +18
   c) (–3) + (–3) + (–3) = –9
   d) (+7) + (+7) + (+7) = +21

2. a) +35 **b)** −60 **c)** −16 **d)** +48
3. −10°C
4. a) Negative **b)** Positive
  c) Negative **d)** Positive
5. a) −63 **b)** +28 **c)** −143
  d) +880 **e)** −17 **f)** 0
6. a) −6 **b)** +10 **c)** 0 **d)** +15
7. (−55) × 6 = −330; 330 mL
8. Answers will vary. For example:
  Jari spends $7 on lunch at school. How much does it cost him to buy lunch for 5 school days? (+5) × (−7) = −35; $35
9. a) +5 **b)** −4 **c)** −3 **d)** +9
10. a) −6 **b)** +7 **c)** −7 **d)** +5
11. a) 13 weeks **b)** (+65) ÷ (+5) = +13
  c) Tyler has $65 to withdraw.
12. a) Negative **b)** Positive
  c) Negative **d)** Positive
13. a) +8 **b)** −8 **c)** −11 **d)** +14
14. a) +9 **b)** −4 **c)** −3 **d)** 0
15. a) −2 **b)** −3 **c)** +5 **d)** +9
16. a) (−63) ÷ (−3) **b)** +21; 21 days
  c) She does not return any of the candies to the jar.
17. Answers may vary. For example:
  Renira had $72 in gift certificates to her favourite book store. She spent all the gift certificates by buying one book each week. What was the average cost of a book? (+72) ÷ (−9) = −8; $8
18. (−21) ÷ (−1) = +21; (−21) ÷ (+1) = −21
  (−21) ÷ (−3) = +7; (−21) ÷ (+3) = −7
  (−21) ÷ (−7) = +3; (−21) ÷ (+7) = −3
  (−21) ÷ (−21) = +1; (−21) ÷ (+21) = −1
19. a) Multiply **b)** Divide
  c) Subtract **d)** Multiply
20. a) 16 **b)** −1 **c)** 2 **d)** 12
21. a) −16 **b)** −1 **c)** 12 **d)** −18
22. a) 3 **b)** −3 **c)** −4
23. a) +1 **b)** +2 **c)** +6
24. a) Corey: +1; Suzanne: −2 **b)** Corey

Unit 2 Practice Test, page 99

1. a) +90 **b)** −66 **c)** −6 **d)** +13
  e) −48 **f)** −4 **g)** +11 **h)** +5
2. a) +98 **b)** −3 **c)** −4 **d)** +21
3. −20°C
4. a) Receive $90 **b)** Spend $105
  c) Receive $63
5. a) Answers may vary. For example:
  −2 and +1, +2 and −1
  b) +3 and +4 **c)** −2 and −1 **d)** +4 and −1
  e) Answers may vary. For example: Find all pairs of integers with a difference of −2.
  +2 and +4, 0 and +2, −2 and 0, +1 and +3, −1 and +1

Unit 2 Unit Problem: Charity Golf Tournament, page 100

1. a) 3(0) + 2(+1) + (−1) + 2(−2) + (+2)
  b) −1 or 1 under par
2. a) 4, 4, 4, 2, 3 **b)** 31 **c)** −1 or 1 under par
3. a) Kyle: 35 **b)** Delaney: 28
  c) Hamid: 26
4. a) Hamid, Hanna, Delaney, Chai Kim, Kyle, Weng Kwong
  b) Hamid; −6
  c) Hanna and Delaney; −5 and −4

Unit 3 Operations with Fractions, page 102

3.1 Using Models to Multiply Fractions and Whole Numbers, page 108

5. a) $\frac{5}{9} \times 45; 45 \times \frac{5}{9}$ **b)** $\frac{3}{8} \times 32; 32 \times \frac{3}{8}$
  c) $\frac{1}{12} \times 36; 36 \times \frac{1}{12}$ **d)** $\frac{4}{5} \times 25; 25 \times \frac{4}{5}$
6. a) $3 \times \frac{1}{4}; \frac{1}{4} \times 3$ **b)** $7 \times \frac{2}{5}; \frac{2}{5} \times 7$
  c) $4 \times \frac{3}{10}; \frac{3}{10} \times 4$
7. a) $\frac{2}{3} + \frac{2}{3} + \frac{2}{3} + \frac{2}{3} + \frac{2}{3} + \frac{2}{3}$ **b)** 4
  c)
  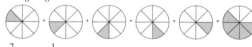
  d) $\frac{2}{3} \times 6 = 4$
8. a) $4 \times \frac{4}{5} = 3\frac{1}{5}$ **b)** $9 \times \frac{1}{2} = 4\frac{1}{2}$
  c) $3 \times \frac{5}{6} = 2\frac{1}{2}$
9. a) $5 \times \frac{1}{2} = \frac{5}{2}$ **b)** $4 \times \frac{3}{4} = 3$
10. a) $4 \times \frac{1}{2} = 2$ **b)** $5 \times \frac{2}{3} = 3\frac{1}{3}$
11. a) $5 \times \frac{1}{8} = \frac{5}{8}$

  b) $\frac{2}{5} \times 3 = 1\frac{1}{5}$

  c) $4 \times \frac{5}{12} = 1\frac{2}{3}$

12. a) 12 **b)** 8 **c)** 6 **d)** 4 **e)** 3 **f)** 2
13. a) 24 **b)** 16 **c)** 18 **d)** 20 **e)** 9 **f)** 10

14. a) $1\frac{5}{7}$

b) $1\frac{1}{3}$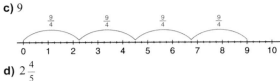

c) 9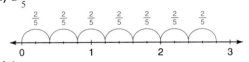

d) $2\frac{4}{5}$

15. a) 4

b) 3

c) 9

d) 6

16. a) $2\frac{2}{5}$ **b)** $3\frac{8}{9}$ **c)** 10 **d)** $2\frac{1}{2}$ **e)** $10\frac{1}{2}$ **f)** $4\frac{1}{2}$

17. $24 \times \frac{2}{3} = 16$; 16 hours

18. a) For example: Jerry ordered 5 pizzas for his birthday party. $\frac{3}{8}$ of each pizza is left over. How much pizza is left over in total?
$5 \times \frac{3}{8} = 1\frac{7}{8}$

b)
c)
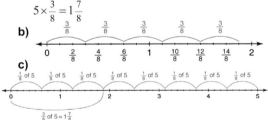

19. For example: Parri only likes black jelly beans. At Halloween she receives 16 packets of jelly beans, each containing 3 black jelly beans and 5 jelly beans of other colours. How many packets of jelly beans will she eat? $\frac{3}{8} \times 16 = 6$

20. $\frac{4}{7} \times 28 = 16$; She spent $16 on rides.

21. a) i) 1 **ii)** 1 **iii)** 1 **iv)** 1

b) The product of a number and its reciprocal is 1.

c) For example: $6 \times \frac{1}{6} = 1; 37 \times \frac{1}{37} = 1$

22. $4\frac{1}{2}$ hours

3.2 Using Models to Multiply Fractions, page 113

5. a) 1 **b)** 3 **c)** 20 **d)** 3 **e)** $\frac{3}{20}$

6. a) $\frac{3}{8}$ **b)** $\frac{1}{2}$ **c)** $\frac{1}{5}$ **d)** $\frac{5}{12}$ **e)** $\frac{21}{40}$ **f)** $\frac{3}{5}$

7. a) $\frac{3}{5}$ **b)** $\frac{4}{9}$ **c)** $\frac{1}{6}$ **d)** $\frac{2}{9}$ **e)** $\frac{5}{12}$ **f)** $\frac{1}{2}$

8. a) $\frac{15}{32}$ **b)** $\frac{8}{45}$ **c)** $\frac{1}{6}$ **d)** $\frac{4}{7}$ **e)** $\frac{2}{9}$ **f)** $\frac{16}{25}$

9. For example:
$\frac{1}{7} \times \frac{2}{5} = \frac{2}{35}; \frac{3}{4} \times \frac{7}{8} = \frac{21}{32}; \frac{2}{7} \times \frac{8}{11} = \frac{16}{77}$

10. a) $\frac{3}{5} \times \frac{1}{2} = \frac{3}{10}$ **b)** $\frac{6}{8} \times \frac{1}{3} = \frac{1}{4}$

c) $\frac{1}{3} \times \frac{3}{4} = \frac{1}{4}$ **d)** $\frac{4}{5} \times \frac{2}{3} = \frac{8}{15}$

11. $\frac{1}{4}$

12. a) i) $\frac{3}{10}$ **ii)** $\frac{3}{10}$ **iii)** $\frac{3}{32}$

iv) $\frac{3}{32}$ **v)** $\frac{2}{5}$ **vi)** $\frac{2}{5}$

b) Switching the numerators of 2 fractions does not change the product.

c) For example: $\frac{1}{7} \times \frac{5}{8} = \frac{5}{56}, \frac{5}{7} \times \frac{1}{8} = \frac{5}{56}$;
$\frac{3}{11} \times \frac{1}{2} = \frac{3}{22}, \frac{3}{2} \times \frac{1}{11} = \frac{3}{22}$

13. a) $\frac{1}{6}$ **b)** $\frac{1}{2}; \frac{1}{6}$ **c)** $\frac{1}{10}; \frac{3}{10}$

14. For example: Gwen conducted an experiment involving $\frac{4}{9}$ of the lab rats at her school. Of the rats she used, $\frac{1}{5}$ were female. What fraction of the rats used were female? $\frac{4}{9} \times \frac{1}{5} = \frac{4}{45}$

15. $\frac{2}{7} \times \frac{7}{8} = \frac{1}{4}$

16. $\frac{5}{8} \times \frac{3}{12} = \frac{5}{32}$ $\frac{3}{8} \times \frac{5}{12} = \frac{5}{32}$

17. a) 3 of 5 parts making up the whole are shaded.

b) To show $\frac{5}{3}$ of $\frac{3}{5}$ you need to shade $\frac{5}{3}$ of the $\frac{3}{5}$ already shaded. Since $\frac{5}{3}$ is all 5 parts of the whole, one whole is $\frac{5}{3}$ of $\frac{3}{5}$.

3.3 Multiplying Fractions, page 118

4. a) 2, 4 **b)** 7 **c)** 2, 4, 8 **d)** 3 **e)** 5 **f)** 2, 3, 6

5. a) $\frac{1}{8}$

b) $\frac{5}{6}$ is about 1, $\frac{3}{20}$ is about 0; so $\frac{5}{6} \times \frac{3}{20}$ is close to 0.

c) Yes; $\frac{1}{8}$ is close to 0.

6. $\frac{3}{16}$

7. a) $\frac{6}{5}$ **b)** $\frac{3}{10}$ **c)** 1 **d)** $\frac{1}{3}$ **e)** $\frac{5}{6}$ **f)** $\frac{3}{2}$

8. a) $\frac{2}{5}$ **b)** $\frac{1}{4}$ **c)** $\frac{1}{24}$ **d)** $\frac{39}{16}$ **e)** $\frac{11}{8}$ **f)** $\frac{49}{24}$

9. a) $\frac{3}{32}$ **b)** $\frac{1}{6}$

10. For example: Amanda ate $\frac{1}{8}$ of a pizza. Her clumsy friend Cody dropped $\frac{1}{2}$ of the remaining pizza on the floor. How much pizza is left? $\frac{7}{8} \times \frac{1}{2} = \frac{7}{16}$

11. $\frac{3}{8}$

12. a) i) 1 **ii)** 1 **iii)** 1 **iv)** 1

b) For example: $\frac{3}{8} \times \frac{8}{3} = 1$, $\frac{8}{9} \times \frac{9}{8} = 1$, $\frac{13}{6} \times \frac{6}{13} = 1$ The product of a fraction and its reciprocal is 1.

13. Answers may vary. For example:

a) i) $\frac{3}{2} \times \frac{4}{3} = 2$ **ii)** $\frac{12}{5} \times \frac{5}{4} = 3$

iii) $\frac{5}{4} \times \frac{16}{5} = 4$ **iv)** $\frac{15}{2} \times \frac{2}{3} = 5$

b) $\frac{7}{20} \times \frac{20}{7} = 1$

i) $\frac{7}{10} \times \frac{20}{7} = 2$ **ii)** $\frac{7}{20} \times \frac{60}{7} = 3$

iii) $\frac{7}{5} \times \frac{20}{7} = 4$ **iv)** $\frac{35}{20} \times \frac{20}{7} = 5$

14. $\frac{1}{3}$ and $\frac{1}{4}$

15. a) $\frac{9}{40}$ **b)** $\frac{6}{13}$ **c)** $\frac{3}{8}$ **d)** $\frac{4}{13}$

16. a) i) $\frac{24}{25} \times \frac{85}{86} = \frac{1 \times 17}{5 \times 4} = \frac{17}{20}$

ii) $\frac{24}{25} \times \frac{85}{96} = \frac{2040}{2400} = \frac{17}{20}$

17. For example: $\frac{1}{2} \times \frac{3}{2}$

20. b; $\frac{4}{7} \times \frac{3}{5} = \frac{12}{35}$

21. a) $\frac{2}{3}$ **b)** $\frac{4}{5}$ **c)** $\frac{2}{3}$ **d)** $\frac{7}{13}$

3.4 Multiplying Mixed Numbers, page 125

4. a) $3\frac{1}{2}; \frac{7}{2}$ **b)** $2\frac{1}{5}; \frac{11}{5}$ **c)** $1\frac{6}{7}; \frac{13}{7}$

5. a) $\frac{23}{10}$ **b)** $\frac{33}{8}$ **c)** $\frac{23}{6}$ **d)** $\frac{5}{3}$ **e)** $\frac{17}{5}$

f) $\frac{11}{2}$ **g)** $\frac{18}{7}$ **h)** $\frac{32}{9}$ **i)** $\frac{20}{3}$

6. a) $3\frac{2}{3}$ **b)** $3\frac{3}{4}$ **c)** $4\frac{1}{5}$ **d)** $1\frac{3}{8}$ **e)** $3\frac{1}{6}$

f) $4\frac{3}{7}$ **g)** $5\frac{1}{2}$ **h)** $4\frac{3}{10}$ **i)** $4\frac{5}{8}$

7. a) 8 **b)** 8 **c)** 21 **d)** 15

8. a) 8 **b)** $\frac{18}{5} \times \frac{20}{9}$ **c)** 8

d) Yes; Estimate and answer are the same.

9. a) $6\frac{3}{4}$ **b)** $8\frac{1}{2}$ **c)** $3\frac{1}{3}$ **d)** $9\frac{3}{5}$

10. a) 2 **b)** $7\frac{1}{3}$ **c)** 4 **d)** $3\frac{3}{5}$

11. a) 5 **b)** $14\frac{1}{6}$ **c)** $3\frac{7}{9}$ **d)** 8 **e)** 6 **f)** $4\frac{1}{5}$

12. a) $4\frac{3}{8}$ **b)** $8\frac{1}{15}$ **c)** $5\frac{15}{32}$

d) $14\frac{1}{16}$ **e)** $3\frac{11}{25}$ **f)** $2\frac{11}{40}$

13. a) $35\frac{1}{4}$ **b)** $35.25

14. $6\frac{5}{12}$ h or 6 h 25 min

15. For example: Josh spends $3\frac{1}{2}$ hours on his phone every week. Mark spends $2\frac{1}{8}$ as much time on the phone as Josh. How much time does Mark spend on the phone in a week? $7\frac{7}{16}$ hours

16. 7 innings

17. a) Layton: 5; Meghan and Josh: $12\frac{1}{2}$

b) $5\frac{5}{12}$ **c)** $13\frac{1}{3}$ **d)** $21\frac{1}{4}$ **e)** 255

18. Least product: a; Greatest product: d

19. a) $16\frac{8}{27}$ **b)** $12\frac{3}{8}$ **c)** $11\frac{13}{16}$

Unit 3 Mid-Unit Review, page 128

1.a) $\frac{1}{8}+\frac{1}{8}+\frac{1}{8}+\frac{1}{8}=\frac{1}{2}$

b) $\frac{3}{5}+\frac{3}{5}+\frac{3}{5}+\frac{3}{5}+\frac{3}{5}+\frac{3}{5}+\frac{3}{5}=4\frac{1}{5}$

c) $\frac{5}{6}+\frac{5}{6}+\frac{5}{6}=2\frac{1}{2}$

d) $\frac{2}{9}+\frac{2}{9}+\frac{2}{9}+\frac{2}{9}+\frac{2}{9}+\frac{2}{9}=1\frac{1}{3}$

2.a) $1\frac{3}{4}$ **b)** 3 **c)** $4\frac{1}{5}$ **d)** $1\frac{1}{4}$

3.a) 14 **b)** 2 **c)** $\frac{1}{8}$

4.a) $\frac{5}{16}$ **b)** $\frac{1}{2}$ **c)** $\frac{2}{5}$ **d)** $\frac{1}{4}$

5.a) $\frac{2}{9}$ **b)** $\frac{4}{15}$ **c)** $\frac{6}{11}$ **d)** $\frac{1}{3}$

6.a) $\frac{1}{3}$ **b)** $\frac{1}{5}$ **c)** $\frac{9}{32}$ **d)** $\frac{10}{27}$

7. $\frac{2}{15}$

8.a) 5 **b)** $8\frac{1}{3}$ **c)** $1\frac{25}{32}$ **d)** $8\frac{1}{4}$

9.a) $8\frac{1}{8}$ **b)** $1\frac{1}{10}$ **c)** $2\frac{4}{5}$ **d)** $14\frac{7}{16}$

11. $4\frac{1}{2}$ h

3.5 Dividing Whole Numbers with Fractions, page 132

3.a) $4\div\frac{1}{3}=12$ **b)** $3\div\frac{1}{6}=18$

c) $4\div\frac{2}{3}=6$ **d)** $3\div\frac{3}{5}=5$

4.a)

b) $4;\frac{4}{6}$ **c)** $\frac{4}{5}$ **d)** $4\div\frac{5}{6}=4\frac{4}{5}$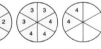

5. 5 subjects

6.a) 4 **b)** 9 **c)** 16 **d)** 12 **e)** 6 **f)** 8

7. Answers may vary. For example:

a) $3\div\frac{1}{4}=12; 3\div\frac{3}{4}=4; 3\div1\frac{1}{2}=2$

b) $2\div\frac{1}{5}=10; 2\div\frac{2}{5}=5; 2\div\frac{10}{5}=2$

c) $4\div\frac{1}{6}=24; 4\div\frac{1}{3}=12; 4\div\frac{4}{3}=3$

8.a) i) 6 **ii)** 3

b) i) 12 **ii)** 6 **iii)** 4

c) i) $\frac{1}{4}$ **ii)** $\frac{1}{8}$ **iii)** $\frac{1}{16}$

9.a) $7\frac{1}{2}$ **b)** $5\frac{1}{3}$ **c)** $\frac{1}{10}$ **d)** $\frac{5}{16}$

10.a) 20 **b)** 9 **c)** 15

11.a) $\frac{3}{20}$ **b)** 12 **c)** $\frac{11}{60}$

12.a) $2\div\frac{4}{6}=3; 2\div\frac{6}{4}=1\frac{1}{3}$

$4\div\frac{2}{6}=12; 4\div\frac{6}{2}=1\frac{1}{3}$

$6\div\frac{2}{4}=12; 6\div\frac{4}{2}=3$

b) $4\div\frac{2}{6}$ and $6\div\frac{2}{4}$ both have the greatest quotient of 12. $2\div\frac{6}{4}$ and $4\div\frac{6}{2}$ both have the least quotient of $1\frac{1}{3}$.

13. No

$\frac{2}{3}\div4=\frac{1}{6}$

$4\div\frac{2}{3}=6$

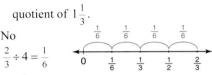
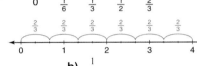

14.a) 24 **b)** $\frac{1}{24}$

c) The quotients are reciprocals.

15. $\frac{25}{4}-5=\frac{25}{4}\div5=1\frac{1}{4}$

$\frac{49}{6}-7=\frac{49}{6}\div7=1\frac{1}{6}$

The numerator is the square of the whole number. The denominator is one less than the whole number.

3.6 Dividing Fractions, page 139

4.a) $\frac{9}{5}$ **b)** $\frac{7}{3}$ **c)** $\frac{8}{7}$ **d)** $\frac{15}{14}$

5.a) 2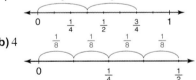

b) 4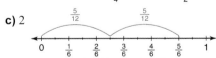

c) 2

6.a) $\frac{10}{9}$ **b)** $\frac{3}{5}\times\frac{10}{9}=\frac{1\times2}{1\times3}=\frac{2}{3}$ **c)** 1

d) Yes; $\frac{2}{3}$ is close to 1.

7.a) $2\frac{1}{2}$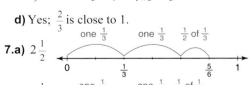

b) $2\frac{1}{4}$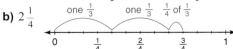

c) $2\frac{1}{3}$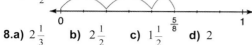

d) $2\frac{1}{2}$

8. a) $2\frac{1}{3}$ b) $2\frac{1}{2}$ c) $1\frac{1}{2}$ d) 2

9. a) $2\frac{2}{15}$ b) $\frac{27}{50}$ c) $2\frac{5}{8}$ d) $\frac{3}{7}$

10. a) $2\frac{1}{3}$ b) $\frac{6}{11}$ c) $7\frac{1}{2}$ d) $\frac{20}{27}$

11. a) $2\frac{7}{9}$ b) 1 c) $\frac{1}{15}$

12. a) $3\frac{2}{3}$ b) $2\frac{3}{4}$ c) $5\frac{1}{2}$ d) $1\frac{5}{6}$

13. a) i) $\frac{6}{5}$ ii) $\frac{5}{6}$ iii) $\frac{35}{24}$
   iv) $\frac{24}{35}$ v) $\frac{25}{12}$ vi) $\frac{12}{25}$

b) When you switch the divisor with the dividend, the quotient is the reciprocal of the original quotient.
$\frac{6}{5} \div \frac{2}{3} = \frac{9}{5}; \frac{2}{3} \div \frac{6}{5} = \frac{5}{9}$
$\frac{4}{7} \div \frac{1}{3} = \frac{12}{7}; \frac{1}{3} \div \frac{4}{7} = \frac{7}{12}$

14. 8

15. a) $1\frac{1}{2}$ b) $2\frac{1}{2}$ c) $1\frac{1}{3}$ d) 5

16. 12

17. a) $\frac{2}{3} \div \frac{4}{5} = \frac{5}{6}; \frac{3}{2} \div \frac{4}{5} = 1\frac{7}{8}; \frac{2}{3} \div \frac{5}{4} = \frac{8}{15}; \frac{3}{2} \div \frac{5}{4} = 1\frac{1}{5}$
$\frac{2}{5} \div \frac{3}{4} = \frac{8}{15}; \frac{5}{2} \div \frac{3}{4} = 3\frac{1}{3}; \frac{2}{3} \div \frac{4}{5} = \frac{3}{10}; \frac{5}{2} \div \frac{4}{3} = 1\frac{7}{8}$
$\frac{2}{4} \div \frac{3}{5} = \frac{5}{6}; \frac{4}{2} \div \frac{3}{5} = 3\frac{1}{3}; \frac{2}{4} \div \frac{5}{3} = \frac{3}{10}; \frac{4}{2} \div \frac{5}{3} = 1\frac{1}{5}$
$\frac{4}{5} \div \frac{2}{3} = 1\frac{1}{5}; \frac{4}{5} \div \frac{3}{2} = \frac{8}{15}; \frac{5}{4} \div \frac{2}{3} = 1\frac{7}{8}; \frac{5}{4} \div \frac{3}{2} = \frac{5}{6}$
$\frac{3}{4} \div \frac{2}{5} = 1\frac{7}{8}; \frac{3}{4} \div \frac{5}{2} = \frac{3}{10}; \frac{4}{3} \div \frac{2}{5} = 3\frac{1}{3}; \frac{4}{3} \div \frac{5}{2} = \frac{8}{15}$
$\frac{3}{5} \div \frac{2}{4} = 1\frac{1}{5}; \frac{3}{5} \div \frac{4}{2} = \frac{3}{10}; \frac{5}{3} \div \frac{2}{4} = 3\frac{1}{3}; \frac{5}{3} \div \frac{4}{2} = \frac{5}{6}$

b) $\frac{5}{2} \div \frac{3}{4}, \frac{4}{2} \div \frac{3}{5}, \frac{4}{3} \div \frac{2}{5}$, and $\frac{5}{3} \div \frac{2}{4}$ all have the greatest quotient of $3\frac{1}{3}$.
$\frac{2}{4} \div \frac{5}{3}, \frac{3}{5} \div \frac{4}{2}, \frac{2}{5} \div \frac{4}{3}$, and $\frac{3}{4} \div \frac{5}{2}$ all have the least quotient of $\frac{3}{10}$.

18. For example: Tahoe has $\frac{7}{8}$ of a bag of chips he wants to share with his friends. Each serving of chips should be $\frac{1}{4}$ of a bag. How many servings are there? $3\frac{1}{2}$

19. a) $\frac{3}{4}$ b) $\frac{5}{4}$ c) $\frac{5}{9}$ d) $\frac{9}{7}$

20. Answers will vary. For example:
$\frac{3}{5} \div \frac{4}{5}; \frac{6}{11} \div \frac{5}{9}; \frac{1}{2} \div \frac{3}{4}$

3.7 Dividing Mixed Numbers, page 145

4. a) $\frac{35}{8}$ b) $\frac{23}{7}$ c) $\frac{37}{6}$ d) $\frac{9}{4}$
   e) $\frac{17}{10}$ f) $\frac{23}{3}$ g) $\frac{23}{9}$ h) $\frac{27}{5}$

5. a) $1\frac{5}{9}$ b) $2\frac{2}{7}$ c) $4\frac{4}{5}$ d) $2\frac{1}{10}$
   e) $2\frac{1}{2}$ f) $3\frac{2}{7}$ g) $5\frac{2}{3}$ h) $2\frac{1}{12}$

6. a) 2 b) 4 c) 1 d) 5

7. a) $\frac{2}{3}$ b) $\frac{9}{5} \div \frac{27}{10}$ c) $\frac{2}{3}$
   d) Yes; The quotient and estimate are both $\frac{2}{3}$.

8. a) $3\frac{1}{3}$ b) $\frac{7}{26}$ c) 1 d) $\frac{3}{5}$

9. a) $1\frac{2}{5}$
   b) $1\frac{1}{2}$

10. a) $\frac{44}{63}$ b) $3\frac{1}{3}$ c) $\frac{13}{36}$ d) 1

11. a) $\frac{57}{80}$ b) $1\frac{5}{28}$ c) $\frac{18}{35}$ d) 1

12. 10

13. $3\frac{1}{2}$ min

14. a) 7 c) $10\frac{5}{8} \div 1\frac{1}{2} = 7\frac{1}{12}$
   d) Amelia can fill 7 planters and $\frac{1}{12}$ of another planter.

15. For example: Sharon has $4\frac{2}{3}$ pounds of cherries with which to make cherry tarts. Each tart requires $\frac{3}{5}$ of a pound of cherries. How many tarts can she make? $7\frac{7}{9}$ tarts

ANSWERS 507

16. Greatest quotient: c; Least quotient: d
17. a) $4\frac{3}{8} \div 3\frac{2}{5}$ is a mixed number since the divisor is smaller than the dividend.
    b) $\frac{175}{136}$; $\frac{136}{175}$ The quotients are reciprocals.
18. Parts a, b, and d have values less than $3\frac{1}{5}$. Parts c, e, and f have values greater than $3\frac{1}{5}$. Part f has a greater value than part e since $\frac{3}{2} > \frac{2}{3}$.
    Calculate c and f:
    c) $4\frac{4}{5}$ f) $4\frac{7}{10}$
    So, c has the greatest value.
19. a) Instead of multiplying, divide by the reciprocal of the second fraction.
    b) Answers may vary. For example: No, since drawing number lines to divide takes too long

3.8 Solving Problems with Fractions, page 151

3. a) Addition   b) Multiplication
   c) Subtraction   d) Multiplication
4. $\frac{11}{12}$ cans; Addition
5. 40 goals; Division
6. a) $\frac{1}{2}$; Subtraction   b) 15; Multiplication
7. $\frac{7}{12}$ h; Subtraction
8. $960; Multiplication
9. 72 cm; Division
10. $\frac{5}{24}$; Subtraction
11. a) $\frac{1}{2}$ cup; Subtraction b) $1\frac{1}{8}$ cups; Multiplication
    c) $1\frac{23}{24}$ cups; Addition d) $\frac{13}{24}$ cup; Subtraction
12. $\frac{1}{12}$; Multiplication
13. $\frac{3}{5}$; Subtraction, then multiplication
14. $\frac{17}{24}$; Division
15. The official was puzzled because the sum of $\frac{3}{8}, \frac{3}{5}$, and $\frac{1}{20}$ is greater than 1.

3.9 Order of Operations with Fractions, page 155

4. a) Subtraction   b) Multiplication
   c) Division   d) Addition
5. Raj; Rena added before she multiplied.
6. a) $\frac{11}{20}$; Multiplication b) $2\frac{1}{3}$; Division
   c) $1\frac{10}{21}$; Division   d) $\frac{1}{48}$; Subtraction
   e) $1\frac{1}{3}$; Division   f) $\frac{8}{9}$; Addition
7. a) $\frac{3}{16}$ b) $1\frac{5}{8}$ c) $1\frac{2}{3}$ d) $1\frac{3}{8}$
8. No; In the first equation you divide first, and in the second equation you multiply first.
9. a) $\frac{2}{5}$ b) $1\frac{1}{5}$ c) $\frac{1}{2}$
10. a) 4   b) $\frac{1}{18}$
11. a) Myra
    b) Robert solved $\left(\frac{3}{4} - \frac{1}{2}\right) + \frac{13}{6} \times \frac{1}{2}$ then multiplied by 4. Joe solved $\left(\frac{3}{4} - \frac{1}{2}\right) + \frac{13}{6}$ before multiplying.
12. a) $2\frac{7}{8}$ b) $1\frac{5}{8}$ c) $5\frac{11}{15}$

Unit 3 Strategies for Success: Checking and Reflecting, page 157

1. a) 3   b) $\frac{2}{5}$ c) $2\frac{11}{12}$ d) $\frac{3}{4}$
2. 12 glasses
3. $\frac{5}{6}$ h

Unit 3 Unit Review, page 159

1. a) $6 \times \frac{2}{5} = 2\frac{2}{5}$ b) $3 \times \frac{6}{7} = 2\frac{4}{7}$
2. a) 1   b) $3\frac{1}{2}$ c) $3\frac{1}{5}$
3. a) 18   b) 4   c) 50   d) $1\frac{1}{2}$
4. a) $\frac{1}{4}$ b) $\frac{6}{25}$ c) $\frac{21}{40}$ d) $\frac{1}{7}$
5. $\frac{3}{20}$
6. a) $\frac{3}{20}$ b) $\frac{3}{40}$ c) $\frac{7}{20}$ d) $\frac{4}{21}$
7. $\frac{3}{10}$
8. For example: $\frac{5}{7}$ of a litter of mice are grey with white patches. The other $\frac{2}{7}$ are black. Of the grey and white mice, $\frac{3}{8}$ are female. What fraction of the litter is grey, white, and female? $\frac{15}{56}$

9. a) $\frac{15}{2}$ b) $\frac{23}{8}$ c) $\frac{107}{10}$

10. a) $3\frac{1}{2}$ b) $7\frac{3}{5}$ c) $\frac{4}{5}$ d) $7\frac{1}{2}$

11. a) $3\frac{1}{6}$ b) $2\frac{13}{16}$ c) $3\frac{3}{20}$ d) $8\frac{2}{3}$

12. $4\frac{1}{12}$ h assuming that he mows at the same rate

13. a) $\frac{1}{10}$ b) 12

14. a) $3\frac{3}{4}$ b) $4\frac{4}{5}$ c) $\frac{3}{20}$ d) $\frac{7}{8}$

15. 16 glasses

16. $13\frac{1}{2}$ people

17. For example: $\frac{3}{4} \div 5 = \frac{3}{20}$

18. a) $1\frac{1}{2}$ b) $1\frac{1}{2}$

19. a) 2 b) $\frac{2}{7}$ c) $1\frac{1}{4}$ d) $\frac{5}{6}$

20. $5\frac{1}{4}$

21. For example: $\frac{3}{5} \div \frac{5}{3} = \frac{9}{25} < 1$

22. a) $\frac{40}{11}$ b) $\frac{31}{6}$ c) $\frac{44}{9}$ d) $\frac{29}{12}$

23. a) $\frac{14}{17}$ b) $1\frac{49}{66}$ c) $2\frac{6}{11}$ d) $\frac{1}{2}$

24. $4\frac{3}{5}$

25. $\frac{1}{8}$

26. 882 tickets

27. a) $\frac{3}{10}$ b) 9 students

28. a) $\frac{3}{5}$; Multiplication b) $2\frac{2}{11}$; Subtraction
    c) $2\frac{2}{5}$; Multiplication d) $\frac{3}{5}$; Division

29. a) $\frac{3}{4}$ b) $1\frac{3}{4}$ c) $\frac{1}{2}$ d) $\frac{5}{36}$

30. Carlton should have written $\frac{14}{5} \div \frac{9}{12} = \frac{14}{5} \times \frac{12}{9}$.
    Correct answer: $3\frac{11}{15}$

Unit 3 Practice Test, page 162

1. 6 2. $\frac{5}{12}$

3. a) 7 b) $\frac{3}{16}$ c) $\frac{5}{12}$ d) $\frac{3}{10}$

4. a) $2\frac{1}{32}$ b) 7 c) $2\frac{4}{7}$ d) $\frac{14}{15}$

5. a) $\frac{1}{5}$
   b) 30 since $\frac{3}{5}$ of 30 is 18 and $\frac{1}{3}$ of 30 is 10, which are both whole numbers

6. a) $1\frac{3}{4}$ b) $1\frac{7}{12}$ c) $3\frac{1}{9}$ d) $3\frac{1}{12}$

7. The product of a fraction and its reciprocal is 1.
   For example: $\frac{7}{8} \times \frac{8}{7} = \frac{56}{56} = 1$

8. a) $\frac{2}{5}$ b) $2\frac{1}{8}$

9. a) About $4\frac{1}{2}$
   b)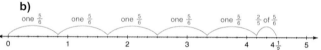
   c) About 5
   d) Each poodle takes the same amount of time to groom.

10. a) $\frac{7}{15}$ b) $\frac{1}{4}$
    c) i) Yes ii) No; $\frac{1}{6}$ cups

11. a) No b) No c) Yes

Cumulative Review Units 1–3, page 167

1. a) i) 1, 2, 3, 4, 6, 7, 12, 14, 21, 42, 84
      ii) 1, 3, 7, 9, 21, 49, 63, 147, 441
      iii) 1, 2, 4, 59, 118, 236
      iv) 1, 2, 3, 4, 5, 6, 9, 10, 12, 15, 18, 20, 25, 30, 36, 45, 50, 60, 75, 90, 100, 150, 180, 225, 300, 450, 900
   b) 441 and 900; Both have an odd number of factors.

2. a) 7.2 b) 7.9 c) 9.5 d) 8.7

3. a) 32.5 cm b) 27.5 cm

4. a) No; $16 + 8 \neq 30$ b) Yes; $16 + 8 = 24$

5. a) No; $2^2 + 5^2 \neq 6^2$ b) Yes; $6^2 + 8^2 = 10^2$
   c) No; $9^2 + 7^2 \neq 12^2$ d) Yes; $18^2 + 24^2 = 30^2$

6. 5 cm

7. 31.1 m

8. a) -72 b) $+112$ c) $+18$ d) -126

9. a) $(+4) \times (-5)$ b) -20

10. For example: Paul withdraws $5 from his account for 8 days. How much has he withdrawn in total? $(+8) \times (-5) = -40$; $40

11. a) -11 b) $+21$ c) $+30$ d) -4 e) -17 f) 0

12. a) -7 b) $+5$ c) -6 d) $+7$

13. a) $(-52) \div (+4)$ b) -13; $13

14. Answers will vary. For example:
    a) i) $(-10) + (+2) = -8$ ii) $(+2) - (+10) = -8$
       iii) $(-4) \times (+2) = -8$ iv) $(-16) \div (+2) = -8$
    b) i) $(-4) + (+2) = -2$ ii) $(+1) - (+3) = -2$

      iii) $(-1) \times (+2) = -2$ **iv)** $(-4) \div (+2) = -2$
c) i) $(-10) + (-2) = -12$ **ii)** $(+2) - (+14) = -12$
      iii) $(-4) \times (+3) = -12$ **iv)** $(+24) \div (-2) = -12$
d) i) $(-7) + (+4) = -3$ **ii)** $(+7) - (+10) = -3$
      iii) $(-1) \times (+3) = -3$ **iv)** $(+15) \div (-5) = -3$

15. a) -24 **b)** -32 **c)** -3

16. a) $2\frac{1}{3}$ **b)** $7\frac{1}{2}$ **c)** $2\frac{5}{8}$ **d)** $3\frac{4}{5}$

17. a) 28 **b)** $17\frac{1}{2}$ **c)** $\frac{25}{36}$

    d) 6 **e)** $4\frac{1}{8}$ **f)** $13\frac{1}{5}$

18. a) 15 **b)** $\frac{5}{18}$ **c)** $2\frac{2}{9}$

    d) $1\frac{1}{4}$ **e)** $1\frac{1}{4}$ **f)** $\frac{9}{14}$

19. $4\frac{8}{9}$ or about 5 bottles

20. a) $\frac{9}{16}$ **b)** $2\frac{7}{24}$ **c)** $1\frac{3}{7}$ **d)** $3\frac{1}{3}$

21. a) $4\frac{1}{4}$; subtraction   **b)** $8\frac{1}{8}$; multiplication

Unit 4 Measuring Prisms and Cylinders, page 168

4.1 Exploring Nets, page 174

4. 6 rectangular faces

5. 6 square faces

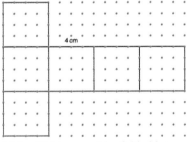

6. Net a; Net b cannot be folded into a rectangular prism.

7. 2 square faces and 4 rectangular faces

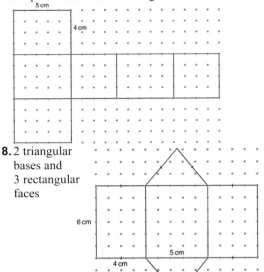

8. 2 triangular bases and 3 rectangular faces

9. a, b) A and F; right hexagonal prism: 2 hexagonal bases and 6 square faces
B and D; pentagonal pyramid: 1 pentagonal base and 5 triangular faces
C and E; right pentagonal prism: 2 pentagonal bases and 5 rectangular faces

10. Net A, Net B, Net C

11. Net C

12. a) Right triangular prism: 2 equilateral triangular bases and 3 rectangular faces
For example:

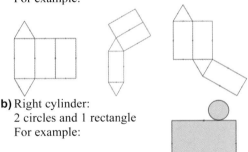

b) Right cylinder: 2 circles and 1 rectangle
For example:

13. a) For example:

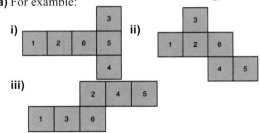

b) For example: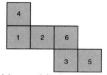

14. a) Square pyramid   **b)** Square prism
  c) Rectangular pyramid
  d) Pentagonal pyramid   **e)** Triangular pyramid
15. Like a net, wrapping paper is folded into the shape of various objects.
Unlike a net, wrapping paper is not folded to make an object.
16. a)
  b)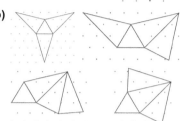

17. b)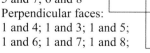
  c) Both nets of a cube with different colour arrangements
  d) Same net, different colour labels

4.2 Creating Objects from Nets, page 180

3. b
4. a) Right triangular prism
  c) 2 congruent triangular bases and 3 rectangular faces
5. a) C: Triangular pyramid
   D: Right triangular prism
   E: Right hexagonal prism
   F: Cube with pyramid on top
6. a) Right prism with 2 congruent L-shaped bases
  c) Yes; 2 congruent L-shaped bases and 6 rectangular faces
  d) Parallel faces: 1 and 2; 3 and 5; 3 and 7; 4 and 6; 4 and 8; 5 and 7; 6 and 8
   Perpendicular faces: 1 and 4; 1 and 3; 1 and 5; 1 and 6; 1 and 7; 1 and 8;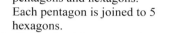
2 and 3; 2 and 4; 2 and 5; 2 and 6; 2 and 7; 2 and 8; 3 and 4; 3 and 6; 3 and 8; 4 and 5; 4 and 7; 5 and 6; 5 and 8; 6 and 7; 7 and 8
7. a) No
  b) There are several ways to correct the diagram.
   For example:
   Move squares A and B to: 1 and 2; 3 and 4; 1 and 3; or 2 and 4.

8. A soccer ball is made of pentagons and hexagons. Each pentagon is joined to 5 hexagons.
9. a) Net of a right rectangular pyramid
  b) Net of a right triangular prism
  c) Not a net   **d)** Not a net
10. a
11. B, C, E, F, H, J
12. a) 8 congruent faces
  b) 8 congruent triangular faces, each with equal angles and side lengths
13. a) Square pyramid   **b)**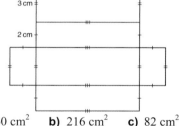

4.3 Surface Area of a Right Rectangular Prism, page 186

4. 174 cm^2; Add the areas of all faces.
5. 92 cm^2
6. a) 160 cm^2 **b)** 216 cm^2 **c)** 82 cm^2
7. a) 164 m^2 **b)** 158 cm^2
9. For example:
  a) 5 cans
  b) The height of the room is 3 m. The ceiling and floor are not being painted.
10. a) 9 cm^2 **b)** 3 cm
11. 2 700 000 Euros
12. 12 000 m^2; Assuming the windows cover one-quarter of 4 sides
13. Greatest surface area: R; Least surface area: Q
14. i) Increases but doesn't double
   ii) Decreases but not by factor of 2
15. a) 1580 cm^2 **b)** 436 cm^2
16. 2 m by 2 m by 5 m
17. 3 cm by 4 cm by 6 cm

4.4 Surface Area of a Right Triangular Prism, page 191

4. 81 cm²; Add the areas of all faces.
5. Triangular bases and rectangular faces on the sides
6. 2100 cm²

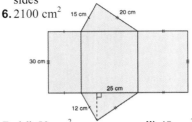

7. a) i) 50 cm²   **ii)** 48 cm²
   b) The surface area of the prism is the same as the area of the net.
8. Prism D: 528.0 cm²; Prism A: 147.7 cm²; Prism C: 117.0 cm²; Prism B: 102.4 cm²
9. a) 336 cm²   **b)** 334.4 m²   **c)** 481.2 mm²
10.

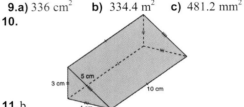

11. b
12. No; The surface area will increase by a factor of 4.
13. 14 400 cm²
14. 1872.7 cm²; Assuming the triangular bases are not covered
15. No, since the surface area of the rectangular prism does not include the side along which the cut was made
16. Between 7.33 cm and 11.5 cm
17. a) About 7.4 cm   **b)** 374 cm²

Unit 4 Mid-Unit Review, page 194

1. a)

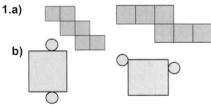

   b)

2. a) A and C are nets
   b) A forms a regular triangular pyramid. C forms a pentagonal pyramid.
3. a) 88 cm²   **b)** 447 cm²
4. 59 318 m²; Assuming the top and bottom are not glass
5. a) 10 752 cm²
   b) About 10 800 cm²; Wrapping paper must overlap a bit

4.5 Volume of a Rectangular Prism, page 198

4. a) 120 cm³   **b)** 729 cm³   **c)** 6000 cm³
5. b) 11 200 cm³
6. a) A: 120 cm³; B: 120 cm³; C: 120 cm³
   b) They are the same.   **c)** No
7. a) 67.5 cm³   **b)** 96 cm³   **c)** 25.2 cm³
9. a) 420 462 cm³, 626 859 cm³
   b) About 1.5 times
10. 10 cm
11. a) 40.6 m³   **b)** 3 trailers
12. a) 2000 m³   **b)** 1800 m³   **c)** 1000 m³
13. a) Possible dimensions: 1 cm by 2 cm by 18 cm; 1 cm by 1 cm by 36 cm; 1 cm by 3 cm by 12 cm; 1 cm by 4 cm by 9 cm; 2 cm by 2 cm by 9 cm; 1 cm by 6 cm by 6 cm; 3 cm by 3 cm by 4 cm; 2 cm by 3 cm by 6 cm
   Decimal solutions are possible as well.
   b) i) Prism with dimensions 1 cm by 1 cm by 36 cm
   ii) Prism with dimensions 3 cm by 3 cm by 4 cm
14. a) 1260 cm³   **b)** 42 cm³
   c) For example: Philip could cut the fudge into 3 columns and 10 rows.
   d) 7 cm by 2 cm by 3 cm
15. a) Doubles   **b)** Increases by a factor of 4
   c) Increases by a factor of 8
   True for all rectangular prisms
16. To double volume, double any one dimension. Surface area will increase but not double.
17. a) 8640 cm³   **b)** 60 cm by 36 cm by 32 cm
   c) A box with dimensions 60 cm by 36 cm by 32 cm has the least surface area. So, the least amount of material is needed to make it.
18. a) Sketches may vary. Possible dimensions: 1 cm by 1 cm by 24 cm; 2 cm by 1 cm by 12 cm; 3 cm by 1 cm by 8 cm

4.6 Volume of a Right Triangular Prism, page 205

3. a) 225 cm³   **b)** 312 cm³
4. a) 21.16 cm³   **b)** 217.5 cm³   **c)** 45 m³
5. a) 955.5 cm³   **b)** 240 m³   **c)** 3.83 m³
6. a) 532 cm³   **b)** 108 cm³
7. 18 cm³
8. 7.5 cm
9. a) i) $A = 5$ cm², $l = 1$ cm; $A = 1$ cm², $l = 5$ cm
   ii) $A = 9$ m², $l = 1$ m; $A = 1$ m², $l = 9$ m; $A = 3$ m², $l = 3$ m
   iii) $A = 8$ m², $l = 1$ m; $A = 4$ m², $l = 2$ m; $A = 2$ m², $l = 4$ m; $A = 1$ m², $l = 8$ m
   iv) $A = 18$ cm², $l = 1$ cm; $A = 9$ cm², $l = 2$ cm; $A = 6$ cm², $l = 3$ cm; $A = 3$ cm², $l = 6$ cm; $A = 2$ cm², $l = 9$ cm; $A = 1$ cm², $l = 18$ cm

b) i) 2   **ii)** 3   **iii)** 4   **iv)** 6
10.a) 120 cm³   **b)** 6
11. 10 m²
12.a) 1.125 m³   **b)** 3.375 m³
13.b) A: 180 cm³; B: 126 cm³
   c) Change the length of 7 cm to 10 cm
14.a) 2250 cm³   **b)** 18 cm   **c)** 60%
15. $l = 11$ cm
   Possible b and h values: 1 cm, 36 cm; 36 cm, 1 cm; 2 cm, 18 cm; 18 cm, 2 cm; 3 cm, 12 cm; 12 cm, 3 cm; 4 cm, 9 cm; 9 cm, 4 cm; 6 cm, 6 cm
16.a) 231.4 cm²; 113.9 cm³
   b) i) $b = 7$ cm, $h = 6.2$ cm, $l = 21$ cm
      $b = 7$ cm, $h = 3.1$ cm, $l = 42$ cm
      $b = 3.5$ cm, $h = 6.2$ cm, $l = 42$ cm
      $b = 14$ cm, $h = 3.1$ cm, $l = 21$ cm
      $b = 3.5$ cm, $h = 12.4$ cm, $l = 21$ cm
      $b = 3.5$ cm, $h = 3.1$ cm, $l = 84$ cm
   ii) Either two of the dimensions are doubled or one is increased by a factor of 4.
17.a) 36 m²; 12 m³   **b)** 63.1 m²; 24 m³
   c) 96 m²; 48 m³   **d)** 144 m²; 96 m³

4.7 Surface Area of a Right Cylinder, page 212

4.a) 88 cm² **b)** 25 cm² **c)** 101 cm²
5.a) A cylinder with radius 2 cm and height 5 cm
   b) A cylinder with radius 1 cm and height 3 cm
   c) A cylinder with radius 2 cm and height 6 cm
6.a) 50 cm² **b)** 94 cm² **c)** 251 m²
8.a) 214 cm² **b)** 19 046 mm² **c)** 4 m²
9. 174 m²   **10.** 12 m²
11.a) 94 cm²   **b)** 4255
12. 191 cm²   **13.** 37 267 cm² or 4 m²
14.a) 14 137 cm²
   b) 1414 cm²; Assuming the heads do not go over the edge of the shell
15. Cylindrical tubes
16.a) 66 cm   **b)** 10.5 cm
   c) 346 cm²   **d)** 1352 cm²
17. 483 cm²

4.8 Volume of a Right Cylinder, page 218

4.a) 785 cm³   **b)** 63 cm³   **c)** 1609 cm³
5.a) 503 cm³   **b)** 8836 mm³   **c)** 328 m³
6. 1571 cm³   **8.** 196 cm³
9. The cylinders have the same volume because they have the same radius and height.
10. Bottle C; It has the greatest radius and height, and so the greatest volume.
11.a) 462 cm³   **b)** 1583 cm³
12. 5301 cm³   **13.** 441 786 cm³
14. 9 521 684 cm³ or about 9.5 m³
15. 404 mL
16. A cylinder with radius 2 m and height 1 m

17.a) 96 m³   **b)** 12 219 m³
   c) 4.58 m by 4.58 m by 4.58 m
18. 28 cm

Unit 4 Strategies for Success: Choosing the Correct Answer, page 220

1.c) 108 m²   **2.b)** 99 cm³

Unit 4 Unit Review, page 223

1. The net of a rectangular prism must have 3 pairs of congruent rectangles.
2.a) Right hexagonal prism   **b)** Cube
   c) Right cylinder   **d)** Pentagonal pyramid
3. Net A; To correct net B, move the rectangle from the top right to the bottom right.
4.a) Pentagonal pyramid **b)** Triangular pyramid
   c) To form a triangular prism:

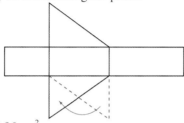

5.a) 96 cm²
   b) Find the area of one face and multiply by 6.
6.a) 72 m²; 36 m³   **b)** 114 cm²; 72 cm³
   c) 15 000 cm²; 125 000 cm³
7.a) 10 rolls and 1 can
   b) Assuming you must buy whole rolls and cans
8.a) A: 1 m by 1 m by 28 m; B: 1 m by 2 m by 14 m
      C: 1 m by 4 m by 7 m; D: 2 m by 2 m by 7 m
   b) A: 114 m²; B: 88 m²; C: 78 m²; D: 64 m²
9.a) 24 m³   **b)** 45 cm³
10.a) 7.2 cm²   **b)** 0.96 cm³
11.a) 14.2 m²   **b)** 2.3 m³
   c) Rotating an object will not change its volume.
12. 6 m³
13.a) Double the height or base of the triangle.
   c) The volume doubles.
14.a) 428 mL
   b) A pocket of air is left in the soup can.
15. 49 m²
16. 12 174 m²

Unit 4 Practice Test, page 226

1.a) A square prism with a square pyramid on top

2. a) Triangular prism: 2 triangular bases   **b)** Cylinder: 2 circles and 1 rectangle

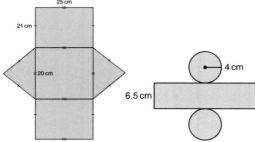

3. a
4. a) 632 cm²   **b)** 200 cm²   **c)** 88 cm²
5. a) 220.5 m³   **b)** 120 m³
6. a) 776 cm²; 1344 cm³   **b)** 17.99 m²; 3.0625 m³
7. a) 166.95 cm³   **b)** 126 m³
8. a) The volume increases by a factor of 9.
  c) 27.5625 m³
9. a) 8.5 m²   **b)** 1.5 m³
10. A piece of paper rolled into a cylinder lengthwise

Unit 5 Percent, Ratio, and Rate, page 232

5.1 Relating Fractions, Decimals, and Percents, page 239

6. a) $\frac{50}{100}$ or $\frac{1}{2}$; 0.50; 50%   **b)** $\frac{36}{100}$ or $\frac{9}{25}$; 0.36; 36%
  c) $\frac{87}{100}$; 0.87; 87%   **d)** $\frac{4}{100}$ or $\frac{1}{25}$; 0.04; 4%
7. a) $\frac{3}{100}$; 0.03   **b)** $\frac{51}{100}$; 0.51
  c) $\frac{98}{100}$ or $\frac{49}{50}$; 0.98   **d)** $\frac{29}{100}$; 0.29
8. a) $\frac{1}{8}$; 0.125; 12.5%   **b)** $\frac{341}{400}$; 0.8525; 85.25%
  c) $\frac{139}{400}$; 0.3475; 34.75%
9. a) $\frac{147}{200}$; 0.735   **b)** $\frac{17}{80}$; 0.2125
  c) $\frac{7}{80}$; 0.875   **d)** $\frac{3}{250}$; 0.012

10. a)   **b)**
  c)   **d)**
11. a)   **b)**
  c)   **d)**

12. a) $\frac{1}{400}$; 0.0025   **b)** $\frac{3}{500}$; 0.006
  c) $\frac{1}{200}$; 0.005   **d)** $\frac{19}{5000}$; 0.0038
13. a) $0.00\overline{6}$; $0.\overline{6}\%$ **b)** 0.045; 4.5%
  c) 0.014; 1.4%   **d)** 0.032; 3.2%
14. a) $\frac{69}{200}$; 34.5%   **b)** $\frac{23}{10\,000}$; 0.23%
  c) $\frac{73}{400}$; 18.25%   **d)** $\frac{7}{1000}$; 0.7%
15. Fiona is correct.
16. Junita; $83.\overline{3}\%$ is greater than 82.5%.
17. Answers may vary. For example:
  a) i) $\frac{5}{8}$ means one whole divided into 8 equal parts, with 5 parts shaded
  ii) A cherry rhubarb pie is divided into 8 equal slices. Laura and her friends eat 3 slices. $\frac{5}{8}$ of the pie is left.
  b) i) $\frac{5}{8}$ means 5 objects divided into 8 equal groups.
  ii) 5 watermelons are shared among 8 people.
18. a) 16 red squares, 12 green squares, and 18 blue squares Shadings may vary. For example:
  b) $\frac{1}{24}$; $0.041\overline{6}$; $4.1\overline{6}\%$

  c) In a 6-cm by 9-cm rectangle, 18 squares will be red, 13.5 squares will be green, and 20.25 squares will be blue.
Answers will vary. For example: Yes, but it would have been more complicated because it would involve part squares;
In a 7-cm by 7-cm square, $16.\overline{3}$ squares will be red, 12.25 squares will be green, and 18.375 squares will be blue.
19. Kyle; 78.6% is greater than $76.\overline{6}\%$.
20. a) No
  b) $\frac{1}{16}$; The student made an error when he converted 6.25% to a fraction.
21. a) =   **b)** >   **c)** >   **d)** <   **e)** =   **f)** =

5.2 Calculating Percents, page 246
4. a)   **b)**

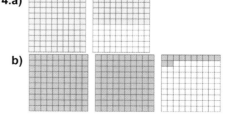

c)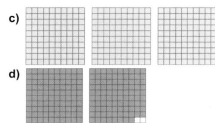

d)

5. a) 1.2 **b)** 2.5 **c)** 4.75
d) 0.003 **e)** 0.0053 **f)** 0.0075

6. a) $\frac{17}{10}$; 170% **b)** $\frac{33}{10}$; 330%
c) $\frac{3}{1000}$; 0.3% **d)** $\frac{56}{10\,000}$; 0.56%

7. $25.20

0% — $9.00 (100%) — 200% — $25.20 (280%) — 300%

8. Answers may vary. It is not possible for one individual to give 110% or to put in more than 100% of their effort.

9. a) Answers may vary. For example:
A charity has a goal for the amount of money they wish to receive in donations. It receives more than 100% of its goal.
A baker has a recipe for 6 dozen cookies. He wants to make 8 dozen, so he must use more than 100% of the ingredients in the original recipe.
b) 5000 tickets were sold in a raffle for a new boat. The chance of winning written as a percent is between 0% and 1%.
If you guess every answer on a 100 question multiple-choice test, your chances of getting 100% are between 0% and 1%.

10. a) i) $33.\overline{3}\%$ **ii)** $66.\overline{6}\%$ **iii)** 100%
iv) $133.\overline{3}\%$ **v)** $166.\overline{6}\%$ **vi)** 200%
b) Each time the numerator increases by 1, the percent increases by $33.\overline{3}\%$.
c) i) $233.\overline{3}\%$ **ii)** $266.\overline{6}\%$ **iii)** 300%
iv) $333.\overline{3}\%$ **v)** $366.\overline{6}\%$ **vi)** 400%

11. a) i) 720 **ii)** 72 **iii)** 7.2 **iv)** 0.72
b) The digits move one place to the right each time.
c) i) 7200 **ii)** 0.072

12. a) 5 runners

13. a) Shapes may vary. For example:

14. About 3140
a) 3139.5 or about 3140 **b)** 3120
c) No, the answers are different. Juan's answer is correct.

15. a) 168 people
16. a) About 20 people **b)** 15 people **c)** 1985
17. $66.\overline{6}\%$ **18.** $110 000
19. 58 cm

5.3 Solving Percent Problems, page 252

3. a) 30 **b)** 16 **c)** 200 **d)** 150
4. a) 20 **b)** 24 **c)** 800 **d)** 40
5. a) 100% **b)** 50%
6. a) 20% **b)** 25%
7. a) $833.\overline{3}$ g **b)** 500 cm **c)** 1500 g
8. a) 7.5% **b)** About 138%
9. a) About 4.55% **b)** 41.2%
10. 169 840 miners
11. About 78.2%
12. a) Jemma: about 1.98 kg; George: 1.95 kg
b) Explanations may vary. Jenna's mass after Week 2 is calculated on her mass after Week 1, which is greater than her birth mass.
13. a) 859 320 people **b)** About 953 845 people
c) About 37.64%
d) No; the overall increase in population is greater than 35%.
14. a) About 13 711 crimes per 100 000 population
b) No; the total decrease from 2004 to 2006 is less than 10%.
15. a) About 167 cm **b)** About 180 cm
Assumptions may vary. For example:
The girls' height increases at the given average rate.
16. Answers will vary, based on gender and current height.
17. No. The original price is $20. It is greater than 120% of the sale price ($19.2).
18. a) 5.6 cm **b)** 44%
c) 31.36 cm^2 **d)** 68.64%
19. 200 marbles
20. Answers may vary. For example:
5 m by 45 m or 15 m by 15 m

5.4 Sales Tax and Discount, page 260

4. a) $1.05 **b)** $0.63 **c)** $1.54
5. a) $5.40 **b)** $1.50 **c)** $1.08
6. a) $0.97 **b)** $4.29 **c)** $3.64
7. a) i) PST: $1.30; GST: $1.56
ii) PST: $7.62; GST: $9.15
b) i) $28.85 **ii)** $169.22
8. a) i) $18.00 **ii)** $54.00
b) i) $71.99 **ii)** $66.00
c) i) $76.31 **ii)** $69.96
9. $389 120
10. Choice A; $35 for 2 DVDs is a better deal than $40 for 2 DVDs.
11. a) About 40% **b)** $12.07

12. No; the total reduction in price was less than 50%.
13. $44.95
14. No; the sale price including taxes is $127.79.
15. a) $86.00   b) $65.69
16. It makes no difference whether the discount is calculated before or after the tax is added. Anika will pay $59.33.
17. Strictly Sports offers the better deal. $41.44 is $0.95 cheaper than $42.39.
18. a) $87.72   b) $12.28
19. a) Yes; the total cost is $22.20.
    b) $17.80
20. The sale price of the skateboard including taxes is $32.47. $39.99 – 14% means a price of $34.39, which is more than the actual sale price of the skateboard.

Unit 5 Mid-Unit Review, page 263

1. a) $\frac{3}{5}$; 0.6   b) $\frac{39}{400}$; 0.0975
   c) $\frac{975}{1000}$ or $\frac{39}{40}$; 0.975

2. a) 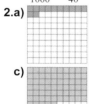   b)
   c)   d)

3. a) $\frac{9}{50}$; 18%   b) $\frac{3}{500}$; 0.6%
   c) $\frac{7}{8}$; 87.5%   d) $\frac{3}{400}$; 0.75%

4. a) 1.45   b) 3.5   c) 0.0044   d) 0.002

5. Answers may vary. For example: No, a mark of 112% is not possible. If Jon answered correctly all the questions on the test, he would score 100%.

6. $45.50
7. a) 20   b) 50
8. 700 tickets   9. 16%   10. $28.93
11. a) $39.99   b) $45.19   c) $31.63   d) $13.56
12. $43.52

5.5 Exploring Ratios, page 266

4. a) $\frac{5}{8}$ b) $\frac{3}{4}$ c) $\frac{4}{9}$ d) $\frac{24}{25}$

5. a) 95%   b) 80%   c) 37.5%   d) $83.\overline{3}\%$

6. a) red candies to green candies
   b) blue candies to green candies
   c) green candies to the total number of candies
   d) red candies to green candies to blue candies
   e) red candies to green and blue candies

7. Answers may vary. For example:
   a) 3 : 15; 20%; $\frac{1}{5}$ b) 1 : 15; $6.\overline{66}\%$; $\frac{1}{15}$
   c) 7 : 4; 7 to 4   d) 7 : 1 : 3; 7 to 1 to 3
8. a) 5 : 7   b) About 71.4%
9. a) i) 9 : 7   ii) 8 : 3   iii) 3 : 1 : 2
   b) i) 9 : 16; $\frac{9}{16}$; 56.25%; 7 : 16; $\frac{7}{16}$; 43.75%
      ii) 8 : 11; $\frac{8}{11}$; $72.\overline{72}\%$; 3 : 11; $\frac{3}{11}$; $27.\overline{27}\%$
      iii) 3 : 6; $\frac{1}{2}$; 50%, 1 : 6; $\frac{1}{6}$; $16.\overline{6}\%$, 2 : 6; $\frac{1}{3}$; $33.\overline{3}\%$

11. a) i) 8 : 3   ii) 5 : 1   iii) 3 : 1 : 5   iv) 8 : 25
    b) i) 5 : 3   ii) 3 : 1   iii) 3 : 1 : 3   iv) 4 : 16

12. Answers may vary. For example:
    a) $\frac{2}{7}$ as a ratio compares 2 out of 7 parts of a group to the total number of parts of the group. It can also be written as 2 : 7.
    b) There are 7 students in the student council: 2 boys and 5 girls. The ratio of boys to the total number of students is 2 : 7 or $\frac{2}{7}$ or about 28.57%.

13. Answers may vary. For example:
    a)
    b)
    c)
    d) The ratio in part a may be a part-to-part or a part-to-whole ratio. The ratios in parts b and c are part-to-part ratios.

14. a) 11 cups
    b) i) 3 : 2   ii) 2 : 3   iii) 2 : 2 : 1
    c) 5 : 11; $\frac{5}{11}$; $45.\overline{45}\%$
    d) i) 2 : 2   ii) 2 : 3   iii) 2 : 2 : 1
       4 : 10; $\frac{2}{5}$; 40%
    e) Answers will vary. For example: Patrick decides to add 1 extra cup of raisins to the recipe. Write the ratio of raisins to the total amount of ingredients.
       (*Answer:* 1 : 12; $\frac{1}{12}$; $8.\overline{3}\%$)

15. Answers may vary. For example: ratio, problem, percent, taxes
16. No. Jeff got $\frac{2}{5}$ of the cranberries.
17. Answers may vary. For example:
    a) i) Squares to triangles: 3 : 5

ii) Green shapes to blue triangles: 2 : 1
   iii) Red squares to red triangles: 2 : 3
   iv) Green squares to all shapes: 1 : 11
   b) Replace a red circle with a green square. The ratio of all red shapes to red triangles is 7 : 3; the ratio of green squares to triangles is 2 : 5.
18. Answers will vary. For example: 3 : 5 squirrel; cinnamon; mushroom

5.6 Equivalent Ratios, page 274

5. Answers may vary. For example:

a)
1st term	2	3	38
2nd term	4	6	76

b)
1st term	4	14	24
2nd term	6	21	36

c)
1st term	5	6	18
2nd term	20	24	72

6. Answers may vary. For example:

a)
1st term	6	9	15
2nd term	8	12	20

b)
1st term	28	98	1274
2nd term	8	28	364

c)
1st term	216	144	312
2nd term	225	150	325

7. Answers may vary. For example:

a)
1st term	2	4	16
2nd term	6	12	48
3rd term	12	24	96

b)
1st term	24	84	120
2nd term	10	35	50
3rd term	14	49	70

c)
1st term	12	168	264
2nd term	2	28	44
3rd term	4	56	88

8. **a)** 1 : 3 **b)** 2 : 3 **c)** 1 : 4 : 6 **d)** 22 : 14 : 3
9. **a)** 4 : 1 **b)** 1 : 3 **c)** 6 : 2 : 1 **d)** 4 : 8 : 3
10. **a)** 8 **b)** 60 **c)** 40 **d)** 7
11. **a)** 2 : 3 : 4 and 6 : 9 : 12; 8 : 5 : 4 and 16 : 10 : 8; 3 : 6 : 9 and 1 : 2 : 3; 3 : 4 : 5 and 9 : 12 : 15
   b) Explanations may vary. For example: In each pair, the first ratio is in simplest form. The second ratio is formed by multiplying or dividing the terms of the first ratio by the same number.
12. Answers may vary. For example:

a)
Fiction	3	21	51
Non-fiction	1	7	17

   b) There is an infinite number of different answers for part a. The ratio in simplest form can be multiplied by any number.
13. 30 cm by 15 cm
14. Drawings may vary. For example:
   a) i) ⬜🟫⬛⬜🟫⬜ 🟫⬜🟫⬜🟫⬜
   ii) △▲△▲△ ▲△▲
   iii) ●○●●○○●●● ○○●○●○●●○● ○○●●○●●

iv)

15. **a)** No **b)** No **c)** Yes **d)** Yes
16. **a)** 12 **b)** 20
17. **a)** 12 **b)** 84 **c)** 12 **d)** 21
18. **a) i)** 8 : 1 **ii)** 8 : 9 **iii)** 1 : 9
    b) i) 17 : 1 **ii)** 17 : 18 **iii)** 1 : 18

Strategies for Success: Explaining Your Thinking, page 277

1. Yes; There are 365 (or 366) days in a year, so only the first 366 students can have a unique birthday. The remaining students will share a birthday with someone else.
2. $\frac{2}{3}$ cup sugar, 500 mL milk, $3\frac{1}{3}$ mL vanilla
3. $0.35 \times 12 = 4.20$
   No; A dozen donuts will cost $4.20 with the coupon. $4.20 is more expensive than the sale price of $3.99.

5.7 Comparing Ratios, page 284

4. **a)** 1 : 4 **b)** 1 : 8 **c)** 1 : 7
   d) 1 : 9 **e)** 1 : 3 **f)** 1 : 6
5. **a)** 4 : 1 **b)** 5 : 1 **c)** 12 : 1
   d) 5 : 1 **e)** 7 : 1 **f)** 12 : 1
6. **a)** 5 cans of white paint and 7 cans of blue paint
   b) 3 cans of white paint and 4 cans of blue paint
7. Tara's
   a) $\frac{21}{35}; \frac{25}{35}$
   b) When the denominators are the same, look at the numerators to compare two fractions.
8. Henhouse B; Assumptions may vary. For example: I assume that the ratio of eggs produced daily by each henhouse is consistent.
9. Mixture A
10. Nadhu; 65 out of 117 is better than 54 out of 117, or $\frac{65}{117} > \frac{54}{117}$.
11. **a)** Calgary Cougars **b)** No
12. Recipe A
13. **a)** Ms. Arbuckle's; 2 more fiction books than Mr. Albright's class
    b) Ms. Arbuckle: $41.\overline{6}\%$
    Mr. Albright: 42.9%
14. **a)** 2 : 1 and 3 : 2
    b) Add 1 more can of concentrate to B.
15. **a)** A: 4 : 12; B: 3 : 15; C: 2 : 3
    b) A: 1 : 3; B: 1 : 5; C: 1 : 1.5
    c) Shade C **d)** Shade B
16. Marcel's reasoning is incorrect.
17. Glider A
18. **a)** 70 **b)** The second box, with a ratio of 3 : 2

19.a) No. The ratios are not equivalent.
  b) 3 scoops for 2 cups of water

5.8 Solving Ratio Problems, page 291

4.a) 36 **b)** 18 **c)** 10 **d)** 63 **e)** 33 **f)** 26
5.a) 12 **b)** 8 **c)** 7 **d)** 3 **e)** 3 **f)** 10
6.a) 9 **b)** 15 **c)** 10 **d)** 7 **e)** 70 **f)** 5
7.a) 27 **b)** 28 **c)** 12 **d)** 16 **e)** 33 **f)** 56
8. 225 shots
9. 148 dentists
10. 0.3 m or 30 cm
11.a) No. There are no measures given.
  b) 15 cm
12. 10 500 000 cm or 105 km
13. 5.14 cm
14.a) 0.15 m or 15 cm **b)** 7.2 m
15. 8 cm
16.a) 10 trees
17. $2.\overline{6}$ m³ cement, $5.\overline{3}$ m³ gravel
18.a) 39 tickets **b)** 26 tickets **c)** $1111.50
19. 1.05 m
20.a) 24 students **b)** 27 students
21. Ratio of my height : height of a flagpole = ratio of length of my shadow : length of shadow of flagpole

5.9 Exploring Rates, page 298

4.a) 60 words/min **b)** 25 m/min
  c) 20 pages/h
5.a) 15 km/h **b)** 24 km/h **c)** 10 km/min
6.a) 55 flyers/h **b)** 60 cupcakes/h
  c) 4.5°C/h
7.a) rate **b)** rate **c)** ratio **d)** ratio
8.a) $1.13/L **b)** $0.25/cob **c)** $0.42/can
9.a) 1.5 goals/game
  b) 53 goals; assuming she continues to score at the same rate
10. 120 beats/min; when you run, your heart rate increases.
11.a) $0.48 **b)** $2.40 **c)** 25 m
12.a) $10.50/h **b)** $367.50
13.a) $1.20 **b)** $3.00 **c)** $12.00 **d)** 1.5 kg
14.a) 30.3 m/s **b)** 13.9 m/s
15.a) 144 km/h **b)** 5.4 km/h
16.a) 25 km **b)** 25 km/h
17.a) $50.00 **b)** £12
18.a) About 56 **b)** About 2
  Assumptions may vary. For example:
  Petra takes no breaks and works at an even rate.
19. 8 min/km
20.a) i) 7 min/km **ii)** 8 min/km **iii)** 8.5 min/km
  b) Answers may vary. For example: 6 h 36 min

5.10 Comparing Rates, page 303

5.a) $133/week **b)** 85 km/h
  c) $0.29/bottle **d)** $0.33/can

6.a) $36.00 in 4 h **b)** $4.50 for 6 muffins
  c) $0.99 for 250 mL
7.a) The 500-mL can is the better buy.
  b) Answers may vary. For example: Delaney may have only needed 110 mL.
  c) The customer may have needed 500 mL of mushroom soup, so she could go with the better deal.
8.a) 8 for $2.99 **b)** 2 L for $4.49
  c) 150 mL for $2.19 **d)** 125 g for $0.79
9.a) 87.5 km
  b) The average speed is the mean distance travelled each hour, or 87.5 km/h.
  c) 8 h
10.b)

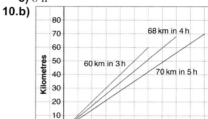

11.a) Petra's job; $9.25/h is better than $9/h.
  b) Answers may vary. For example: Yes, I would choose to work as a lifeguard. It pays more, and you get to work more hours per week.
12.a) $12\frac{2}{3}$ **b)** 304 points
13. About 4.92 cm
14.a) Brand B
  b) Brand A: $3.61/kg; Brand B: $2.21/kg
  c) Brand B is the better buy.
  d) Becky may not have enough money or room to store it.
15.a) i) $16\frac{2}{3}$ min **ii)** About 31 min
  b) i) Swimming; 52 min
    ii) Cycling and walking
16. $520.8\overline{3}$ kg
17. Answers may vary. For example:
  a) A rate compares 2 quantities with different units: a bike travelled 40 m in 5 s.
  b) A unit rate compares a quantity to a unit: a snail crawls 1.75 km in 100 h or 0.0175 km/h.
18.a) Toyota Echo **b)** 14 L
19. $8.73
20.a) i) About 3 people/km²
    ii) About 137 people/km²
    iii) About 338 people/km²
21.a) 8 km **b)** 12 km/h faster

Unit 5 Unit Review, page 308

1. a) $\frac{13}{20}$; 65% **b)** $\frac{69}{10\,000}$; 0.69%

c) $\frac{3}{80}$; 3.75% **d)** $\frac{393}{400}$; 98.25%

2. Conner: 87.5% > 83.$\overline{3}$%

3. a) $\frac{19}{50}$; 0.38 **b)** $\frac{15}{16}$; 0.9375

c) $\frac{79}{10\,000}$; 0.0079 **d)** $\frac{1}{500}$; 0.002

4. a) 1.6 **b)** 3.1 **c)** 0.0027 **d)** 0.009
5. a) 44 800 people
6. 25 cards   **7.** 65
8. a) Both mandrills had the same mass at the end of month 2.
  b) No; by the end of month 2, Amy gained more than 45% of her original mass.
9. 112.5 cm or 1.125 m
10. a) 205.8 cm by 235.2 cm **b)** 3.96%
11. a) $69.99 **b)** 28.6%
12. $76.27
13. $36.11; the cost would be the same.
14. a) i) 2 : 3 **ii)** 5 : 3 **iii)** 5 : 10 = $\frac{1}{2}$ = 50%
  iv) 5 : 2 : 3
  b) i) 1 : 2 **ii)** 4 : 2 **iii)** 4 : 7 = $\frac{4}{7}$ = 57.1%
  iv) 4 : 1 : 2
15. a) 7 : 4 **b)** 3 : 4 **c)** 3 : 11
16. a), **b)**, **c)** [diagrams]
17. a) i) 1 : 3 **ii)** 3 : 1 **iii)** 3 : 8 **iv)** 1 : 6 : 3
  b) i) Purple to blue
   ii) Yellow to red or purple to green
   iii) Yellow to blue to red
18. a) 12 girls **b)** 3 : 2
19. Explanations may vary. For example: Divide terms by a common factor, such as 5 (5 : 2 : 6) or, multiply terms by the same number, such as 3 (75 : 30 : 90)
20. a) 30 **b)** 40 **c)** 35 **d)** 108
21. a) 8 : 1 **b)** $\frac{11}{12}$: 1 **c)** $1\frac{1}{2}$: 1 **d)** $2\frac{3}{4}$: 1
22. a) Stronger **b)** Weaker
23. a) Ms. Beveridge's class
  b) Answers will vary. For example: No; I used equivalent ratios. The ratios given are part-to-part ratios and I need part-to-whole ratios to use percents.
24. 180 pike   **25.** About 19
26. a) 400 mL
  b) 15 people: about 2.14 L pop; 857.14 mL orange juice
   20 people: about 2.86 L pop; 1.14 L orange juice
27. a) 40 km/h **b)** 250 m/min **c)** $8.00/h
28. a) The wild horse is faster. 936 m/min; 800 m/min
  b) Cougar to wild horse: 117 : 100
29. a) 16.$\overline{6}$ m/s **b)** 11.$\overline{1}$ m/s
30. a) $9.50/h **b)** $237.50
31. a) i) $1.07/L **ii)** $4.46/kg **iii)** $0.44/100 g
32. a) 8.5 L for $7.31 **b)** 12 candles for $5.99
  c) 5 kg of grass seed for $2.79
33. Jevon
34. Aaron's job as a ticket seller pays more.

Unit 5 Practice Test, page 312

1. a) 132 **b)** 14 **c)** 2 **d)** 17.85
2. a) 39 **b)** 7 **c)** 28 **d)** 18
3. a) $5.16/h **b)** 12 min/puzzle **c)** $1\frac{1}{3}$ km/min
4. 350 boxes
5. a) $57.37 **b)** $63.68
6. No; the house is less expensive at the end of 2006.
7. a) 24 batteries for $9.29
  b) 100 g of iced tea mix for $0.29
8. a) The Tigers **b)** The Leos
9. No; the medium size of orange juice is the best buy. It costs $2.17/L, compared to $2.79/L and $2.26/L.
10. Answers may vary. For example:
  a) The Jessup family ate 6 of 8 slices of the sugar pie, or $\frac{3}{4}$ of it.
  b) Carly has 6 yellow gumballs and 2 purple ones. The ratio of yellow gumballs to all gumballs is 6 : 8 or 3 : 4 or $\frac{3}{4}$.
  c) 4 sisters split 3 brownies and got $\frac{3}{4}$ of a brownie each.
  d) The ladybug flew 3 km in 4 h.

Unit 6 Linear Equations and Graphing, page 316

6.1 Solving Equations Using Models, page 324

5. a) $s = 4$ **b)** $t = -3$ **c)** $a = 3$ **d)** $b = -6$
6. a) $x = 2$ **b)** $s = 3$ **c)** $c = 1$ **d)** $m = -2$
7. a) $6n + 3 = 21$; $n = 3$
8. a) $6n - 3 = 21$; $n = 4$
9. a) $3n + 4 = 22$; $n = 6$

10. a) No; Curtis' model should subtract 2 unit tiles.
  b) Change the two yellow unit tiles on the left side of the model to red; $x = 5$
11. a) $x = 3$ **b)** $x = -5$ **c)** $x = 6$ **d)** $x = -3$
12. a) Breanna added an extra a-mass on the left side of the scale.
  b) The balance scales should show 3 identical a masses in one pan and three 8-g masses in the other pan; $a = 8$
13. a) $x = -5$ **b)** $x = 5$ **c)** $x = -5$ **d)** $x = -8$
14. a) $4a + 2 = 34$; $a = 8$
15. a) $n = 1$ **b)** $n = 5$ **c)** $n = 20$
16. Answers will vary. For example:
  a) All equations that have positive values and an integer solution; for example: $3b + 2 = 11$; $b = 3$
  b) All equations that have at least one negative value but an integer solution. For example: $4f - 2 = 10$; $f = 3$
  c) All equations that can be solved with balance scales can be solved with algebra tiles.
17. a) One heart, one star, and two smiley faces equal one heart, one smiley face, and three stars.
  b) 22 g; it is not possible to determine the mass of a heart.

6.2 Solving Equations Using Algebra, page 331
5. a) $x = 4$ **b)** $a = 3$ **c)** $m = 2$
  d) $x = 3$ **e)** $x = 4$ **f)** $x = 6$
6. a) $x = -4$ **b)** $x = -4$ **c)** $x = -3$ **d)** $x = -5$
7. a) $x = -5$ **b)** $n = 3$ **c)** $t = \frac{1}{2}$
  d) There are no mistakes.
8. a) $x = -6$ **b)** $x = 3$ **c)** $x = -2$ **d)** $x = -4$
9. a) $72 + 24w = 288$ **b)** $w = 9$
10. a) $85 + 2s = 197$ **b)** $s = 56$
11. a) $x = -6$ **b)** $c = -5$ **c)** $b = -4$
  d) $a = -5$ **e)** $f = -3$ **f)** $d = 11$
12. a) $n = \frac{1}{3}$ **b)** $x = \frac{3}{2}$ **c)** $p = \frac{4}{5}$
  d) $p = \frac{1}{5}$ **e)** $e = \frac{3}{4}$ **f)** $g = \frac{4}{5}$
13. a) $2n + 7 = -3$ **b)** $h = -5$; -5°C
14. Answers may vary. For example:
  a) A family wants to spend a day fishing. They rent 2 fishing boats, and some rods for $720. How many rods did they rent?
  b) $2 \times 300 + 20r = 720$; $r = 6$
  c) Use guess and test.
15. Answers may vary. For example:
  b) A basement is flooded with 316 L of water. After how many minutes of pumping is there 1 L of water left?
  b) $316 = 15m + 1$; $m = 21$; 21 min

6.3 Solving Equations Involving Fractions, page 336
3. a) $t = 30$ **b)** $a = 56$ **c)** $b = 18$ **d)** $c = 27$
4. a) $d = -20$ **b)** $f = -40$ **c)** $k = -36$ **d)** $m = 35$
5. a) $\frac{b}{4} = 8$ **b)** $b = 32$; 32 golf balls
6. a) $\frac{n}{6} = 9$; $n = 54$ **b)** $\frac{n}{-4} = -3$; $n = 12$
  c) $\frac{n}{-5} = 7$; $n = -35$
7. a) $n = 28$ **b)** $m = 33$ **c)** $x = 24$ **d)** $s = 22$
8. a) $f = 18$ **b)** $t = -36$ **c)** $w = -25$ **d)** $e = -63$
9. a) $\frac{n}{-3} + 1 = 6$; $n = -15$ **b)** $3 - \frac{n}{9} = 0$; $n = 27$
  c) $\frac{n}{-2} + 4 = -3$; $n = 14$
10. a) $\frac{s}{2} - 11 = 12$ **b)** $s = 46$; 46 baseballs
11. a) Yes; each student gets $\frac{n}{5}$ of the bag, then gives 1 treat to the teacher, and is left with 9 treats.
  b) $n = 50$; 50 treats
12. a) $\frac{s}{3} + 5 = 41$ **b)** $s = 108$; 108 students
13. a) Correct **b)** $t = 48$ **c)** $r = -40$
14. b) $n = 105$

6.4 The Distributive Property, page 342
4. a) i) 77 **ii)** 77 **b) i)** 25 **ii)** 25 **c) i)** -10 **ii)** -10
  The expressions in each pair are equivalent.
5. Five groups of 1 positive x-tile and 2 positive unit tiles are equivalent to 5 x-tiles and 10 unit tiles. Both use the same tiles, only the tiles are grouped differently.
6.

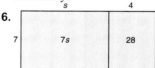

7. a) $2x + 20$ **b)** $5a + 5$ **c)** $10f + 20$
  d) $72 + 6g$ **e)** $64 + 8y$ **f)** $5s + 30$ **g)** $27 + 3p$
  h) $44 + 4r$ **i)** $7g + 105$ **j)** $63 + 9h$
8. a) $3x - 21$ **b)** $4a - 12$ **c)** $9h - 45$ **d)** $56 - 7f$
  e) $5 - 5s$ **f)** $6p - 12$ **g)** $88 - 8t$ **h)** $30 - 20v$
  i) $10b - 80$ **j)** $11c - 44$
9. $P = 2(b + h)$; $P = 2 \times b + 2 \times h$
10. Answers will vary. For example: In multiplication, the order of the terms does not matter. The area of a rectangle with $b = 2$ cm and $h = 1$ cm is $b \times h = h \times b = 2$ cm^2.
11. The expression in Part a: $54 - 9t$ is equal to $9(6 - t)$.
12. a) $-6c - 24$ **b)** $-8a + 40$ **c)** $10f - 70$
  d) $-24 - 3g$ **e)** $-64 + 8y$ **f)** $2s - 10$
  g) $5t + 40$ **e)** $-81 + 9w$
13. Expressions in parts c and d are equivalent.

c) $2(t + 3) = 2t + 6$ or $6 + 2t$
d) $9 + x = x + 9$
14.a) $15(25 + 14)$ or $15 × 25 + 15 × 14$
b) $585; Answers may vary. For example: I prefer the unexpanded expression. It is easier to solve.
15.a) $5(9 + 8)$ or $5 × 9 + 5 × 8$
b) $85; Answers may vary. For example: I prefer the unexpanded expression. It is easier to solve.
16.a) iv   b) ii   c) iii   d) i
17. $2(3(m + 2)) = 6(m + 2) = 6m + 12$
18.a) i) $7y + 21$ ii) $3t - 15$
    iii) $32 - 8s$ iv) $-12p - 36$
b) i) $7(5 + y - 2)$ $7(5 + y) - 7(2)$
    $7 × 5 + 7y - 7 × 2$
    $7 × 5 + 7(y - 2)$ $7(5 - 2 + y)$
19.a) $14 + 2b + 2c$ b) $-66 + 11e - 11f$
c) $r - s + 8$ d) $60 + 10y + 10w$
e) $5j - 75 - 5k$ f) $4g - 48 + 4h$

6.5 Solving Equations Involving the Distributive Property, page 347

4.a) $x = 7$ b) $p = 15$ c) $y = 3$ d) $a = -5$
5.a) $a = -13$ b) $r = 14$ c) $y = -2$ d) $c = 16$
6.a) $2(c + 3) = 20$ b) $c = 7$
7. The equation doubles the cards Marc had before he was given 3 more cards.
8.a) $2(w + 8) = 26$ b) $w = 5$; 5 cm
9.a) $6(x - 5) = 90$ b) $x = 20$; $20
10.a) $8(6 + m) = 264$ b) $m = 27$; $27
11.a) $-5(n + 9) = 15$ b) $n = -12$
12.a) $-4(n - 7) = 36$ b) $n = -2$
13.a) No
b) Kirsten divided the right side of the equation by +8 instead of –8. The correct solution is $x = 2$.
14.a) $t = 0$ b) $p = \frac{13}{2}$, or 6.5
c) $r = \frac{3}{4}$, or 0.75   d) $s = -12$
15.a) $1500 = 25(n + 40)$ b) $n = 20$; 20 guests
16. Answers may vary. For example:
a) Glenn and Lisa won a radio contest for a free roller-skating party for themselves and five other friends. The cost per person includes a $2.00 skate rental deposit that is returned to the renter when the skates are returned. If the total value of the roller-skating party works out to be $42, what is the cost per person, before and after the skates are returned?
b) $n = 8$; The cost per person is $8 before the skates are returned and $6 after.
17.a) $p = 5$ b) $x = \frac{3}{8}$, or 0.375
c) $s = \frac{1}{2}$, or 0.5

Unit 6 Mid-Unit Review, page 350

1.a) $x = -9$ b) $x = -9$ c) $x = 3$ d) $x = -4$
2.a) $3g + 4 = 13$ b) $g = 3$; 3 granola bars
3.a) $x = -9$ b) $x = -3$ c) $x = \frac{1}{3}$ d) $x = \frac{5}{6}$
4.a) $125 + 12p = 545$ b) $p = 35$
5.a) $n = -32$ b) $m = 15$
c) $b = -18$ d) $f = -32$
6.a) $\frac{n}{-7} = 4; n = -28$ b) $\frac{k}{-9} = -3; k = 27$
c) $\frac{m}{-2} + 5 = 0; m = 10$

7.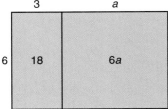

8.a) $3x + 33$ b) $60 + 5y$ c) $28 - 7a$ d) $12t - 72$
9.a) $x = 5$ b) $p = 7$ c) $r = -1$ d) $s = -7$
10.a) $2(n + 6) = 26$ b) $n = 7$; 7 points

6.6 Creating a Table of Values, page 356

4.a)
x	y
1	2
2	3
3	4
4	5
5	6

b)
x	y
1	4
2	5
3	6
4	7
5	8

c)
x	y
1	2
2	4
3	6
4	8
5	10

5.a)
x	y
1	3
2	5
3	7
4	9
5	11

b)
x	y
1	1
2	3
3	5
4	7
5	9

c)
x	y
1	-1
2	-3
3	-5
4	-7
5	-9

6. (2, 11), (4, 29), (5, 38)

7.a)
h	w
1	7
2	14
3	21
4	28
5	35

b) 15 h   c) $168

8.a)
x	y
-3	-1
-2	0
-1	1
0	2
1	3
2	4
3	5

b)
x	y
-3	-6
-2	-5
-1	-4
0	-3
1	-2
2	-1
3	0

c)
x	y
-3	1
-2	2
-1	3
0	4
1	5
2	6
3	7

9.

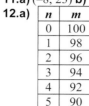

a)
x	y
−3	9
−2	7
−1	5
0	3
1	1
2	−1
3	−3

b)
x	y
−3	11
−2	6
−1	1
0	−4
1	−9
2	−14
3	−19

c)
x	y
−3	−27
−2	−19
−1	−11
0	−3
1	5
2	13
3	21

10. (1, 2), (5, −10), (7, −16)
11. a) (−8, 23) **b)** (12, −17) **c)** (−12, 31) **d)** (15, −23)
12. a)
n	m
0	100
1	98
2	96
3	94
4	92
5	90

b) 20 months   **c)** 86 kg

13. a) m represents the product, t represents the number 9 is being multiplied by.
b)
t	m
0	0
1	9
2	18
3	27
4	36
5	45

c) Patterns may vary. For example: The tens digit in the product increases by 1 each time and the ones digit decreases by 1 each time. The value of m starts at 9 and increases by 9 each time. The sum of the digits of m is equal to 9.
d) Yes; 126 is divisible by 9 because its digits add up to 9.
e) 153

14. a) (−4, −14) **b)** (6, 26) **c)** (3, 14) **d)** (−1, −2)
15. a) (−2, −18) **b)** (−8, −48) **c)** (6, 22) **d)** (1, −3)

6.7 Graphing Linear Equations, page 363

4. a) When x increases by 1, y increases by 4. The points lie on a line that goes up to the right.
b) When x increases by 1, y decreases by 3. The points lie on a line that goes down to the right.

5. a) 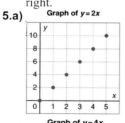 **b)**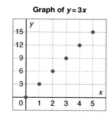

c) 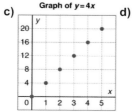 **d)**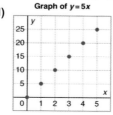

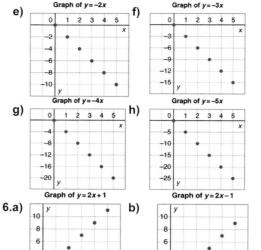

6.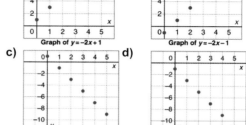

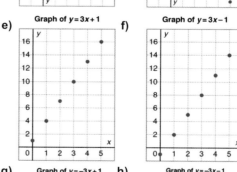

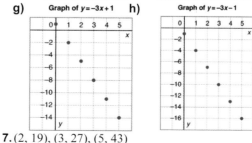

7. (2, 19), (3, 27), (5, 43)
To find a missing number, substitute the given number into the equation and then solve for the unknown.
8. (−3, 13), (−2, 7), (2, −17), (3, −23)
To find the missing numbers, substitute the given numbers into the equation and then solve for the remaining unknown.

9. a) Graph of $c = 11 + 2n$

b) The points go up and to the right. As the x value increases by 1, the y value increases by 2.

c) (6, 23)

10. a) Graph of $m = 100 - 2n$

b) The points go down and to the right. As the x value increases by 2, the y value decreases by 4.

c) (7, 86)

11. a)

x	y
1	20
2	28
3	36
4	44
5	52
6	60
7	68
8	76
9	84
10	92
11	100

b) 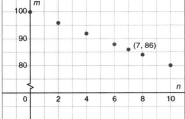 Graph of $m = 8n + 12$

c) The points lie on a line that goes up and to the right. As the x value increases by 1, the y value increases by 8.

d) Yes

12. a) Graph of $y = 8x + 2$ **b)** Graph of $y = -8x - 2$

c) Graph of $y = -7x + 4$ **d)** Graph of $y = 5x - 4$

13. a)

n	p
10	−12 000
20	−9 000
30	−6 000
40	−3 000
50	0
60	3 000
70	6 000
80	9 000

The negative values of p represent a loss of money.

b) 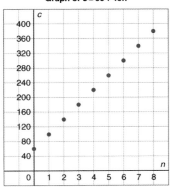 Graph of $p = 300n - 15\,000$

c) As the value of n increases by 10, the value of p increases by 3000. The points lie on a line that goes up and to the right.

d) $7500; I used the graph.

14. a) Graph of $c = 60 + 40n$

b) When n increases by 1, c increases by 40. The points lie on a line that goes up and to the right.

c) No; answers may vary. For example: You cannot work −1 h.

15. a)

i) Graph of $y = -9x + 4$
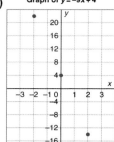

ii) Graph of $y = 6x - 3$
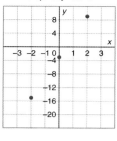

iii) Graph of $y = -7x - 2$
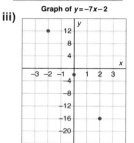

iv) Graph of $y = 4x + 11$
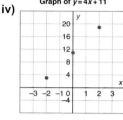

v) Graph of $y = 7x + 5$
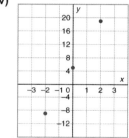

vi) Graph of $y = 3x - 8$
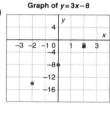

vii) Graph of $y = -9x - 6$

viii) Graph of $y = -8x + 7$
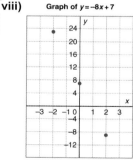

b) Graphs ii, iv, v, and vi go up and to the right. Graphs i, iii, vii, and viii go down and to the right.

c) If the x term is positive, the graph goes up to the right; if it is negative, the graph goes down to the right.

Unit 6 Technology: Using Spreadsheets to Graph Linear Relations, page 367

1. When the input increases by 1, the output increases by 2.

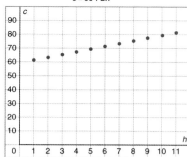

$c = 60 + 2h$

2. If Chris is planning on using the ATV for under 5 h, using the "Rambler" is cheaper. For more than 5 h, the "Northern" is cheaper; for exactly 5 h, the price is the same.

Unit 6 Strategies for Success: Choosing a Strategy, page 369

1. 152 fence posts **2.** 16 teams
3. Ivan has $25.00 and Marsha has $35.00.
4. a) $7.95   **b)** $22.95
5. The order of the beads on a necklace is reversed when a necklace is flipped over.
  a) There are 3 different necklaces which can be made: Green, Yellow, Red; Yellow, Red, Green; Red, Green, Yellow
  b) There are 6 different necklaces which can be made: Green, Yellow, Red, Red; Yellow, Green, Red, Red; Yellow, Red, Green, Red; Green, Red, Yellow, Red; Red, Green, Yellow, Red; Green, Red, Red, Yellow

Unit 6 Unit Review, page 371

1. a) $7c = 56$ **b)** $c = 8$; 8 coins
2. a) $x = 5$ **b)** $x = -2$ **c)** $x = 4$
   d) $x = -4$ **e)** $x = 6$ **f)** $x = -6$
3. a) $8 + 3g = 29$ **b)** $g = 7$; 7 gardens
4. a) $x = 4$ **b)** $x = 3$ **c)** $x = \frac{2}{3}$
   d) $x = -7$ **e)** $x = \frac{1}{3}$ **f)** $x = 2$
5. a) $3h + 6 = 3$ **b)** $h = -1$; $-1°C$
6. a) $p = 12$ **b)** $t = -90$
   c) $w = -54$ **d)** $e = -88$
7. a) $h = 14$
8. a) $\frac{f}{5} = 52$ **b)** $f = 260$; 260 fish
9.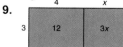
10. a) $6x + 54$ **b)** $33 - 12c$
   c) $-35s + 25$ **d)** $-12a + 8$
11. b) $5t - 20$
12. a) $x = 3$ **b)** $b = 12$ **c)** $f = 17$

d) $s = \frac{24}{5}$, or 4.8

13. a) $-4(x - 7) = 36$ **b)** $x = -2$

14. a) No
b) $c = -9$; Chas should have written +10 after multiplying –2 and –5.

15. a)

x	y
–3	–11
–2	–10
–1	–9
0	–8
1	–7
2	–6
3	–5

b)

x	y
–3	8
–2	7
–1	6
0	5
1	4
2	3
3	2

16. a)

n	s
1	6
2	12
3	18
4	24
5	30
6	36
7	42
8	48

b) 42
c) 11
d) 6 packages

17. a)

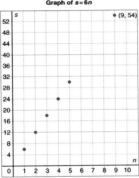

b) When n increases by 1, s increases by 6.
c) (9, 54)

18. a) 18   **b)** 3   **c)** –52   **d)** 0

19. a)

n	p
0	200
1	240
2	280
3	320
4	360
5	400
6	440
7	480

b) $560
c) 7 memberships

20. a)

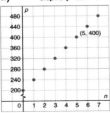

b) When n increases by 1, p increases by 40.
c) (5, 400)

21. a)

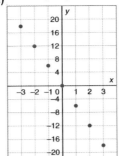

b)

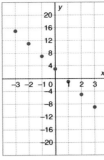

c)

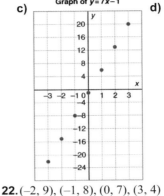

d)

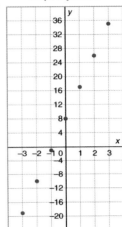

22. (–2, 9), (–1, 8), (0, 7), (3, 4)

Unit 6 Practice Test, page 374

1. b) $s = 2$
2. a) $3 - 3r = -6$; $3 - 3r + 3r = -6 + 3r$; $3 + 6 = -6 + 6 + 3r$; $9 = 3r$; $3 = r$

3. a)

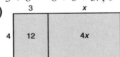

b) No, you could not draw a picture to show that $-4(x + 3)$ and $-4x - 12$ are the same, because it is not possible to draw a rectangle with negative side lengths and areas.

4. a) $x = 12$ **b)** $n = -21$ **c)** $p = 42$ **d)** $x = 8$
5. a) $14p + 200 = 424$ **b)** $p = 16$; 16 people

6. a)

x	y
–3	19
–2	13
–1	7
0	1
1	–5
2	–11
3	–17

b)

x	y
–3	–25
–2	–18
–1	–11
0	–4
1	3
2	10
3	17

7. a)

a	g
10	0
9	1
8	2
7	3
6	4
5	5
4	6
3	7
2	8
1	9
0	10

b) Graph of $g = 10 - a$

c) When a decreases by 1, g increases by 1.
d) (5, 5) because both boys get an equal share

8. a) 0   b) 22   c) 3   d) −38

Unit 6 Unit Problem: Planning a Ski Trip, page 376

1. You should choose company A: Company A charges $5525; company B charges $7225.
2. $c = 23p$; $p = 37$; 37 students
3. a) $I = -15$; −15°C   b) $c = -8$; −8°C
4. a)

p	h
0	1500
10	1800
20	2100
30	2400
40	2700
50	3000
60	3300
70	3600
80	3900
90	4200
100	4500

b) Graph of $h = 1500 + 30p$

c) When p increases by 10, h increases by 300.
d) $4350; $45.79/person

Cumulative Review Units 1–6, page 378

1. a) 1   b) 256   c) 6.6   d) 121
2. a) 20 cm²; $\sqrt{20}$ cm   b) 45 cm²; $\sqrt{45}$ cm
3. $(+24) \times (-3) = -72$; 725
4. a) −60   b) +80   c) −19   d) −19
5. a) $\frac{5}{12}$ b) $1\frac{3}{5}$ c) $5\frac{3}{5}$ d) $4\frac{19}{24}$
6. a) 5   b) 4
   c) The hygienist sees 4 patients and has $\frac{1}{2}$ of $1\frac{1}{6}$ h free time.
7. c) A box with base 2 units by 4 units and back taller than front and sides.
8. 832 cm²
9. a) 600 cm³   b)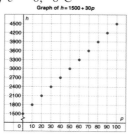

c) 660 cm²
10. 12%
11. a) i) 9 : 1   b) 1 : 2   c) 5 : 6   d) 2 : 1 : 9
    b) 27
12. a) 12 rolls of paper towels for $5.59
    b) 500 mL of mouthwash for $3.99
13. a) $46.65   b) $38.84
14. a) $x = -5$ b) $s = -7$ c) $t = 96$ d) $f = -40$
15. $A = 8(6 + x)$; $A = 48 + 8x$
16. a) $-2(x + 11) = -4$ b) $x = -9$
17. a) (2, −13)   b) (0, 5)
    c) (−3, 32)   d) (4, −31)
18. a) Answers may vary. For example:

n	p
10	80
20	110
30	140

b) $185   c) 70

Unit 7 Data Analysis and Probability, page 380

7.1 Choosing an Appropriate Graph, page 387

3. a) Answers will vary. For example:
   Adult women watch TV 4 h more than adult men.
   Children watch 15 h of TV a week.
   Teens watch 2 h less of TV than children.
   b) Answers will vary. For example:
   Adult women watch more TV than adult men.
   Teens watch the least amount of TV.
   Adult women watch more TV than any other age group.
   c) The bar graph; It is easier to determine the number of hours of television each group watches.

4. a), b)

	Strengths	Limitations
Bar Graph	The heights of the bars can be used to compare the responses. It is easy to read.	The percent of students who chose each response cannot be read from the graph.
Pictograph	The key is one symbol equals 3 students. The key can be used to calculate the total number of students who littered.	Some people may find this graph difficult to read accurately because not all of the symbols are complete.

c) Answers may vary. For example: the bar graph; it is easier to read than the pictograph with partial pictures.

d) A line graph would not be appropriate because the data were not collected over time. A circle graph would be appropriate if you wanted to know percents.

5.a) Answers will vary. For example:
7 people got an A.
1 more person got a B than a C.
3 people got a D.
b) Answers will vary. For example:
More people got a B than any other grade.
More than half the class got an A or B.
About 75% did not get an A.
c) The bar graph; you can see the exact number of students.
d) Bar graphs, since they show the actual number of students, and his class has more students who got As and Bs
e) Ms. Taylor's class; it had a higher percent of students who got As and Bs

6.a) Both compare the winning times of the men's and women's 400-m hurdles at six Olympic games.
b) The line graph makes it easier to compare women's or men's times from one Olympic game to the next.
The bar graph makes it easier to compare women's and men's times for a particular year.
c) The line graph makes it difficult to accurately read the times.
The double bar graph makes it difficult to compare from one year to the next.
d) i) The double line graph; it clearly shows the change from one year to the next.
ii) The bar graph; the difference in bar heights for any year is easy to see.

7. Answers will vary. For example:
a) Practice run times for 2 months
b) Student shoe sizes
c) Average heights of boys and girls for several grades
d) Number of apples eaten in a week
e) How a student spends her time during summer vacation

8.a) Yearly sales; it shows a trend over time.
b) Answers may vary. For example: Bar graph; it would break down the number of pairs by size.

9.a) The line graph shows better the general decrease in attendance. The bar graph shows better the actual mean attendance numbers.
b) For both graphs, it is difficult to accurately read the numbers that are not directly on a grid line.
c) Answers may vary. For example: The line graph; it shows best the change over time.
d) No; we are not interested in mean attendance numbers as parts of a whole.

10.a)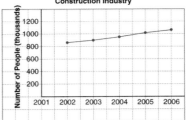
b) Allows you to see a trend in data which are collected over a period of time
Difficult to read accurately because none of the points are on a grid line

11.a)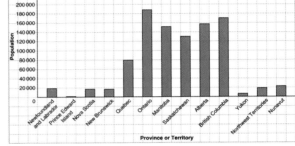
b) Allows for direct comparison of populations in different regions
Difficult to read accurately the exact populations
Does not show populations as percents

12. Answers may vary. For example:
i) A line graph to show the trend over time
ii) A double bar graph to compare her parents
iii) A circle graph to show the percent of time her brother spent on each activity
iv) A line graph to show the change in height over time

13.a)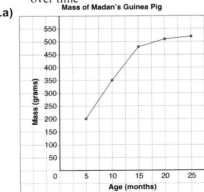
b) Allows you to see a trend in data collected over a long period of time
Difficult to read accurately the data because not all of the points are on grid lines
c) 8 months: about 290 g; 30 months: about 525 g

ANSWERS **527**

Unit 7 Technology: Using Spreadsheets to Record and Graph Data, page 392

1. a)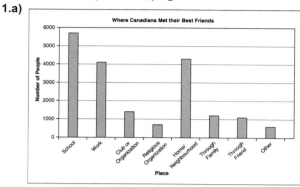

b) Answers may vary.

2. a)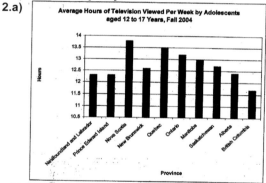

b) Circle graph; since each piece of data is independent and should not be represented as part of a whole
Line graph since data are not collected over time

3. a)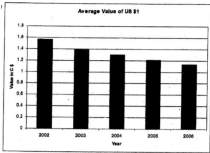

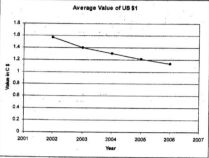

b) The line graph; the data are collected over a period of time.

c) A circle graph, because the data do not represent parts of a whole.

7.2 Misrepresenting Data, page 399

3. Graph B is misleading because its vertical axis does not start at zero, making it appear that Russ sold at least twice as much as the others.

4. a) Graph A gives the impression that half the students want to go to Stanley Park.
Graph B gives the impression that the votes are roughly equal.

b) Graph A is misleading because its horizontal axis doesn't start at zero.
Graph B is misleading because the scale on its horizontal axis is too large.

c) The creator of Graph A would most likely like to go to Stanley Park.

d) Answers may vary. For example: The creator of Graph B probably doesn't want to go anywhere.

e) Start the horizontal axis at 0, and let 1 grid square represent 1 student.

5. Conclusions a and b are incorrect.
Both conclusions were based on Graph A, whose vertical axis doesn't start at 0.

6. a) No; Nick thought the larger symbols represented a greater number of animals.

b) All the fish symbols should be drawn the same size.

c) Bird; The symbols for birds are so large it looks like most students have birds.

7. a) No; the drop is only 5% from 3rd to 4th term.

b) The vertical axis should start at zero.

8. a) Manufacturer A has a lot more trucks still on the road.
  b) 97.5, 96.5, and 95.5
  c) Manufacturer A's trucks are about as dependable as those of the other three manufacturers; the four companies have nearly the same number of trucks on the road after 10 years.
  d) Start the vertical axis at zero.

9. a) Kathy's plant
  b) The symbols are drawn in different sizes; it looks like Arlene's plant is the tallest.
  c) Make all symbols the same size.
  d) No; Arlene probably drew it to make her plant look the tallest.

10. a) More girls participate in sports than boys.
  b), c) The girls' bars are thicker and the boys' scale larger.
  d) Use the same scale and the same bar width for both.
  e) A double bar graph

11. a) The graph is misleading because the angle it is drawn on makes the closest bars appear taller.
  b) Do not draw the graph on an angle.

13. By making the symbols of a pictograph different sizes
By altering where an axis begins its numbering
By slightly removing a sector from a circle graph

14. a) No, because you don't know how much money either of them has to spend, only the percent they spend on each activity
  b) A double bar graph

15. a) Draw a bar graph with a large scale that starts at $50 000.
  b) Draw a bar graph with a small scale.
  c) Draw a bar graph. Start the vertical axis at 0 and let 1 grid square represent $25 000.

16. Answers may vary. For example:
  a) Draw a 3-D bar graph on an angle to make the Pizza Bar appear larger.
  b) Draw a bar graph with a large scale.
  c) Draw a bar graph and start the vertical scale at 195.

Unit 7 Technology: Using Spreadsheets to Investigate Formatting, page 405

1. a) This graph makes it appear that Thornton made 3 times as many points as some of the players.

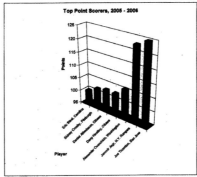

b) This graph makes it look as though Jagr scored about as well as any of the other players.

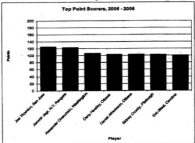

2. Answers may vary. For example:
  a) This double bar graph accurately shows the number of men and women in the Olympics because it has an appropriate scale, no enlarged bars, and the vertical axis starts at zero.

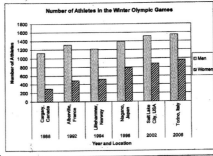

b)

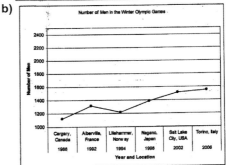

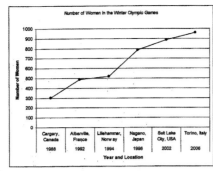
c)
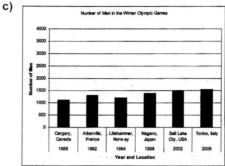
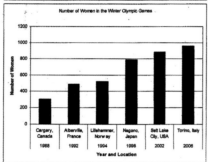

Unit 7 Mid-Unit Review, page 406

1. a) The bar graph shows the number of endangered species of each type of animal. The circle graph shows the percent of endangered species of each type of animal.
  b) The bar graph allows for comparison of numbers. The circle graph shows percents or parts of a whole.
  c) In the circle graph, the original numbers are lost. The bar graph does not show percents.
  d) The bar graph; Percents of endangered species are less relevant than actual numbers.
  e) No; The data were not collected over a period of time.
2. a) No; The Hawks have scored more points per game and are improving faster.
  b) The smaller scale on the Ravens' graph and the fact that their vertical axis doesn't start at 0.
  c) Have the same scales and start both vertical axes at the same number

7.3 Probability of Independent Events, page 411

3. a) $\frac{1}{4}$ **b)** $\frac{1}{2}$

4. a) $\frac{4}{9}$ **b)** $\frac{2}{9}$ **c)** $\frac{4}{9}$ **d)** $\frac{2}{3}$

5. a) $\frac{1}{12}$ **b)** $\frac{1}{4}$ **c)** $\frac{1}{4}$

6. a) i) $\frac{3}{100}$ **ii)** $\frac{3}{20}$ **iii)** $\frac{3}{100}$ **iv)** $\frac{6}{25}$

7. a) $\frac{1}{36}$ **b)** $\frac{1}{36}$ **c)** $\frac{5}{12}$ **d)** $\frac{1}{4}$ **e)** $\frac{1}{4}$

8. a) i) $\frac{1}{24}$ **ii)** $\frac{5}{78}$

  c) $\frac{1}{312}$; The rule is much faster than using a tree diagram.

9. The probability of rolling the same colour twice is $\frac{1}{16}$. However, Marcus forgets that there are 4 different colours. So, a person actually has a $\frac{1}{4}$ chance of winning.

10. a) $\frac{1}{5}$
  b), c) Answers may vary depending on assumptions. For example: $\frac{1}{25}$; I assumed he places the first pair of socks back in the drawer before trying again.

11. a) $\frac{1}{4}$ **b)** For example:

	B	G
B	BB	GB
G	BG	GG

12. a) i) $\frac{1}{12}$ **ii)** $\frac{1}{9}$ **iii)** $\frac{1}{9}$
  b) No; the colour of the first marble will affect which marbles are left and their probabilities for being drawn.

13. a) $\frac{1}{5}$ **b)** $\frac{9}{10}$ **c)** $\frac{1}{10}$ **d)** 0

14. $\frac{1}{16}$

15. Answers will vary. For example:
  a) Rolling a number less than 7 and tossing heads
  b) Rolling a 2 or 5 and tossing tails
  c) Rolling a 1, 3, 4, or 6 and tossing heads

7.4 Solving Problems Involving Independent Events, page 420

4. a) $\frac{1}{8}$ **b)** $\frac{1}{8}$ **c)** $\frac{1}{8}$

5. a) $\frac{1}{216}$ b) $\frac{1}{36}$

6. a) $\frac{1}{32}$ b) $\frac{3}{16}$ c) $\frac{1}{8}$

7. $\frac{1}{10\,000}$

8. a) $\frac{1}{1000}$ b) $\frac{1}{10}$ c) $\frac{6561}{10\,000}$

9. a) $\frac{1}{64}$ b) $\frac{1}{32}$ c) $\frac{3}{104}$

10. a) $\frac{1}{1024}$ b) $\frac{9}{1024}$ c) $\frac{243}{1024}$

11. $\frac{1}{17\,576}$

12. a) $\frac{27}{512}$ b) $\frac{9}{1024}$ c) $\frac{5}{512}$

13. a) About 0.013%
    b) $\frac{1}{6}$; The events are independent.
    c) No; the probability of drawing the white marble 6 times in a row looks at 6 events and their outcomes altogether, so it is much less likely to happen.

14. a) $\frac{5}{64}$ b) $\frac{7}{192}$ c) 0

15. a) 30% b) 21% c) 49%

16. a) $\frac{1}{216}$ b) $\frac{215}{216}$
    c) The sum is 1 because rolling 6s and not rolling 6s account for all possibilities.

Unit 7 Technology: Using Technology to Investigate Probability, page 423

1. b) i) $\frac{1}{8}$ or 12.5% ii) $\frac{1}{4}$ or 25%
   c) The results are very similar.

Unit 7 Unit Review, page 424

1. a) Line graph:
   The number of barrels of oil produced increased every year except for 1999 and 2005.
   The lowest production year was 1999.
   The highest production year was 2004.
   Bar graph:
   The amount of oil produced never exceeded 2.5 million barrels.
   The amount of oil produced never dropped below 1.5 million barrels.
   Oil production was fairly steady from 1998 to 2001.
   b) The line graph is more appropriate to display the data because it clearly shows the change from year to year.

2. a) A circle graph shows the percent of students who chose each response. A pictograph is more visually appealing and makes it very easy to compare how many students chose each response.
   b) For example, a company who produces snack food might want to know which products are most popular.
   c) For example, Sarah is throwing a birthday party and she's inviting the entire class. She wants to know how many people prefer each kind of snack food.

3. a) The number of awards each group of dogs won.
   b) Answers may vary. For example: The bar graph; it isn't important what percent of the awards each type of dog won, but how many awards each group won.
   c) No; the data does not change over time
   d) Yes; however, you would have to use symbols or divide a symbol into many smaller pieces.

4. a)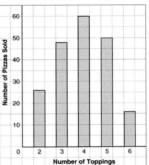

   b) Allows for immediate comparison between the different responses
   Does not show each number of pizzas sold as a percent of the total number of pizzas sold.
   May be difficult to read the value of the bars that do not end on grid lines.

5. a) Yes; she drew the same number of symbols next to each category
   b) Carrots
   c) The symbols for carrots are enlarged to make them appear more popular.

6. The "Not Ready" sections are not drawn to scale; they are drawn shorter to make the percents appear smaller.

7. The second graph; because it does not start the scale on the horizontal axis at 0, but rather at 10; it appears that Party Pizza has at least 10 times as many wrong orders as Pizza Place

8. No; the graph is misleading because its vertical axis does not start at 0. This shortens the bar for the 11-year-olds to look as though very few of them have cell phones

9. a)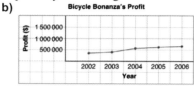

I created the impression of a large growth in profit by not starting the vertical axis scale at 0.

b)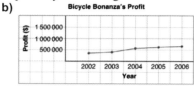

I created the impression of very little growth in profit by using a very large scale on the vertical axis.

10. a) $\frac{1}{6}$

11. a) $\frac{1}{3}$ b) $\frac{1}{3}$ c) $\frac{1}{9}$ d) $\frac{1}{9}$ e) 0

12. a) $\frac{1}{3}$ b) $\frac{1}{27}$ c) $\frac{8}{27}$

13. a) $\frac{1}{2197}$ b) $\frac{1}{140\,608}$ c) $\frac{9}{676}$

14. $\frac{1}{1024}$

15. a) $\frac{1}{64}$ b) $\frac{1}{64}$ c) $\frac{1}{64}$ d) $\frac{3}{64}$

Unit 7 Practice Test, page 428

1. a) Answers may vary. For example:
   i) A bar graph does not allow data to be displayed as percents or parts of a whole.
   ii) In a circle graph, the original data are lost, only the percents are shown.
   b) A pictograph would likely be a bad choice; the data would require a large number of symbols. A line graph could not be used because the data were not collected over a period of time.
   c) A bar graph; the actual number of students would be shown

2. a), b) The first graph gives the impression that sales have not increased substantially by using a large scale that goes far beyond the data. The second graph gives the impression that sales have increased dramatically because the vertical axis does not start at 0 and the scale is very small.
   c) Use an appropriate scale that starts at 0.

3. a) $\frac{1}{4}$ b) $\frac{1}{18}$ c) $\frac{1}{12}$ d) $\frac{1}{6}$

4. a) $\frac{3}{4}$ b) $\frac{9}{64}$ c) $\frac{7}{64}$

Unit 8 Geometry, page 432

8.1 Sketching Views of Objects, page 437

4. a) 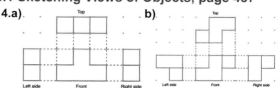 b)

c)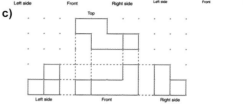

5. A: N, P; B: L, N, Q; C: L, Q; D: J, L, Q; E: K, M, Q

8. 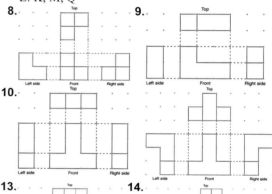 9.

10.

13. 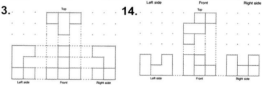 14.

15. 7 possible objects

16.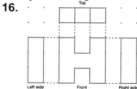

17. a) No
    b) Answers may vary. For example: Front, back, and side views the same
    c) Answers may vary. For example: Front and side views the same

d) Answers may vary. For example: No views the same

8.2 Drawing Views of Rotated Objects, page 444

3. a) Horizontally 90° counterclockwise
  b) Horizontally 180°
  c) Horizontally 90° clockwise
4. a) Front view: B; top view: E; left side view: A; right side view: A
  b) Front view: G; top view: C; left side view: F; right side view: D
5. Part a
6. Students' answers should show the rectangle representing the vertical side piece.
7. i), iii) **ii)**
8. a) **b)**
9. i) **ii)**
iii)
11. **12.**

8.3 Building Objects from Their Views, page 450

4. Object C
5.
6.
7.
8.
9. b) A staircase **c)** Yes
  d) Yes; build the same object but without the middle cube on its back.
10.
11. Answers will vary. For example:
13. a) **b)** 112 cm^2; 56 cm^3
14. a) **b)**
15.
16. a) 9 **b)** 7 **c)** 4

Unit 8 Mid-Unit Review, page 455

1.
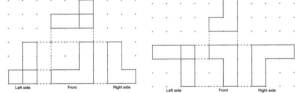
2. a) **b)**
3. a) **b)**

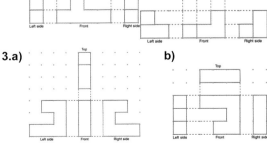

ANSWERS **533**

4. Same views as when rotated 90° clockwise
6. a) b) b)

8.4 Identifying Transformations, page 460

5. a) Reflection b) Rotation of 180° or reflection
   c) Translation 1 unit left
   d) Translation 1 unit right
6. a) B   b) D   c) A   d) C   e) F
7. a) Rotation of 90° counterclockwise
   b) Translation 2 units up
   c) Rotation of 90° counterclockwise
   d) Rotation of 180°
8.
9. A could be a translation image of D or E, or a rotation image of C.
   B could be a reflection image of D, or a rotation image of D.
   C could be a rotation image of A, D, or E.
   D could be a reflection image of B, a translation image of A or E, or a rotation image of C.
   E could be a translation image of A or D, or a rotation image of C.
10. a) The shape should have at least 2 axes of symmetry.
    b) The shape should have 1 axis of symmetry.
    c) The shape should not have any axes of symmetry.
12. Translation: 3 up and 1.5 to the right
    Rotation: 120° counterclockwise or 240° clockwise
    Reflection

8.5 Constructing Tessellations, page 467

6. a) Designs i and iii are tessellations since there are no gaps between shapes.
   b) Designs ii and iv are not tessellations since there are gaps between shapes.
7. a) Yes   b) Yes   c) No
   d) No   e) Yes   f) No
8. a) 90° + 90° + 90° + 90° = 360°
   60° + 60° + 60° + 60° + 60° + 60° = 360°
   108° + 108° + 108° = 324° < 360°
   150° + 150° = 300° < 360°
   120° + 120° + 120° = 360°
   144° + 144° = 288° < 360°
   b) Triangle, square, hexagon

9. Answers will vary. For example:
   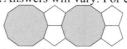
10. No; multiples of 90° and 120° do not add up to 360°.
11. a) Yes; 45° + 45° + 90° + 90° + 45° + 45° = 360°
    b) Yes; 120° + 120° + 60° + 60° = 360°
    c) No; Even though the sum of some angle measures may be 360°, gaps remain among shapes.
    d) Yes; 60° + 60° + 60° + 60° + 60° + 60° = 360° and 120° + 120° + 120° = 360°
    e) No; Even though the sum of some angle measures may be 360°, gaps remain among shapes.
    f) No; Even though the sum of some angle measures may be 360°, gaps remain among shapes.
12. a) Answers will vary. For example:
    Shapes e and f combine to form a hexagon that tessellates.
13. No; 135° + 135° = 270° < 360°
14. Regular octagon and square; The shape tessellates because the sum of the angles at each point is 90° + 135° + 135° = 360°.
16. For example: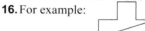
17. a) No   c) For example:
18. a) No   c) For example:

8.6 Identifying Transformations in Tessellations, page 476

3. Answers will vary. For example:
   i) a) Shape C; translation to the right
      b) Shape F; line of reflection is the shared side.
      c) Shape E; rotation 120° counterclockwise about shared vertex
   ii) a) Shape B; translation to the right
       b) Shape B; line of reflection is the shared side.
       c) Shape D; rotation 180° about shared vertex
   iii) a) Shape H; translation down
        b) Shape E; line of reflection is the shared side.
        c) Shape J; rotation 180° about shared vertex
4. Answers will vary. For example:
   i) a) Shape E; Translation down and to the right
      b) Shape D; Line of reflection is the shared side.

c) Shape D; Rotation 60° counterclockwise about shared vertex
ii) a) Shape C; Translation to the right
b) Shape B; Line of reflection is the shared side.
c) Shape E; Rotation 180°
5. Answers will vary. For example:
a) Translate 1 unit down, then rotate 180° about top right vertex. Repeat.
b) Reflect across shared side, then reflect across shared vertex. Repeat.
c) Rotate 90° clockwise about centre vertex. Repeat.
6. a) Answers will vary. For example:
Translate A 4 units to the right to get C.
Translate B 4 units to the right to get D.
Translate E and F 3.5 units down and 1 unit to the left to get G and H.
Translate E and F 3.5 units down and 3 units to the right to get I and J.
b) Answers will vary. For example:
Reflect A across right side to get B.
Reflect B across right side to get C.
Reflect C across right side to get D.
Reflect E across right side to get F.
Reflect G across right side to get H.
Reflect I across right side to get J.

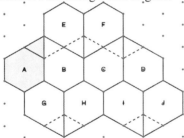

c) The shapes are congruent, so area is conserved.
7. Translations, reflections followed by rotations
8. Answers will vary. For example:
A tessellation can be created by translating the shape 2 units up and 1 unit to the right, then repeating with the new shape.
9. Translations, reflections, rotations of 90°, 180°, and 270°
12. Answers will vary. Students' answers should describe translations, rotations, and reflections.
13. Divide each square into 4 identical squares. Rotate the smaller square in the top left corner 90°, 180°, and 270° to get the other three smaller squares.
14. c) Label the shapes A, B, C, D, and E starting from the top left and going clockwise.
Translate A 8 units right to get C.
Translate A 4 units down and 4 units right to get D.
Translate B 4 units down and 4 units left to get E.

Unit 8 Strategies for Success: Explaining Your Answer, page 481
1. $39.96
2. 4 tables of 8 people and 9 tables of 10 people;
9 tables of 8 people and 5 tables of 10 people;
14 tables of 8 people and 1 table of 10 people

Unit 8 Unit Review, page 483
1. a)

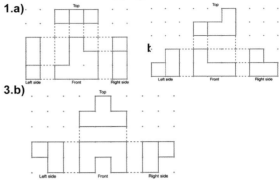

3. b)

4. a) 90° clockwise
b) 90° clockwise or 90° counterclockwise
c) 180° rotation

5.

7. a)   b)

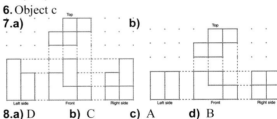

8. a) D b) C c) A d) B
9. Each image is the same.
10. a) Answers may vary. For example:

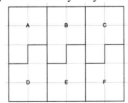

b) Translate A and D to the right to obtain the other shapes.
11. a) No b) Yes c) Yes
12. Answers will vary. For example:

13. a) A and B
15. Translations and rotations

Unit 8 Practice Test, page 486

1.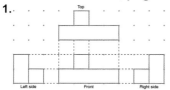

2. a) 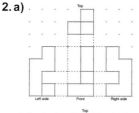 **b)**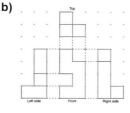

c)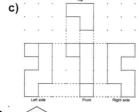

3.

4. A: Reflection across the red line
B: 180° rotation about P
D: Reflection across the blue line then the red line

5. a) No; 108° + 108° + 108° = 324° < 360°
b) Yes; 135° + 45° + 90° + 90° = 360°
c) Yes; 120° + 120° + 120° = 360°

6. Answers will vary. For example:
A and D can either be translated 2 units to the right or reflected across their right sides to obtain B and C. Since all shapes are congruent, area is conserved.

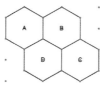

7. Answers may vary. For example:

8. Answers may vary. For example:
Start from the shaded shape and rotate 60° clockwise to obtain the next shape. Rotate the new shape 60° clockwise.

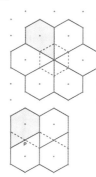

Repeat until tessellation is complete.
Translate shaded shape 2 units to the right to get the next shape over. Rotate the shaded shape 180° about P to get the shape below.

Cumulative Review Units 1–8, page 492

1. a) Logan **b)** 7.3 m
2. a) False; $\sqrt{5}+\sqrt{2} \doteq 3.65 > 2.65 \doteq \sqrt{7}$
  b) True; $\sqrt{46} \doteq 6.78$
  c) True; $\sqrt{36}+\sqrt{64} = 6+8 = 14$
3. $20
4. a) –6 **b)** 2 **c)** 2 **d)** –1
5. $3\frac{3}{4}$
6. a) $\frac{7}{12}$ h **b)** 7 batches
7. Cube
8. a) Triangular prism **b)** Cylinder
  c) Hexagonal prism
9. a) 1 unit by 1 unit by 60 units; least like cube, 242 square units
  b) 3 units by 4 units by 5 units; most like cube, 94 square units
10. a) $\frac{253}{400}$, 0.6325 **b)** $\frac{9}{800}$, 0.01125
  c) $\frac{7}{2500}$, 0.0028 **d)** $\frac{7}{1000}$, 0.007
11. a) $7\frac{1}{2}$% of $2.00 is $0.15; cost in 2007: $2.15
  12% of $2.15 is $0.26; cost in 2008: $2.41
  b) No; $119\frac{1}{2}$% of $2.00 is $2.39
12. a) i) 3 : 8 ii) 5 : 12
  b) i) $\frac{3}{8}$ ii) $\frac{5}{12}$
  c) 2 : 5
13. a) 36 **b)** About 36 min
14. a) 8 apples/min **b)** 12 fence posts/h
  c) 12 km/h
15. a) 52 + 12d **b)** –35 + 42c
  c) –72d + 56 **d)** 48e – 6
16. a) Adding 7 to both sides of the equation, instead of subtracting 5 from both sides
  b) Yes; Substitute –4 for x into the original equation.
  c) It did not affect the solution because Felix always did the same operations on both sides of the equations.
17. a)

x	y
–2	6
–1	3
0	0
1	–3
2	–6

b)

x	y
–2	5
–1	4
0	3
1	2
2	1

18. a) Answers may vary. For example:

n	c
3	24
6	48
9	72

b) 96   **b)** 18

d) Answers may vary. For example:

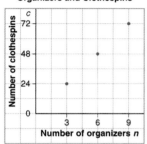

e) Linear   **f)** (6, 48)

19. a) Answers may vary. For example:
The bar graph allows you to determine easily the difference in percents among various uses of water.
The circle graph shows the different water uses as parts of a whole.

b) Answers may vary. For example: In both graphs, we cannot tell how many people were surveyed.

c) Circle graph; because various water uses are parts of a whole

d) No; because the data was not collected over a period of time

20. a) The first graph gives the impression that government funding has not increased much. The second graph gives the impression that government funding has increased dramatically.

b) The first graph uses a large scale on the vertical axis. The second graph starts its vertical axis at 155 000.

c) Groups advocating for more government funding would use the first graph to show the government is not giving enough money. The government would use the second graph to show it is providing a lot more money.

21. a) $\frac{49}{400}$ **b)** $\frac{3}{100}$ **c)** $\frac{21}{200}$ **d)** $\frac{39}{400}$

22. $\frac{77}{1000}$ or 7.7%

23. a) $\frac{1}{64}$

24.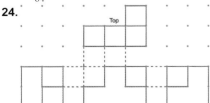

25.

26. Answers will vary. For example:
From A to B, translate 4 units right.
From A to C, translate 8 units right.
From A to D, translate 2 units right and 2 units up.
From A to E, translate 6 units right and 2 units up.
Since all shapes are congruent, area is conserved.

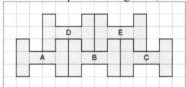

27. Yes

Illustrated Glossary

acute angle: an angle measuring less than 90°

acute triangle: a triangle with three acute angles

algebraic expression: a mathematical expression containing a variable: for example, $6x - 4$ is an algebraic expression

approximate: a number close to the exact value of an expression; the symbol $\doteq$ means "is approximately equal to"

area: the number of square units needed to cover a region

array: an arrangement in rows and columns

average: a single number that represents a set of numbers (see *mean*, *median*, and *mode*)

axis of rotation: the straight line around which an object is turned

bar graph: a graph that displays data by using horizontal or vertical bars

bar notation: the use of a horizontal bar over a decimal digit to indicate that it repeats; for example, $1.\overline{3}$ means 1.333 333 …

base: the side of a polygon or the face of an object from which the height is measured

bisector: a line that divides a line segment or an angle into two equal parts

capacity: the amount a container can hold

Cartesian Plane: another name for a coordinate grid (see *coordinate grid*)

central angle: the angle between the two radii that form a sector of a circle; also called sector angle

certain event: an event with probability 1, or 100%

chance: a description of a probability expressed as a percent

circle graph: a diagram that uses sectors of a circle to display data

circumference: the distance around a circle, also known as the perimeter of the circle

common denominator: a number that is a multiple of each of the given denominators; for example, 12 is a common denominator for the fractions $\frac{1}{3}, \frac{5}{4}, \frac{7}{12}$

common factor: a number that is a factor of each of the given numbers; for example, 3 is a common factor of 15, 9, and 21

commutative property: the property of addition and multiplication that states that numbers can be added or multiplied in any order; for example, $3 + 5 = 5 + 3$; $3 \times 5 = 5 \times 3$

composite number: a number with three or more factors; for example, 8 is a composite number because its factors are 1, 2, 4, and 8

composite shape: the result of combining one or more shapes to make a new shape

concave polygon: has at least one angle greater than 180°

congruent: shapes that match exactly, but do not necessarily have the same orientation

consecutive numbers: integers that come one after the other without any integers missing; for example, 34, 35, 36 are consecutive numbers, so are $-2, -1, 0,$ and 1

conservation of area: under a transformation, the area of a shape does not change

constant term: the number in an expression or equation that does not change; for example, in the expression $4x + 3$, 3 is the constant term

convex polygon: has all angles less than 180°

coordinate axes: the horizontal and vertical axes on a grid

coordinate grid: a two-dimensional surface on which a coordinate system has been set up

coordinates: the numbers in an ordered pair that locate a point on the grid (see *ordered pair*)

cube: an object with six congruent square faces

cubic units: units that measure volume

cylinder: an object with two parallel, congruent, circular bases

database: an organized collection of facts or information, often stored on a computer

denominator: the term below the line in a fraction

diagonal: a line segment that joins two vertices of a shape, but is not a side

diameter: the distance across a circle, measured through its centre

digit: any of the symbols used to write numerals; for example, 0, 1, 2, 3, 4, 5, 6, 7, 8, and 9

dimensions: measurements, such as length, width, and height

discount: the amount by which a price is reduced

discrete data: data that can be counted

distributive property: the property stating that a product can be written as a sum or difference of two products; for example, $a(b + c) = ab + ac$, $a(b - c) = ab - ac$

dividend: the number that is divided

divisor: the number that divides into another number

double bar graph: a bar graph that shows two sets of data

equation: a mathematical statement that two expressions are equal

equilateral triangle: a triangle with three equal sides

equivalent: having the same value; for example, $\frac{2}{3}$ and $\frac{6}{9}$; 3:4 and 9:12

estimate: a reasoned guess that is close to the actual value, without calculating it exactly

evaluate: to substitute a value for each variable in an expression

even number: a number that has 2 as a factor; for example, 2, 4, 6

event: any set of outcomes of an experiment

experimental probability: the probability of an event calculated from experimental results

expression: a mathematical phrase made up of numbers and/or variables connected by operations

factor: to factor means to write as a product; for example, $20 = 2 \times 2 \times 5$

formula: a rule that is expressed as an equation

fraction: an indicated quotient of two quantities

frequency: the number of times a particular number occurs in a set of data

greatest common factor (GCF): the greatest number that divides into each number in a set; for example, 5 is the greatest common factor of 10 and 15

height: the perpendicular distance from the base of a shape to the opposite side or vertex; the perpendicular distance from the base of an object to the opposite face or vertex

hexagon: a six-sided polygon

horizontal axis: the horizontal number line on a coordinate grid

hypotenuse: the side opposite the right angle in a right triangle

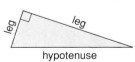

image: the shape that results from a transformation

impossible event: an event that will never occur; an event with probability 0, or 0%

improper fraction: a fraction with the numerator greater than the denominator; for example, both $\frac{6}{5}$ and $\frac{5}{3}$ are improper fractions

independent events: two events in which the result of one event does not depend on the result of the other event

inspection: solving an equation by finding the value of the variable by using addition, subtraction, multiplication, and division facts

integers: the set of numbers
… $-3, -2, -1, 0, +1, +2, +3, …$

inverse operation: an operation that reverses the result of another operation; for example, subtraction is the inverse of addition, and division is the inverse of multiplication

irrational number: a number that cannot be represented as a terminating or repeating decimal; for example, π

irregular polygon: a polygon that does not have all sides equal or all angles equal

isometric: equal measure; on isometric dot paper, the line segments joining 2 adjacent dots in any direction are equal

isometric drawing: a representation of an object as it would appear in three dimensions

isosceles triangle: a triangle with two equal sides

legend: part of a circle graph that shows what category each sector represents

legs: the sides of a right triangle that form the right angle; (see *hypotenuse*)

linear relation: a relation that has a straight-line graph

line graph: a graph that displays data by using points joined by line segments

line segment: the part of a line between two points on the line

line symmetry: a shape that can be divided into 2 congruent parts, so that the parts coincide when the shape is folded along a line of symmetry

lowest common multiple (LCM): the lowest multiple that is the same for two numbers; for example, the lowest common multiple of 12 and 21 is 84

mass: the amount of matter in an object

mean: the sum of a set of numbers divided by the number of numbers in the set

measure of central tendency: a single number that represents a set of numbers (see *mean, median,* and *mode*)

median: the middle number when data are arranged in numerical order; if there is an even number of data, the median is the mean of the two middle numbers

midpoint: the point that divides a line segment into two equal parts

mixed number: a number consisting of a whole number and a fraction; for example, $1\frac{1}{18}$ is a mixed number

mode: the number that occurs most often in a set of numbers

multiple: the product of a given number and a natural number; for example, some multiples of 8 are 8, 16, 24, …

natural numbers: the set of numbers 1, 2, 3, 4, 5, …

negative number: a number less than 0

net: a pattern that can be folded to make an object

numerator: the term above the line in a fraction

numerical coefficient: the number by which a variable is multiplied; for example, in the expression $4x + 3$, 4 is the numerical coefficient

obtuse angle: an angle whose measure is greater than 90° and less than 180°

obtuse triangle: a triangle with one angle greater than 90°

octagon: an eight-sided polygon

odd number: a number that does not have 2 as a factor; for example, 1, 3, 7

operation: a mathematical process or action such as addition, subtraction, multiplication, or division

opposite integers: two integers with a sum of 0; for example, +3 and −3 are opposite integers

ordered pair: two numbers in order, for example, (2, 4); on a coordinate grid, the first number is the horizontal coordinate of a point, and the second number is the vertical coordinate of the point

order of operations: the rules that are followed when simplifying or evaluating an expression

origin: the point where the *x*-axis and the *y*-axis intersect

outcome: a possible result of an experiment or a possible answer to a survey question

parallel lines: lines on the same flat surface that do not intersect

parallelogram: a quadrilateral with both pairs of opposite sides parallel

part-to-part ratio: a ratio that compares a part of the whole to another part of the whole

part-to-whole ratio: a ratio that compares a part of the whole to the whole

pentagon: a five-sided polygon

percent: the number of parts per 100; the numerator of a fraction with denominator 100

percent decrease: to calculate a percent decrease, divide the decrease by the original amount, then write the quotient as a percent
Percent decrease (%) = $\frac{\text{Decrease}}{\text{Original amount}} \times 100$

percent increase: to calculate a percent increase, divide the increase by the original amount, then write the quotient as a percent
Percent increase (%) = $\frac{\text{Increase}}{\text{Original amount}} \times 100$

perfect square: a number that is the square of a whole number; for example, 16 is a perfect square because $16 = 4^2$

perimeter: the distance around a closed shape

perpendicular lines: intersect at 90°

pictograph: a graph that uses a symbol to represent a certain number, and repetitions of the symbol illustrate the data (see page 384)

plane: a flat surface with the property that a line segment joining any two points lies completely on its surface

polygon: a closed shape that consists of line segments; for example, triangles and quadrilaterals are polygons

polyhedron (*plural,* **polyhedra**)**:** an object with faces that are polygons

population: the set of all things or people being considered

prediction: a statement of what you think will happen

prime number: a whole number with exactly two factors, itself and 1; for example, 2, 3, 5, 7, 11, 29, 31, and 43

prism: an object that has two congruent and parallel faces (the *bases*), and other faces that are parallelograms

probability: the likelihood of a particular outcome; the number of times a particular outcome occurs, written as a fraction of the total number of outcomes

product: the result when two or more numbers are multiplied

proper fraction: a fraction with the numerator less than the denominator; for example, $\frac{5}{6}$

proportion: a statement that two ratios are equal; for example, $r:24 = 3:4$

pyramid: an object that has one face that is a polygon (the *base*), and other faces that are triangles with a common vertex

Pythagorean Theorem: the rule that states that, for any right triangle, the area of the square on the hypotenuse is equal to the sum of the areas of the squares on the legs

Pythagorean triple: three whole-number side lengths of a right triangle

quadrant: one of four regions into which coordinate axes divide a plane

quadrilateral: a four-sided polygon

quotient: the result when one number is divided by another

radius (*plural,* **radii**)**:** the distance from the centre of a circle to any point on the circle

range: the difference between the greatest and least numbers in a set of data

rate: a comparison of two quantities measured in different units

ratio: a comparison of two or more quantities with the same unit

reciprocals: two numbers whose product is 1; for example, $\frac{2}{3}$ and $\frac{3}{2}$

rectangle: a quadrilateral that has four right angles

rectangular prism: a prism that has rectangular faces

rectangular pyramid: a pyramid with a rectangular base

reflection: a transformation that is illustrated by a shape and its image in a line of reflection

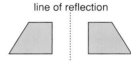

reflex angle: an angle between 180° and 360°

regular polygon: a polygon that has all sides equal and all angles equal

regular prism: a prism with regular polygons as bases; for example, a cube

regular pyramid: a pyramid with a regular polygon as its base; for example, a tetrahedron

related denominators: two fractions where the denominator of one fraction is a factor of the other

relation: a rule that associates two terms

repeating decimal: a decimal with a repeating pattern in the digits to the right of the decimal point; it is written with a bar above the repeating digits; for example, $\frac{1}{15} = 0.0\overline{6}$

rhombus: a parallelogram with four equal sides

right angle: a 90° angle

right triangle: a triangle that has one right angle

rotation: a transformation in which a shape is turned about a fixed point

scale: the numbers on the axes of a graph

scalene triangle: a triangle with all sides different

sector: part of a circle between two radii and the included arc

sector angle: see *central angle*

simplest form: a ratio with terms that have no common factors, other than 1; a fraction with numerator and denominator that have no common factors, other than 1

spreadsheet: a computer-generated arrangement of data in rows and columns, where a change in one value results in appropriate calculated changes in the other values

square: a rectangle with four equal sides

square number: the product of a number multiplied by itself; for example, 25 is the square of 5

square root: a number which, when multiplied by itself, results in a given number; for example, 5 is a square root of 25

statistics: the branch of mathematics that deals with the collection, organization, and interpretation of data

straight angle: an angle measuring 180°

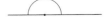

surface area: the total area of the surface of an object

symmetrical: possessing symmetry (see *line symmetry*)

systematic trial: solving an equation by choosing a value for the variable, then checking by substituting

term: (of a fraction) the numerator or the denominator of the fraction

(of a ratio) each of the quantities constituting a ratio; for example, in the ratio 4:5, 4 and 5 are both terms

terminating decimal: a decimal with a certain number of digits after the decimal point; for example, $\frac{1}{8} = 0.125$

tessellate: to use congruent copies of a shape to cover a plane with no overlaps or gaps

tetrahedron: an object with four equal triangular faces; a regular triangular pyramid

theoretical probability: the number of favourable outcomes written as a fraction of the total number of possible outcomes

three-dimensional: having length, width, and depth or height

three-term ratio: a comparison of three quantities with the same unit

transformation: a translation, rotation, or reflection

translation: a transformation that moves a point or a shape in a straight line to another position on the same flat surface

trapezoid: a quadrilateral that has one pair of parallel sides

triangle: a three-sided polygon

two-dimensional: having length and width, but no thickness, height, or depth

two-term ratio: a comparison of two quantities with the same unit

unit fraction: a fraction that has a numerator of 1

unit price: the price of one item, or the price of a particular mass or volume of an item

unit rate: a quantity associated with a single unit of another quantity; for example, 6 m in 1 s is a unit rate; it is written as 6 m/s

variable: a letter or symbol representing a quantity that can vary

vertex (*plural,* **vertices):** the point where 2 sides of a shape meet, or the point where 3 or more edges of an object meet

vertical axis: the vertical number line on a coordinate grid

volume: the amount of space occupied by an object

whole numbers: the set of numbers 0, 1, 2, 3, …

x-axis: the horizontal number line on a coordinate grid

y-axis: the vertical number line on a coordinate grid

zero pair: two opposite numbers whose sum is equal to zero

zero property: the property of addition that states adding 0 to a number does not change the number; for example, $3 + 0 = 3$; for multiplication, multiplying a number by 0 results in the product 0; for example, $3 \times 0 = 0$

Index

A
acute triangle, 39, 40, 464
algebra,
    solving equations involving fractions with, 334, 335
    solving equations with, 327–330
algebra tiles,
    modelling the distributive property with, 340
    solving equations with, 320–323, 328, 329
area,
    of a circle, 209
    of a rectangle, 6, 105
    of a square, 7, 17–19, 22–24, 40, 41, 47, 48
    of a triangle, 18
area model,
    multiplying fractions with, 115, 116
    multiplying mixed numbers with, 122, 123
ascending order, 14
average speed, 296, 297
axis of rotation, 441–444

B
balance-scales model,
    solving equations with, 319, 320, 327
"bank" model,
    for dividing integers, 77, 80
    for multiplying integers, 65, 66
bar graph, 382, 384, 385
    constructing with spreadsheet software, 391
    formatting on spreadsheet software, 403
    misrepresented data on, 395, 396
base of a net, 171
base of a prism *vs.* base of a polygon, 184
brackets, 85, 90, 91, 154

C
calculators,
    investigating square roots with, 29
circle,
    area of, 209
    circumference of, 209
circle graph, 382, 384–386
    constructing with spreadsheet software, 392
    misrepresented data on, 396
circumference, 209

clockwise rotation, 441–444, 457, 459, 474
coloured tiles,
    dividing integers with, 80
    multiplying integers with, 65, 66
    solving equations with, 318
common denominators, 123
    dividing fractions with, 136–138
    dividing mixed numbers with, 142
    subtracting fractions with, 154
common factors, 124
    in multiplying fractions, 115–117
commutative property, 71
composite shape, 467
concave quadrilateral, 464
congruent bases, 171
congruent circles, 171, 210
congruent number, 6
congruent shapes, 457, 458, 475
conservation of area, 473
convex quadrilateral, 464
counterclockwise rotation, 441–444, 457, 459, 473
counters,
    multiplying fractions by whole numbers with, 106
    multiplying fractions with, 111
cube, 171
cubic centimetres (cm^3), 196, 197, 199 *Practice*
curved brackets, 85, 90, 91
cylinder, 171, 214 *Math Link*
    surface area of, 209–211
    volume of, 215–217

D
data,
    formatting on spreadsheet software, 403, 404
    misrepresented, 394–398
decagons, 474
decimal percent, 235
decimals,
    relating to fractions and percents, 234–238, 242–245, 248–252, 282, 283
descending order, 14
diorama, 228 *Unit Problem*
discount, 257, 258
discrete data, 360, 383
distributive property, 71, 72, 338–341
    solving equations with, 344–346
dividend, 12, 85
division equations, 131, 136
division facts, 13, 85

divisor, 12, 85
double bar graph, 383
double line graph, 404

E
edge of a net, 171
equations,
    involving distributive property, 344–346
    involving fractions, 333–335
    models for solving with, 318–323
    of a linear relation, 351–355, 359–362
    solving using algebra, 327–330
equivalent fractions, 116
equivalent ratios, 269–273, 288–290
    comparing ratios with, 280, 281
events, 407–410, 417–419
expand, 341
expanded form of a number, 72
experimental probability, 423

F
faces of a net, 171
factors, 11–14, 115
fractal, 164 *Unit Problem*
fraction circles,
    dividing whole numbers by fractions with, 132
    multiplying fractions by whole numbers with, 107
fractions,
    as ratios, 265, 266
    dividing into whole numbers, 129–132
    division of, 135–138
    equations involving, 333–335
    in math problems, 147–150
    in mixed numbers, 121–124
    multiplication models for, 110–112
    multiplying as factors, 115–118
    multiplying by whole numbers, 104–107
    order of operations for, 153, 154
    relating to decimals and percents, 234–238, 242–245, 248–252, 282, 283

G
Game:
    Empty the Rectangles, 416
    Fitting In, 28
    Largest Box Problem, 201
    Make the Number, 349
    Spinning Fractions, 127

Target Tessellations, 470
Triple Play, 278
What's My Product?, 76
geometry software,
    creating tessellations with, 479
    viewing objects with, 440
goods and services tax (GST), 256–259
graphs, 382–386
    constructing with spreadsheet software, 366, 367, 391, 392
    finding unit rates with, 296
    for linear relations, 352, 359–362
    formatting on spreadsheet software, 403, 404
greatest common factor, 271
grouping symbol, 85, 91
grouping terms, 90

H

harmonized sales tax (HST) (*see* sales tax)
hexagon, 463
hexagonal pyramid, 171
horizontal rotation, 441–444
hundredths chart for
    representing percents with, 235, 244
hypotenuse, 31–33, 46–48
hypothesis, 314 *Unit Problem*

I

improper fractions, 122, 143, 144
independent events,
    probability of, 408–410
    solving problems with, 417–419
integers,
    division models of, 77–80
    division rules for, 84–86
    multiplication models of, 64–67
    multiplication rules for, 70–72
    order of operations with, 90, 91
interactive isometric drawing tool, 454
inverse operations, 12, 129
irrational number, 19
isometric, 435
isometric drawing, 435, 436
isosceles right triangle, 31

L

legs, 31–33, 46–48
length of a line segment, 17–19
line graph, 383
    constructing with spreadsheet software, 391
    formatting on spreadsheet software, 404

misrepresented data on, 397, 398
line segment measurements, 17–19
linear relations,
    creating tables of values for, 352–355
    graphing, 359–362
    graphing with spreadsheet software, 366, 367
linking cubes,
    building objects from their views, 447–449
    viewing objects with, 435, 436, 440
    viewing rotated objects with, 442–444
    viewing with an interactive isometric drawing tool, 454

M

Math Link:
    Art, 453
    History, 42, 214
    Science, 358
    Sports, 67
    Your World, 134, 301, 422
misrepresented data, 394–398
mixed numbers,
    dividing, 141–144
    multiplying, 121–124
multiplication,
    dividing fractions with, 136, 137
    dividing mixed numbers with, 143
multiplication equations, 105, 136
multiplicative identity, 71

N

negative integers, 64–67, 71, 72, 77–80, 84–86, 94, 95
nets, 170–173
    creating objects from, 177–179
    of a rectangular prism, 182–185
    of a triangular prism, 188
number line,
    dividing fractions by fractions with, 135, 138
    dividing integers with, 78, 79
    dividing mixed numbers with, 142
    dividing whole numbers by fractions (and vice versa) with, 130, 131
    estimating square roots with, 24
    multiplying fractions by whole numbers with, 105, 106
    multiplying integers with, 64, 67
    showing percents on, 243–245, 249–252

O

objects,
    building from their views, 447–449
    identifying transformations in tessellations, 471, 475
    rotated views of, 441–444
    sketching views of, 434–437
    sketching with geometry software, 440
    tessellations of, 462–467
    transformations of, 456–459
    viewing with an interactive isometric drawing tool, 454
obtuse triangle, 39, 40, 464
octagonal pyramid, 179
order of operations, 90, 91
    for expressions as a fraction, 91
    with fractions, 153, 154
ordered pair, 353, 355
outcomes, 407–410, 417–419

P

palindromic number, 16 *Practice*
part-to-part ratio, 265, 266, 281
part-to-whole ratio, 265, 266, 282, 283
Pattern Blocks,
    multiplying fractions with, 111
pentagonal prism, 171
percent decrease, 251, 252
percent increase, 251
percents,
    as ratios, 265, 266
    calculating sales taxes in, 256–259
    comparing ratios with, 282, 283
    relating to decimals and fractions, 234–238, 242–245, 248–252, 282, 283
perfect square (*see* square numbers)
perimeter,
    of a square, 7, 24
phonograph cylinder, 214 *Math Link*
pictograph, 384, 385
    comparing ratios with, 280
    finding unit rates with, 295
    misrepresented data on, 396
    multiplying fractions by whole numbers with, 105
pie charts, 392
plane, 462–467
plus/minus statistics, 67 *Math Link*
polygons, 171, 179
    tessellations of, 463

polyhedron, 171, 179
positive integers, 64–67, 71, 72, 77–80, 84–86, 94, 95
power of 10, 235
pressure, 358 *Math Link*
prisms, 434
    nets of, 171–173
    surface area of, 183–185, 188–190, 209–211
    volume of, 195–197, 202–204, 215–217
probability,
    of independent events, 407–410, 417–419
    simulated on a Web site virtual manipulative, 423
proper fractions, 113 *Practice*, 122
proportion, 287–290
provincial sales tax (PST), 256–259
pyramid, 171
Pythagoras, 32
Pythagorean Theorem, 4, 31–33, 39–42, 46–48
    verifying with geometry software, 37, 38
Pythagorean triple, 41, 42

Q

quadrilateral, 6
    tessellation of, 464
quotient, 12, 77, 85

R

rates, 294–297, 300–303
ratios, 264–266
    comparing, 279–283
    equivalent, 269–273, 288–290
    solving math problems with, 287–290
rectangle, 6
    area of, 105
    multiplying fractions with models of, 112
    multiplying mixed numbers with models of, 122, 123
    tessellation of, 466, 467
    rectangular prism, 434
    surface area of, 183–185
    volume of, 195–197
reflection, 457, 458, 472, 473
regular dodecagon, 175 *Practice*
regular hexagonal pyramid, 179
regular polygon, 171
regular pyramid, 171
remainders, 131, 132, 138
repeated addition, 64, 65, 105
repeating decimal, 237
rhombus,
    transformations in, 473

right angles, 6
right cylinders, 171
right prisms, 171
right triangle, 31–33, 37, 38, 40
rotation, 441–444, 457–459, 473

S

sale price, 257–259
sales tax, 256–259
scalene right triangle, 31
scatter plot, 367
selling price, 257-259
side length of a square, 17, 22–24, 47, 48
Sierpinski Triangle, 164 *Unit Problem*
simplest form, 235, 236, 271
    of a product, 123
simulation, 490 *Investigation*
speed, 294, 296, 297
spreadsheet software,
    constructing graphs with, 391, 392
    formatting graphs on, 403, 404
    graphing linear relations with, 366, 367
square, 6
    area of, 17–19, 22–24, 40, 41, 47, 48
    side length of, 17, 22–24, 47, 48
square brackets, 90, 91
square dot paper, 435
square numbers (*also* perfect square), 6, 7, 11–14
square prism, 171, 172
square pyramid, 179
square roots, 12–14
    as decimals, 22–24
    estimating, 22–24
    investigating with calculators, 29
square units, 13, 18, 19, 23
surface area,
    of a cylinder, 209–211
    of a rectangular prism, 183–185
    of a triangular prism, 188–190

T

tables,
    constructing with spreadsheet software, 366, 367
    creating values of linear relations in, 351–355
    finding unit rates with, 296
tessellate, 463–467
tessellations, 462–467
    creating with geometry software, 479
    transformations in, 471–475

tetrahedron, 171
Theodorus, 36 *Practice*
theoretical probability, 423
three-term ratio, 265, 266
"Traffic Light" strategy, 157
transformation, 456–459
    in tessellations, 471–475
translation, 457, 458, 473
tree diagram, 409, 418
triangle,
    area of, 18
    creating tessellations with geometry software, 479
    tessellation of, 464, 472
triangular prism, 172, 179
    surface area of, 188–190
    volume of, 202–204
triangular pyramid, 171
two-term ratio, 265, 266

U

unit rates, 295–297, 301–303

V

variables, 318–323, 328–330, 334, 335, 361, 362
    in writing formulas, 196, 203
vertex of a net, 171
vertical rotation, 441–444
volume,
    of a cylinder, 215–217
    of a rectangular prism, 195–197
    of a triangular prism, 202–204

W

Web site virtual manipulatives, simulating probability on, 423
Wheel of Theodorus, 36 *Practice*
whole numbers,
    dividing by fractions, 129–132
    multiplying by fractions, 104–107
    order of operations, 90
    properties of, 71
word problems,
    key word operations in, 148–150

Z

zero pairs, 66, 322, 323
zero property, 71
zero value, 77

Acknowledgments

Pearson Education would like to thank the Bank of Canada and the Royal Canadian Mint for the illustrative use of Canadian bills and coins in this textbook. In addition, the publisher wishes to thank the following sources for photographs, illustrations, and other materials used in this book. Care has been taken to determine and locate ownership of copyright material in this text. We will gladly receive information enabling us to rectify any errors or omissions in credits.

Photography

Cover: John Guistina/Imagestate/firstlight.ca
p. 3 Ian Crysler; p. 4 (top) Arthur S. Aubry/Photodisc Collection/Getty Images; p. 4 (centre) bridge Corel Collection Southwestern U.S.; p. 5 (top left) Royalty-Free/CORBIS; p. 5 (top right) Martin Bond/Photo Researchers, Inc.; p. 5 (bottom) Vision/Cordelli/Digital Vision; p. 8 Ian Crysler; p. 10 Linda Bucklin/Shutterstock; p. 11 Ian Crysler; p. 17 Ian Crysler; p. 22 Ray Boudreau; p. 26 Angela Wyant/Stone/Getty Images; p. 32 SEF/Art Resource N.Y.; p. 38 Creatas/First Light; p. 50 Burke/Triolo/Brand X/Getty Images; p. 51 Clifford Skarstedt/CP Photo; p. 56 Jeff Greenberg/Photo Edit; pp. 62–63 (top) Corel Collections, *Lakes and Rivers*; p. 62 (top) imagesource/firstlight.ca; p. 63 (middle left) Reuters/CORBIS; p. 63 (middle right) Canadian Press/Jonathan Hayward; p. 63 (bottom) Lawson Wood/CORBIS; pp. 76–77 Ian Crysler; p. 79 Al Grillo/CP Photo; p. 81 Steve Kaufman/A.G.E. fotostock/First Light; p. 84 Ian Crysler; p. 86 DreamPictures/Stone/Getty Images; p. 88 Sandy Grant/CP Photo; p. 89 Ron Hofman/White Rainbow Photography; p. 99 Stewart Cohen/Pam Ostrow/Getty Images; p. 100 Ray Boudreau; p. 104 Ian Crysler; p. 109 Larry Macdougal/First Light; p. 110 Ian Crysler; p. 118 B&Y Photography/Alamy; p. 121 Ian Crysler; p. 134 Wendell Webber/FoodPix/Jupiter Images; p. 141 Ian Crysler; p. 147 Keith Levit/Alamy; p. 151 Blend Images/Alamy; p. 152 The Stock Asylum, LLC/Alamy; p. 153 Jeff Greenberg/Alamy; pp. 168–169 CD case Photodisc; Timberland® shoes Cindy Charles/Photo Edit Inc.; golf balls Bill Ivy/Ivy Images; tennis balls Danilo Calilung/Corbis; all others Ian Crysler; p. 170 Ian Crysler; p. 175 Courtesy of Jason Taylor; p. 176 (top) Andy Crawford/Dorling Kindersley; p. 176 (bottom) Klaus Hackenberg/zefa/Corbis; p. 177 Michael Newman/Photo Edit; p. 181 Jules Frazier; p. 183 Ian Crysler; p. 188 Ian Crysler; p. 195 Ian Crysler; p. 199 Graham Tomlin/Shutterstock; p. 206 Prentice Hall, Inc.; p. 209 Ray Boudreau; p. 213 (left) Katherine Fawssett/The Image Bank/Getty Images; p. 213 (right) ME967X.45, McCord Museum, Montreal; p. 214 Terrence Mendoza/Shutterstock; p. 215 Ian Crysler; p. 217 Reuters/CORBIS; p. 228 (left) BYPhoto/Alamy; p. 228 (centre) Bill Brooks/Alamy; p. 228 (right) Hemera/Photo.com/Jupiter Images; p. 229 Ian Crysler; p. 231 Ray Boudreau; p. 232 Dorling Kindersley Media Library; p. 233 (top to bottom) Dorling Kindersley Media Library; Noella Ballenger/Alamy; Photodisc/Getty Images; Don Mason/CORBIS/MAGMA; p. 248 Ian Crysler; p. 251 Aaron Haupt/

Photoresearchers/First Light; p. 253 Courtesy of Rayven Moon; p. 254 Rubens Abboud/Alamy; p. 256 Ian Crysler; p. 261 Alaska Stock LLC/Alamy; p. 262 Photo Objects.Net Images/Jupiter Unlimited; p. 274 Canadian Press/Don Denton; p. 279 Ian Crysler; p. 282 BlueMoon Stock/Alamy; p. 286 Johner/Getty Images; p. 287 Image Source Pink/Jupiter Images; p. 290 Photodisc Collection/Getty Images; p. 293 Tony Freeman/Photo Edit; p. 294 (top) Michael Probst/CP Photo; p. 294 (centre) Iuri/Shutterstock; p. 294 (bottom) Ian Crysler; p. 297 (top) Tim O'Hara/Corbis; p. 297 (bottom) Jack Cox – Images of Nature/Alamy; p. 298 Larry MacDougal/CP Photo; p. 301 Photodisc/First Light; p. 304 Courtesy of BC Curios Ltd.; p. 308 (left) Ken Gigliotti/CP Photo; p. 308 (right) Kitch Bain/Alamy; p. 310 Natalie Fobes/CORBIS; p. 311 Flip Nicklin/Minden Pictures/Getty Images; p. 314 (top) Vlade Shestakov/Shutterstock; p. 314 (centre) Eric Gevaert/Shutterstock; p. 314 (bottom) Geri Lavrov/Lifesize/Shutterstock; p. 316 (top) Royalty-Free/CORBIS; p. 316 (bottom) Emmanuel Faure/Taxi/Getty Images; p. 317 (top) Bernd Fuchs/firstlight.ca; p. 317 (bottom) Courtesy of Raf Komierowski; p. 318 Ian Crysler; p. 327 Jonathan Hayward/CP Photo; p. 332 Chuck Savage/CORBIS; p. 338 John Connell/Index Stock Imagery; p. 344 Ian Crysler; p. 345 Paul Felix Photography/Alamy; p. 347 Courtesy of Diana Mej; p. 353 Sergey Shandin/Shutterstock; p. 365 Ryan McVey/Stone/Getty Images; p. 367 Brian Bailey/Taxi/Getty Images; p. 376 isifa Image Service s.r.o./Alamy; pp. 380–381 Ian Crysler; p. 393 Digital Vision; p. 416 Ian Crysler; p. 418 Richard I'Anson/Lonely Planet Images/Getty Images; p. 420 Graca Victoria/Shutterstock; p. 421 Stockbyte/Getty Images; p. 422 Loredo Rucchin/iStockphoto; pp. 430–431 Ian Crysler; pp. 432–433 Courtesy of Don Yeomans; p. 433 (top) "The Benefit" Courtesy of Don Yeomans; p. 433 (bottom) "Gunarh and the Whale" Courtesy of Don Yeomans; p. 434–436 Ian Crysler; p. 437 (top) Ian Crysler; p. 437 (bottom) Ray Boudreau; p. 438 (top) Ray Boudreau; p. 438 (bottom) Masterfile Royalty-Free; p. 439 (top) AbleStock.com Images/Jupiter Unlimited; p. 439 (centre) Hemera/MaXx Images; p. 439 (bottom) Masterfile Royalty-Free; p. 440 Ray Boudreau; pp. 441–442 Ian Crysler; p. 443 (top) Ray Boudreau; p. 443 (middle) Ian Crysler; p. 444 (top, middle) Ian Crysler; p. 444 (bottom) Ray Boudreau; p. 445 Ray Boudreau; p. 446 Ian Crysler; pp. 448–450 Ian Crysler; p. 453 Courtesy of IMAX Corporation; p. 455 (top) Ronen/Shutterstock; p. 455 (centre) Ray Boudreau; p. 455 (bottom) Ian Crysler; p.461 "Haida Frog" by Bill Reid, courtesy of Martine Reid; p. 462 (left) Ryan McVay/Photodisc/Getty Images; p. 462 (left inner) Brigitte Sporrer/zefa/Corbis; p. 462 (centre) Peter Baxter/Shutterstock; p. 462 (right inner) oksanaperkins/Shutterstock; p. 462 (right) nolie/Shutterstock; p. 471 Ian Crysler; p. 477 (left) Courtesy of M.C. Escher Company; p. 477 (right) Bonnie Kamin/Photo Edit; p. 478 Philip and Karen Smith/Photodisc/Getty Images; pp. 482–484 Ian Crysler; p. 485 (top) Roman Soumar/CORBIS; p. 485 (bottom) M.C. Escher's Reptiles © 2004 The M.C. Escher Company–Baarn, Holland. All Rights Reserved; p. 488 M.C. Escher's Fish © 2004 The M.C. Escher Company–Baarn, Holland. All Rights Reserved; pp. 489–490 Ian Crysler

Illustrations

Steve Attoe, Pierre Berthiaume, Philippe Germain, Brian Hughes, Paul McCusker, Dusan Petriçic, Michel Rabagliati, Neil Stewart/NSV Productions, Craig Terlson